Proceedings of the 1st International Conference on

Experimental Implementation of Quantum Computation

Sydney, Australia, 16 – 19 January 2001

Editor:

R G Clark

Rinton Press

12 Castleton Road
Princeton, New Jersey 08540
USA
rinton@rinton.com

Published by Rinton Press, Inc.

Printed in the United State of America

ISBN: 1-58949-013-4

PREFACE

The first International Conference on Experimental Implementation of Quantum Computation was held on 16-19 January 2001 at the National Maritime Museum in Sydney, Australia. The conference provided an international forum to review the latest developments in the experimental implementation of quantum computer proposals worldwide. This also included theoretical investigations relevant to the various computer architectures, implementation strategies and computer operation. Additionally the conference addressed applications of few-qubit quantum computers that might be achievable in the medium term. A range of invited presentations provided a framework that set out both the challenges of the various experimental proposals together with progress that has been made in overcoming these challenges.

The IQC01 scientific program comprised 80 presentations from the 175 registered conference delegates, with 13 countries represented. This included oral presentations from 33 distinguished invited speakers, and a further 22 oral presentations selected from contributed papers. A dedicated poster session, in which 25 additional presentations at the forefront of international research in quantum computation were given, reinforced the oral program.

The conference was jointly organized by the Australian Research Council Special Research Centre for Quantum Computer Technology and Los Alamos National Laboratory USA. The Australian and US organizers wish to express their gratitude to the institutions and corporations that provided financial sponsorship.

IQC01 was officially opened by Senator Richard Alston, Minister for Communications, Information Technology and the Arts, Australian Government.

Continuation of this conference will be in the format of a Gordon Research Conference series and will next take place in the US in January 2003. The Chair and Vice Chair of IQC in 2003 will be Professor David Awschalom, University of California, Santa Barbara and Professor Birgitta Whaley, University of California, Berkeley.

I would especially like to thank my colleague Dr Alex Hamilton for the immense assistance and expertise he has willingly provided in the technical production and editorial process of the Proceedings. I would also like to thank the Australian Centre for Quantum Computer Technology's Executive Officer Karen Jury for her invaluable help in the coordination and planning of IQC01. I would finally like to thank my colleagues at the University of New South Wales and Los Alamos National Laboratory for their assistance in the organisation and running of IQC01. It was a wonderful team effort. I am also grateful to Rinton Press, USA for their expert production of these Proceedings.

R.G. Clark

ORGANIZING COMMITTEE

R.G. Clark, Australian Chair	Sydney, Australia
G. Milburn	Brisbane, Australia
S. Prawer	Melbourne, Australia
M. Simmons	Sydney, Australia
A. Dzurak	Sydney, Australia
A. Hamilton	Sydney, Australia
P.C. Hammel, US Chair	Los Alamos, USA
R. Hughes	Los Alamos, USA
D. James	Los Alamos, USA
M. Hawley	Los Alamos, USA
K. Jury, Secretariat	Sydney, Australia

PROGRAM COMMITTEE

R. Hughes, Chair	Los Alamos, USA
G. Milburn	Brisbane, Australia
D. Di Vincenzo	New York, USA
S. Williams	Palo Alto, USA
R. Clark	Sydney, Australia
P.C. Hammel	Los Alamos, USA
D. Awschalom	Santa Barbara, USA
P. Kwiat	Los Alamos, USA

INTERNATIONAL ADVISORY COMMITTEE

B. Kane	Maryland, USA
E. Yablanovitch	Los Angeles, USA
A. Ekert	Oxford, UK
C. Marcus	Cambridge, USA
H. Mooij	Delft, Netherlands
D. Wineland	Boulder, USA
J. Kimble	Los Angeles, USA
P. Zoller	Innsbruck, Austria
P. Knight	London, UK
S. Haroche	Paris, France
A. Zeilinger	Vienna, Austria
H. Walther	Garching, Germany
H. Everitt	Washington, USA
R. Webb	Maryland, USA
A. Steane	Oxford, UK
Y. Yamamoto	Tokyo, Japan
C. Caves	Alberqueque, USA
W. Zurek	Los Alamos, USA
J. S. Tsai	Tsukuba, Japan
J. Lukens	New York, USA
I. Cheung	New York, USA
M. Roukes	Caltech, USA
M. Dykman	Michigan, USA
L. Sham	San Diego, USA
G. Schon	Karlsruhe, Germany

SESSION CHAIRPERSONS

Richard Hughes	Los Alamos, USA
Chris Hammel	Los Alamos, USA
Gerard Miburn	Brisbane, Australia
Daniel James	Los Alamos, USA
Birgitta Whaley	Berkeley, USA

David DiVincenzo....New York, USA
Alex Hamilton....Sydney, Australia
Michelle Simmons....Sydney, Australia
Robert Clark....Sydney, Australia
Halina Rubinsztein-Dunlop....Brisbane, Australia
Sam Braunstein....Bangor, UK
Don Parkin....Los Alamos, USA

INVITED SPEAKERS

Spintronics

D.D. Awschalom....Santa Barbara, USA
D. Loss....Basel, Switzerland
C.H.W. Barnes....Cambridge, UK

New Implantations

Y. Yamamoto....Stanford, USA
M.S. Shahriar....Cambridge, USA
M.I. Dykman....Michigan, USA

Semiconductor Quantum Dots

D.P. DiVincenzo....New York, USA
Y. Hirayama....Saitama, Japan
T. King....Santa Barbara, USA
X. Hu....Maryland, USA

Ion Trap I

D. Kielpinski....Boulder, USA
D.M. Lucas....Oxford, UK

Ion Trap II

R. Blatt....Innsbruck, Austria
R.J. Hughes....Los Alamos, USA

Superconducting Implementations DFS

J.S. Tsai....Tsukuba, Japan
S.L. Braunstein....Bangor, UK
J.P. Dowling....Los Angeles, USA
K. Berggren....Maryland, USA

Single Electron Devices

W.H. Huber....Maryland, USA
Y. Takahashi....Kanagawa, Japan
K. Schwab....Maryland, USA

Silicon-Based QC I, Arrays

K.L. Wang....Los Angeles, USA
A.S. Dzurak.... Sydney, Australia
J.R. Tucker....Illinois, USA

Silicon-Based QC II

P.C. Hammel....Los Alamos, USA
K.B. Whaley....Berkeley, USA

NMR, Industry Perspective, Nanofabrication

P.J. Kuekes....Palo Alto, USA
I. Chuang....San Jose, USA
D.M. Parkin....Los Alamos, USA

Optics/Cavity QED

H. Mabuchi....Los Angeles, USA
G. Milburn....Brisbane, Australia
H. Bachor....Canberra, Australia

Neutral Atoms, Algorithms

P.S. Jessen....Arizona, USA
S.L. Rolston....Maryland, USA
I.H. Deutsch....Albuquerque, USA

Group shot of IQC01 delegates on HMAS Vampire, Australian National Maritime Museum, Sydney Australia.

CONTENTS

4. Superconducting Implementations

5. Silicon Based Quantum Computing

6. Single Electron Devices

7. NMR Quantum Computing

8. Optics / Cavity QED

9. Neutral Atoms

10. DFS and Algorithms

11. Poster Presentations

QUANTUM COMPUTATION USING ELECTRONS TRAPPED BY SURFACE ACOUSTIC WAVES

C. H. W. BARNES, J. M. SHILTON AND A. M. ROBINSON

Cavendish Laboratory, University of Cambridge, Madingley Road, Cambridge CB3 0HE, United Kingdom

We outline a set of ideas for implementing a quantum processor based on technology used in surface acoustic wave (SAW) single-electron transport devices. These devices allow single electrons to be captured from a two-dimensional electron gas by a SAW. We discuss how these devices can be adapted to capture electrons in pure spin states and how both single and two-qubit gates can be constructed. We give designs for readout gates and discuss possible sources of error and decoherence.

1 Introduction

Recent experiments have shown that it is possible to capture a single electron in each minimum of a surface acoustic wave (SAW) as it passes from a two-dimensional electron gas (2DEG) into a quasi-one-dimensional channel (Q1DC) [1]. Our proposal for a quantum processor [2] is based on the assumption that by placing a series of N Q1DCs in parallel, a SAW can be made to capture N electrons in each of its minima, with one electron in each Q1DC. A single quantum computation in such a device would be performed by the N electrons in a single SAW minimum as they are dragged through a pattern of magnetic and non-magnetic surface gates, which carry out single and two-qubit operations. At the end of the computation the electrons would then be channeled into a set of N readout gates that would determine the final spin state of each electron. Each SAW minimum would contain an identical number of electrons prepared in an identical manner and would therefore perform the same computation. This would allow the readout gates to produce a measurable current representing the output of the computation. We discuss these ideas based on using electron spin to encode qubits.

2 Production of single and multiple qubits

2.1 Production of a quantised current using SAWs

Figure 1(a) shows a schematic diagram of a SAW-based quantised-current device [1]. It consists of a GaAs/AlGaAs heterostructure with NiCr/Al interdigitated transducers patterned either side of a central etch-defined mesa which contains a 2DEG. When a high frequency a.c. signal is applied to one of these transducers it produces a SAW, via the piezoelectric effect, which can be detected with the other transducer. As a SAW propagates across the mesa region, the traveling periodic electrostatic potential it produces drags electrons from the 2DEG along with it. For a typical SAW frequency of 3 GHz and an applied power of 10 dBm this produces a measurable current in the nanoAmp range. For the quantised current devices, the mesa is patterned so that it is split into two 2DEG regions (source and drain) connected by a narrow depleted Q1DC. The Q1DC can be formed either by surface Schottky gates, or by an etching technique [3] as shown in Fig 1(a). It has been shown experimentally [1] that over certain ranges of SAW power and gate voltage, the current which is produced

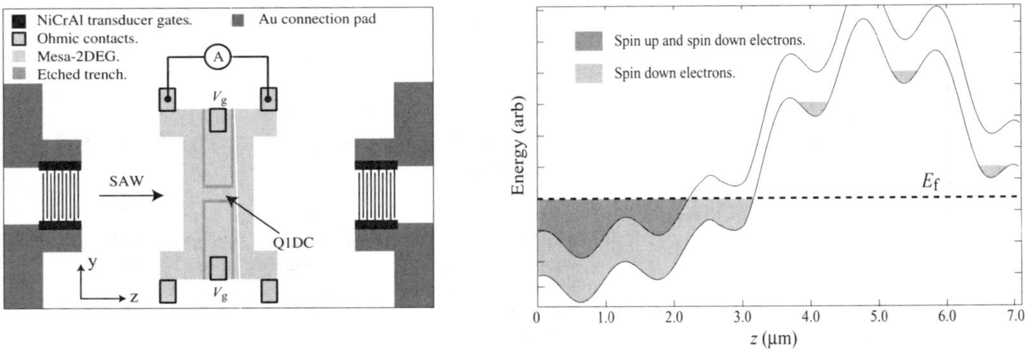

Figure 1. (a) Schematic diagram showing an experimental device which produces a quantised acoustoelectric current.(b) Schematic diagram showing the spin up and spin down occupations of the 1D SAW potential in a finite magnetic field.

by the SAW becomes quantised in units of ef, where e is the elementary electronic charge and f is the SAW frequency. The lowest quantised value observed represents the transport of a single electron from the source to the drain reservoir in each SAW minimum through the Q1DC. The electrons travel through the Q1DC with a mean speed equal to the SAW velocity, which is 2700ms^{-1} on the (001) plane of GaAs. Recent experimental results [4] have demonstrated this effect with an accuracy of five parts in 10^4.

2.2 Discussion of capture process

As a SAW minimum approaches the entrance to a single Q1DC, it narrows and forms into an isolated quantum dot. At the point that the dot forms, it will be occupied by many electrons. As it shrinks in passing through the entrance region to the Q1DC, electrons will be forced out. Electrons that are left in the dot are taken far from equilibrium with the source and drain reservoirs as they pass along the Q1DC. The distances and barrier heights between adjacent quantum dots are sufficiently large that the dots in the Q1DC can be considered to be independent as far as quantum-mechanical tunneling is concerned.

From a classical perspective, the electrons that are most likely to be captured and transported through the Q1DC by the SAW will be those that are least energetic in the rest frame of the SAW. Quantum mechanically, the SAW minima will contain electrons in localised states and the SAW would not be expected to eject these electrons in favour of capturing those in higher energy states. This is borne out by classical simulations and by evaluations of the 1D single-particle time-dependent Schrödinger equation [2].

For quantum computation it is necessary to provide electrons in pure states and this can be achieved by applying an external magnetic field to the device. The simple application of an external magnetic field to an N particle mixed state does not cause it to become a pure state unless it has time to relax through decoherence: the field merely precesses the spin of each electron. However, it has a significant effect on which electrons are captured from the 2DEG by the SAW. The application of an external magnetic field polarises the electrons in the source 2DEG so that the low energy electrons all have the same spin. The

Lande g-factor in GaAs is 0.44 and the amplitude of the SAW electrostatic potential is estimated from experiments to be approximately 1meV [1]. Therefore, the application of a magnetic field with strength $\sim$1T would ensure that all electrons captured would have the same spin. This is illustrated schematically in Fig. 1(b). Once the split-gate voltage is increased to the point where only one electron is captured per cycle these polorised electrons will be in pure spin states even though their orbital motion may be mixed. These electrons can therefore be used as spin qubits. In principle a large number of such qubits may be prepared simultaneously using this method by patterning large numbers of parallel Q1DCs with surface gates [5] or by etching.

3 Qubit operations

A quantum processor performs two kinds of operation on qubits. The first, the single-qubit operation, acts to rotate the spin of a single qubit about an arbitrary direction in space. The second, the two-qubit operation, entangles the spin wave functions of two adjacent qubits by altering the exchange coupling between them.

3.1 Single qubit gate

We can perform a single-qubit operation on a SAW spin qubit by having it pass through the magnetic field from a local static magnet. During the time a qubit is in the field, its wave function will evolve according to the Zeeman term in Schrödinger's equation. For a constant external magnetic field this evolution is a precession of the qubit spin about the direction of the field.

In principle, a local static magnet with a magnetic field that points in an arbitrary direction can be provided by a magnetic force microscope or by controlled ion beam deposition of a ferromagnetic material. However, for universal computation it is only necessary to make single qubit rotations about two independent axes and therefore it is sufficient to provide local magnetic fields that point in two orthogonal directions. As we have mentioned above, in order to provide pure states for the input, the device must be placed in an external field. This field may be used for one of the directions. The magnetic field for the other direction can be provided by a local magnetic split gate. If this is fabricated as a split ring, then stray fields will be reduced. A material such as permalloy could be used since it has a large surface magnetisation ($\sim$1T) and exhibits shape anisotropy which can be exploited in setting the direction of the magnetisation.

If the application of a global external magnetic field proved to be inconvenient for a particular processor design, the input magnetic field and both directions of magnetic field for single qubit operations could be provided by magnetic surface gates. A variable magnetic field could be provided by pulsing a current through a wire adjacent to a Q1DC.

3.2 Two qubit gate

A two-qubit operation can be achieved with surface patterning which causes two adjacent qubits, captured in the same SAW minimum, to be forced into tunnel contact. Suitable gate patterning is shown schematically as the left had side of Fig. 2(a). It consists of etched trenches defining an 'X shape' with a central metallic surface gate which splits this shape into two channels when a negative bias is applied to it. The tunneling coefficient between

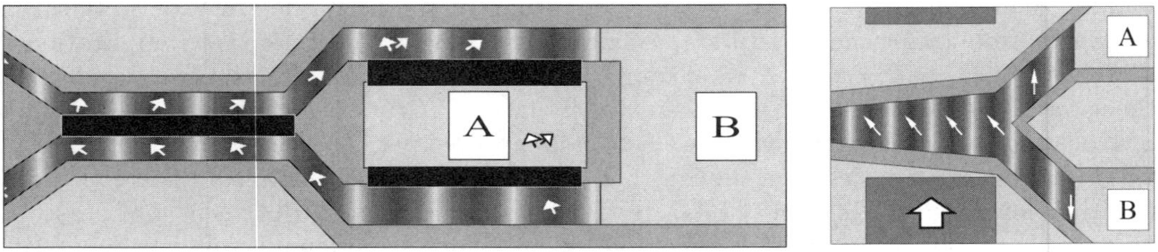

Figure 2. (a)Spin comparison readout gate. (b) Stern-Gerlach readout gate.

adjacent dots is fully controllable using the bias on the central surface gate. For a low bias, the tunneling coefficient is essentially unity, and at a large negative bias it will be zero. For a finite transmission coefficient the wave functions of the two electrons in the two dots entangle via tunneling and the Coulomb interaction and then decouple leaving just one electron in each quantum dot. This represents a two-particle unitary transformation on the initial qubit states of the trapped electrons which can be modelled as an exchange interaction [6,2]. By applying a suitable potential to the central metallic surface gate it is possible to perform a square root of swap operation [6]. This gate in combination with single-qubit operations allow universal quantum computation.

4 Readout

Many proposals for quantum processors rely on single electron transistors for reading the state of the output qubits [7]. Such techniques are also possible here and would be necessary if for a particular algorithm the output from individual computations was needed. With our device design though we can also construct readout gates which measure the ratio $\langle|\alpha|^2\rangle : \langle|\beta|^2\rangle$ for the qubits passing out of each Q1DC. Here α and β are the amplitudes for spin up and spin down and the brackets indicate an average of these qubit spin components over $\sim 3 \times 10^9$ quantum computations per second. Since each computation from each SAW minimum is nominally identical in our scheme, this ratio should be the same as that for the individual qubits exiting from each Q1DC. Such averaging may even compensate for decoherence and errors from random events in a given network. We present two ways for measuring these averages here.

4.1 Double occupation readout

Fig. 2(a) shows a readout gate which is an adaptation of the two-qubit gate. It operates by comparing a test qubit (upper channel) with an unknown qubit (lower channel).

In the central region there is a gold surface gate to produce an adjustable tunnel barrier between the upper and lower channels. This enables entanglement of incoming qubit wave functions. If the test qubit has the same spin orientation as the unknown qubit then no tunneling will occur between the two channels in the central region and the two incoming electrons will simply pass into ohmic contacts B. However, if they have opposite spin, tunneling can occur and there is a finite probability that the exiting dots will be doubly occupied either in the upper or lower channel. The gate is designed so that if this occurs one of these electrons can further tunnel across a narrow barrier to ohmic contact A. Thus the ratio of the currents passing into A and B will reflect the spin of the unknown electron relative to that of the known test electron.

4.2 Stern-Gerlach readout gate

Figure 2(b) shows a readout gate which is based on the Stern-Gerlach effect. The offset split magnetic gate produces a magnetic field which is spatially varying in the y-direction. Through the Zeeman term, this then produces an equal and opposite force F_{SG} on the different spin components of the incoming qubits so that the ratio of the currents flowing into ohmic contatcs A and B should be representative of their spin components along the y-direction.

Since electrons are charged particles, this readout gate must be designed to eliminate, or compensate for, the effect of the Lorentz force F_{Lor}. It is nominally significantly larger $F_{Lor}/F_{SG} = ev_{SAW}B/\frac{1}{2}g\mu_B\partial B_y/\partial y \sim 10^8 B/\partial B_y/\partial y$. The Lorentz force deriving from the magnetic field in the y-direction acts in the direction perpendicular to the 2DEG plane, the x-direction, and is therefore compensated for by the 2DEG quantum well confinement. The Lorentz force deriving from any resultant component of the magnetic field in the x-direction will produce a force in the y-direction and will therefore alter the relative currents into A and B. This could be compensated for by applying a d.c. bias V_{dc} between the two poles of the split magnetic gate to balance the force: $F = 0 = E + v_{SAW}B_R$, where $E = V_{dc}/s$ is the electric field between the pole pieces (separation s) and B_R is the resultant magnetic field in the x-direction. For a separation between pole pieces of s =5μm a resultant field of $B_R = 0.1$T could be compensated by a d.c. bias $V_{dc} = 1.5$mV. The device would need to be calibrated by comparing the currents for polarized and unpolarized currents.

5 Errors and decoherence

The SAW quantum-gate network consists of a set of static metal or etched gates. Errors in their patterning will lead to a quantum-gate network performing a different unitary transformation from the one intended. In the design we have outlined here, this problem is partially mitigated by the built-in variability of the gates: the current through a wire producing a magnetic field can be changed; the value of the external field can be changed;and the potential on two-qubit center gates can be changed. Compensation for static or weakly bound impurities present a more difficult challenge and it will be necessary to implement these devices in the best available MBE grown heterostructures.

Decoherence is not an intrinsic limiting factor to quantum computation since quantum error correction codes can be used to correct for its effects [8]. For our design, implementing these codes would simply result in the necessity to use more gates for the same computation. Semiconductor systems suffer from a large number of possible sources of decoherence. Any coupling between any line of N qubits and any other system of particles could reduce its coherence. The sources of decoherence for our system would include at least: coupling to other lines of N qubits in neighbouring SAW minima, and to other electrons, in surface gates, on donor impurities and in the adjacent 2DEGs; piezoelectric coupling to phonon modes; coupling to nuclear spins; and coupling to radio-frequency photons.

6 Summary

We have suggested a way in which SAW devices can produce a quantum processor. The idea involves the capture of electrons in pure spin states from a 2DEG by a SAW to form qubits. Single-qubit operations are performed using local magnetic fields produced by magnetic surface gate patterning. Two-qubit operations are performed by allowing exchange coupling between qubits through a controllable tunnel barrier. A quantum computation with one of these devices would consist of a single SAW minimum dragging a line of N qubits through a patterned array of single and t wo-qubit gates. Each SAW minimum would perform the same computation. A t wo schemes for reading out the final state of qubits have been proposed: double occupation in t wo-qubit gates; and the Stern-Gerlach effect. The major problem problem for this kind of implementation will be decoherence arising from coupling with: other electrons; radio frequency photons; phonons; and nuclear spins. However, these problems are generic to any semiconductor quantum computation proposal and the repetition (three billion time per second) which our implementation allows could be exploited to counteract these problems.

Acknowledgments

We are grateful to C. J. B. Ford, D. P. DiVincenzo, D. Loss, G. Milburn and G. Burkard for instructive discussions and encouragement in this work. CHWB acknowledges an Advanced Fellowship from the EPSRC.

References

1. J. M. Shilton, V. I. Talyanskii, M. Pepper, D. A. Ritchie, J. E. F. Frost, C. J. B. Ford, C. G. Smith and G. A. C. Jones, J. Phys.: Condens. Matter **8**, L531 (1996).
2. C. H. W. Barnes, J. M. Shilton and A. M. Robinson. Phys. Rev. B **62**, 8410 (2000).
3. A. Kristensen, J. B. Jensen, M. Zaffalon, C. B. Sorensen, S. M. Reimann, M. Michel and A. Forchel, J. Appl. Phys **83** 607 (1998).
4. J. Cunningham, V. I. Talyanskii, J. M. Shilton, M. Pepper, M. Y. Simmons, D. A. Ritchie Phys. Rev. B **60** 4850 (1999).
5. A. G. C. Haubrich, D. A. Wharam, H. Kriegelstein, S. Manus, A. Lorke, J. P. Kotthaus and A. C. Gossard, Appl. Phys. Lett. **70** 3251 (1997).
6. G. Burkard, D. Loss and D. P. Divincenzo, Phys. Rev. B **59** 2070 (1999).
7. A. Shnirman and G. Sch¨ onPhys. Rev. B **57** 15400 (1998).
8. P. W. Shor Phys. Rev. A **52** 2493 (1995).

CRYSTAL LATTICE QUANTUM COMPUTATION

YOSHIHISA YAMAMOTO*, THADDEUS DEWEY LADD, JONATHAN RUSSELL GOLDMAN, AYKUTLU DANA AND FUMIKO YAMAGUCHI

*Quantum Entanglement Project, ICORP, JST, Edward L. Ginzton Laboratory, Stanford University, Stanford, CA 94305-4085, USA *Also at NTT Basic Research Laboratories, Atsugishi, Kanagawa 243-0198, Japan*

A scalable, solid-state implementation of a nuclear spin quantum computer is proposed. Qubits are fluorine nuclear spins in a solid crystal of fluorapatite [$Ca_5F(PO_4)_3$], in which the spin resonant frequencies are separated by a large magnetic field gradient. Quantum logic is accomplished using nuclear-nuclear dipolar couplings with decoupling and selective recoupling RF pulse sequences. Magnetic resonance force microscopy is used for readout. This proposal takes advantage of many of the successful aspects of solution NMR quantum computation, including ensemble measurement and long T_1, but it allows for more qubits and the potential for initialization.

1 Introduction

An increasing number of theoretical developments have recently motivated the construction of a quantum computer with a large number of qubits. The potential for efficient simulation of quantum systems, [1] the discoveries of algorithms for fast factorization [2] and database searching, [3] and the development of quantum error correction [4] have triggered a large number of proposals for experimental implementations of a quantum computer. However, any physical implementation of a quantum computer must battle the fact that well-isolated quantum systems are difficult to couple and to measure, whereas the introduction of necessary couplings and probes leads to the devastating effects of decoherence.

To date, the most successful experimental realization of a multi-qubit quantum computer has been in room-temperature solution NMR. [5] Here, the spin-states of molecular nuclei in a solution are well isolated from an environment, as demonstrated by long thermal relaxation times (T_1) of many seconds. The nuclei of each molecule are weakly coupled by scalar couplings. Measurement without substantial decoherence is made possible by the weak coupling between the nuclear spins and the pick-up coil, which however requires a large ensemble of $\sim 10^{22}$ uncoupled, identical molecules in the solution to achieve a reasonable signal-to-noise ratio. There is much debate, however, as to whether solution NMR currently does or ever will exhibit the signatures of truly *quantum* information processing. [6] The presence of a large, mixed ensemble renders quantum entanglement and wave-function collapse of individual spins unmeasurable. Certain manipulations allow the processing of an "effective pure state" in which the ensemble behaves nearly identically to a single quantum system. [5,7] However, it has been shown that the evolution of effective pure states of existing solution NMR computers can be described without any entanglement, and either more qubits or much larger nuclear polarizations will be needed before entanglement is demonstrable. [8] Even if we neglect the debate of what is "quantum" and what is not, the principal limitation due to the small polarizations in solution NMR quantum computation is that the usable signal decreases exponentially with the number of qubits, leaving this method unlikely to exceed the 10 qubit level without extensive modification. [9]

In this paper, we propose an implementation using solid-state NMR. In a solid crystal

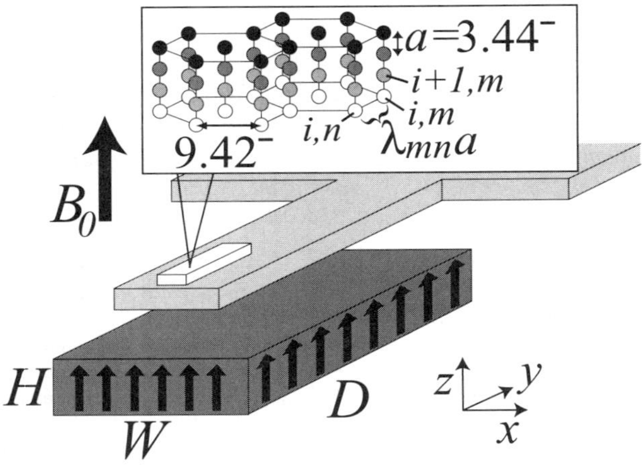

Figure 1. The fluorapatite crystal, of length 10 μm and width roughly 1 μm, is mounted near the end of a cantilever. A dysprosium micromagnet with dimensions $W = 10$ μm, $H = 4$ μm, $D = 400$ μm is aligned parallel to the crystal planes. A field gradient of 1.4 T/μm is produced at the crystal, a distance of 2.07 μm above the magnet. The insert shows the ^{19}F sites of the crystal structure. The different shades indicate different resonant frequencies.

there exists the potential for initialization of the nuclear spins by simple cooling of the lattice to extremely low temperatures or transferred optical pumping. [10] The thermal relaxation time T_1 can be dramatically lengthened when moving from a solution to an insulating, nonmagnetic crystal, and the nuclear-nuclear dipole coupling can be faster. Hence, a solid crystal has great potential for quantum computation in terms of timescales and scalability. [11]

2 Basic Scheme

These advantages, however, come at a cost: several difficult and related problems arise in implementing a solid-crystal NMR computer. One problem is the complicated network of dipolar couplings in a solid, [12] which must be suppressed to increase a T_2 decoherence time but selectively retained for logic. We address this problem with both "selective averaging" radio frequency (RF) pulse sequences [13] and a well-chosen crystal structure. [14] Another problem arises from the need to distinguish and detect nuclei in the periodic lattice of a crystal; for this we establish a very large, static, one-dimensional magnetic field gradient with a microfabricated, high-magnetization ferromagnet. [15] Only a small ensemble of nuclei may fit into the area over which the gradient field is homogeneous, and therefore a sensitive means than standard inductive pick-up is needed to measure the nuclear spin states. We propose using magnetic resonance force microscopy (MRFM). [14]

A schematic for the solid-crystal NMR quantum computer we propose appears in Fig. 1. The quantum computer is an ensemble of N one-dimensional chains of n spin-1/2 nuclei. Due to the field gradient, the resonant frequencies of the nuclear spins within a chain are different. Hence the secular component of the dipolar Hamiltonian which couples the ith

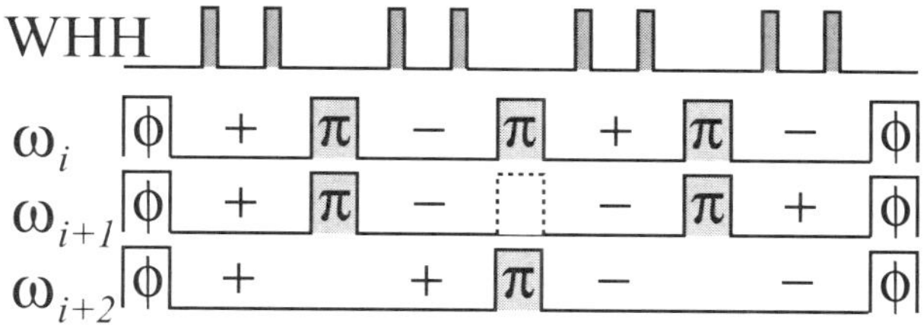

Figure 2. An example RF pulse sequence. The first line (WHH) shows broadband decoupling of qubit copies; the narrow, shaded pulses are appropriately phased $\pi/2$ pulses, [17] made sufficiently narrow (broadband) to affect all nuclei in the crystal. The subsequent three lines show selective pulses to decouple qubits; these wider pulses are sufficiently long (narrowband) to affect only individual planes. Single spin rotations, denoted by ϕ, may be performed in between cycles. Multi-bit gates such as controlled-NOT may be performed by combining single spin rotations with selective recoupling. For example, qubits i and $i+1$ may be recoupled by adding a π pulse at the time noted by the dotted line. [13]

spin to the jth spin within the same mth chain is written [16]

$$\begin{aligned}\hat{\mathcal{H}}_{ijmm} &= \frac{\mu_0}{4\pi}\gamma^2\hbar^2\frac{1-3\cos^2\phi}{[|j-i|a]^3}\hat{I}^z_{im}\hat{I}^z_{jm}\\ &\equiv \hbar\delta\omega_{ij}\hat{I}^z_{im}\hat{I}^z_{jm},\end{aligned} \tag{1}$$

where γ is the gyromagnetic ratio ($2\pi \times 40$ MHz/T for ^{19}F), a is the distance between spins in a chain, and ϕ is the angle between the chain of spins and the large applied magnetic field, which is taken to be in the z-direction. Other terms of the dipolar Hamiltonian average to zero on a timescale of $1/\Delta\omega_{ij}$ or faster, where $\Delta\omega_{ij} = \gamma a|\nabla B^z_{ij}|$ is the separation of the ith and jth resonant frequencies caused by the field gradient ∇B^z_{ij} between them. The Hamiltonian of Eq. (1) may be "switched off" by applying a periodic succession of narrow band π pulses at, for instance, the ith resonant frequency. [16] Simultaneous decoupling of more than two qubits may be accomplished by timing the selective π pulses according to the entries of an appropriately sized Hadamard matrix; a pair of qubits may be selectively recoupled in order to implement two-bit gates. [13]

Weak and undisturbing measurement is possible in this scheme because, orthogonal to the chain direction, a nucleus of resonant frequency ω_i has a large plane of copies with equal frequency. The coupling between copies has a different form; the nth and mth nucleus of identical resonant frequency ω_i are coupled by the Hamiltonian [16]

$$\hat{\mathcal{H}}_{iimn} = \frac{1}{2}\frac{\mu_0}{4\pi}\gamma^2\hbar^2\frac{1-3\cos^2\theta_{mn}}{(a\lambda_{mn})^3}\left(3\hat{I}^z_{im}\hat{I}^z_{in} - \hat{\mathbf{I}}_{im}\cdot\hat{\mathbf{I}}_{in}\right), \tag{2}$$

where λ_{mn} is the distance between the two nuclei in units of a, and θ_{mn} is the angle between the vector which connects them and the direction of the applied field. This coupling between copies must be "switched off" all the time to prevent interference between ensemble members. Decoupling these spins may be done by any number of pulse sequences which have been developed over the past 40 years in solid-state NMR, the simplest of which is the WAHUHA pulse sequence. Such sequences can reduce dipolar broadening by more than three orders of magnitude. [17]

Fortunately, the different forms of $\hat{\mathcal{H}}_{ijmm}$ and $\hat{\mathcal{H}}_{iimn}$ allow constant decoupling of copies without adverse effect on the manipulations of couplings within individual computers (one-dimensional chain). The qubit-qubit coupling of Eq. (1) may in general be rescaled and rotated by the broadband WAHUHA-type sequence, but it may still be controlled for logic gates with suitably phased π pulses. The broadband WAHUHA-type pulse sequence which decouples all copies is unaffected by the addition of selective π-pulses, insofar as those π-pulses are sufficiently short in comparison to the pulse spacing. This criterion slows a computation process, however, since the π pulses must last a time of about $1/\Delta\omega_{i,i+1}$ to be frequency selective. A simple example to illustrate the decoupling scheme is shown in Fig. 2. This example is limited by the adverse effect of the finite π pulse-length on the WAHUHA decoupling, but more sophisticated sequences can be designed with standard techniques to compensate for this and other limiting effects. [17]

3 Quasi-One-Dimensional Crystal

The only problematic dipolar couplings which remain after applying the pulse sequences described above are those between qubits with different resonant frequencies ($i \neq j$) and in different chains ($m \neq n$). These couplings only arise when two planes of qubits are selectively recoupled during a two-bit gate operation. The needed time duration for a logic operation may be minimized by maximizing the desired qubit-qubit coupling – this is accomplished by first letting a be the smallest nuclear distance in the crystal. Second, we set $\phi = 0$, hence putting the gradient parallel to the applied field. Third, only nearest-neighbor qubit couplings are used ($j = i + 1$), with longer distance couplings accomplished by bit-swapping. [18] After making these choices, we have $\delta\omega \equiv \delta\omega_{i,i+1} = -\mu_0\gamma^2\hbar/2\pi a^3$, and the target qubit of frequency ω_i in the mth chain is coupled to control qubit copies at ω_{i+1} in the other chains by the Hamiltonian

$$\begin{aligned}\hat{\mathcal{H}}^*_{im} &= -\hbar\delta\omega \sum_n \frac{\lambda^2_{mn} - 2}{2(1+\lambda^2_{mn})^{5/2}} \hat{I}^z_{im}\hat{I}^z_{i+1,n} \\ &\equiv -\hbar\delta\omega \sum_n b_{mn}\hat{I}^z_{im}\hat{I}^z_{i+1,n}. \end{aligned} \tag{3}$$

The decoherence caused by this undesired Hamiltonian is minimized if the coupling constants are sufficiently small to obey

$$\frac{\sigma}{\delta\omega} = \frac{1}{2}\sqrt{\sum_n b^2_{mn}} \ll 1, \tag{4}$$

where σ (the square root of the second moment [17]) is the effective linewidth of the qubit ensemble during recoupling, which we demand to be much smaller than the dipolar coupling $\delta\omega$ use for a logic operation. This inequality is best satisfied by a crystal whose nuclei couple only in isolated chains.

A crystal which approximates this condition is fluorapatite, $Ca_5F(PO_4)_3$, whose ^{19}F structure is shown in Fig. 1. In fluorapatite, we find $\sigma/\delta\omega \approx 1/58$, roughly six times smaller than in a simple cubic crystal such as CaF_2. The one-dimensional nuclear structure of fluorapatite has been recognized in several NMR experiments. [19] Decoherence timescales in this crystal are analagous to the well-known case of CaF_2. The T_2 time is limited by

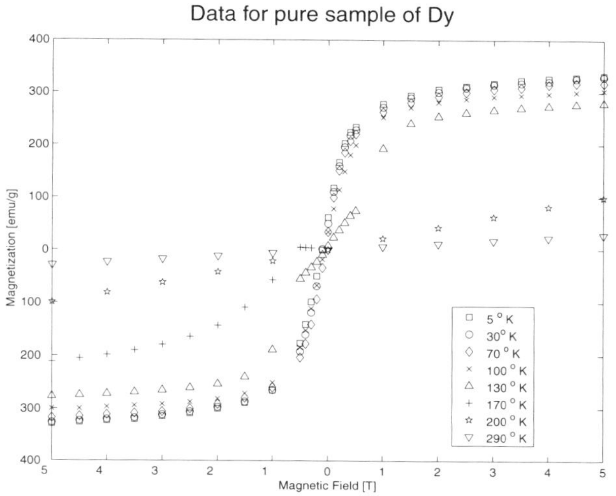

Figure 3. The magnetization c haracteristicsof a Dy micro-magnet.

dipolar broadening; if dipolar couplings are perfectly controlled, the timescale for internal decoherence is pushed toward the spin-lattice relaxation time constant T_1, which is limited by thermal fluctuations of paramagnetic impurities (rare-earth substitutions, for example) and can easily be several hours for reasonable crystal purities and temperatures. [20]

4 Micro-Magnet and Magnetic Resonance Force Microscope

The field gradient is accomplished by a micron-sized ferromagnetic parallelopiped. Calculations similar to those in Ref. [15] show that such a magnet, made of dysprosium and placed in a 7 T external field as shown in Fig. 1, can produce a field gradient of 1.4 T/μm at a distance of 2.07 μm above the magnet, which leads to $\Delta\omega \equiv \Delta\omega_{i,i+1} = 2\pi \times 19.2$ kHz. A 1 μm by 10 μm fluorapatite crystal above the magnet contains $N = 10^7$ equivalent-frequency qubit copies in each xy-plane. The field variation along the y-direction is negligible since the magnet is much longer than the crystal. The field variation along the x-axis is sufficiently small that all of the equivalent-frequency nuclei lie within a bandwidth much smaller than $\Delta\omega$. Figure 3 shows the magnetization characteristics of a Dy micromagnet fabricated on a Si substrate. We also note that inhomogeneous broadening is constantly refocused in this scheme by the narrow-band π pulse sequences during both decoupling and selective recoupling.

The presence of a large magnetic field gradient provides a natural means for performing MRFM on a magnetization M^z, since this technique is sensitive to the gradient force given by $F = M^z \nabla B^z$. [21] The crystal is mounted on a microfabricated cantilever which oscillates in the z-direction, as in Fig. 1. The long axis of the crystal-cantilever heterostructure and magnet are aligned. The experiment is performed in high vacuum ($< 10^{-5}$ torr) and at low temperatures. A coil is used to generate the RF pulses for logic operations and decoupling sequences; it also generates the continuous-wave radiation for readout. An optical

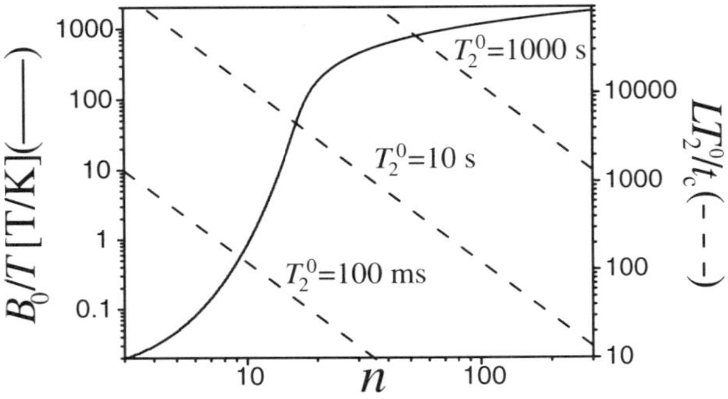

Figure 4. A plot showing the scalability of the present scheme. The solid curve, corresponding to the left axis, shows the magnetic field (in T) divided by temperature (in K) needed in order for a number of qubits n to be measurable. The dashed lines, corresponding to the right axis, plot the number of logic gates (decoherence time T_2 divided by the pulse sequence cycle time t_c) times the length of a "block" of the decoupling sequence L against n for several values of T_2^0.

fiber-based displacement sensor is used to monitor deflection of the cantilever using interferometry. Sub-Ångstrom oscillations can be detected; larger oscillations can be damped with active feedback which avoids additional broadening while maintaining high sensitivity. [22]

A calculation using Redfield theory [16] with the thermal noise statistics of a high-sensitivity cantilever [23] estimates the T_2 timescale due to the cantilever's thermal drift as $T_2 = k\omega_c Q a^2/\Delta\omega^2 k_B T \approx 200$ ms, where $k \approx 10^{-3}$ N/m is the spring constant, $\omega_c/2\pi \approx 10$ kHz is the resonant frequency, and $Q \approx 20,000$ is the quality factor of the cantilever. Active feedback stabilization is expected to increase this timescale by as many as four orders of magnitude, bringing it close to the lengthy T_1 timescale.

Readout is performed using cyclic adiabatic inversion, [24] which modulates the magnetization of a plane of nuclei at a frequency near or on resonance with the cantilever. The spins of resonant frequency ω_i are irradiated with the RF field $B^x = 2B_1 \cos\{\omega_i t - (\Omega/\omega_m)\cos(\omega_m t)\}$, where ω_m is the modulation frequency chosen to be near the resonance of the cantilever, and Ω is the frequency excursion, which should be much smaller than $\Delta\omega$. [25] Simultaneous detection of signal from multiple planes is possible if the different planes to be measured are driven at distinct modulation frequencies ω_m.

The force resolution for MRFM is limited by thermal fluctuations of the cantilever. [23] Force resolutions of 5.6×10^{-18} N/$\sqrt{\text{Hz}}$ have been reported for single crystal cantilevers at 4 K. [26] To estimate the force involved in a measurement following a quantum computation, we consider the case in which initialization is imperfect, so that the temperature T is non-zero and an effective pure state must be used. [5,7] The usable magnetization is estimated [9] as

$$M^z = \gamma\hbar \frac{N}{2^n} \frac{\sinh(n\hbar\gamma B_0/2k_B T)}{\cosh^n(\hbar\gamma B_0/2k_B T)}. \tag{5}$$

In the high temperature limit, this magnetization approximates to $(\gamma^2\hbar^2 B_0/2k_B T)Nn2^{-n}$ showing the exponential downscaling which plagues solution NMR. Even in this limit, with the relatively small N of 10^7 and using the field gradient described above in a 7 T field at 4 K, the force which needs to be measured works out to about $10^{-15}n2^{-n}$ newtons, allowing $n \sim 10$ qubits. As seen in Fig. 4, the situation is improved in larger fields and at lower temperatures, where M^z improves exponentially in B_0/T. In the realistic laboratory extreme $B_0 = 20$ T and $T = 10$ mK, this design may scale to $n \sim 300$ qubits, even with a standard MRFM. [26] The transferred optical pumping NMR technique [10] should achieve a larger number of qubits in a standard experimental condition.

5 Scalability

A primary limitation to the long-term scalability of this scheme is the number of logic operations which can fit into the decoherence time when there is a large number of qubits. The decoupling/recoupling pulse scheme based on Hadamard matrices has a cycle time $t_c(n)$ of $Ln^2/\Delta\omega$, where the parameter L depends on the length of the homonuclear decoupling subsequence and only weakly on the number of qubits n. [13] Since single spin rotations must occur between cycles, the clock speed is set by this cycle time. The number of logic operations we may perform is approximately $t_c(n)$ divided into the remaining decoherence time T_2^0. This timescale T_2^0 is due to residual homonuclear couplings, cantilever thermal drift, and thermal relaxation resulting from magnetic impurities; these contributions are limited by the experimental performance of the homonuclear decoupling pulse sequence, cantilever feedback stabilization, and crystal growth technique, respectively. Figure 4 shows the number of logic operations possible for several values of T_2^0. The bottom dashed trace ($T_2^0 = 100$ ms) corresponds to the case in which homonuclear couplings are perfectly controlled but cantilever drift is completely unsuppressed, while the middle ($T_2^0 = 10$ s) and top ($T_2^0 = 1000$ s) traces assume suppression by 20 and 40 dB with a negative feedback circuit. The limited number of logic operations at high numbers of qubits is a remaining problem which may be addressed through error correcting codes, [4] more sophisticated pulse sequences, or the use of a more one-dimensional system.

In a perfectly one-dimensional system with high n, the Hadamard decoupling scheme may be truncated to decouple sets of only $n_d < n$ nearest neighbors and then repeated periodically along a chain. The remaining couplings result in decoherence with a timescale T_2^{qq} which scales as n_d^3, and since $t_c \sim n_d^2$, T_2/t_c may be made arbitrarily large with large enough T_2^0.

6 Conclusion

A remaining challenge in this proposed scheme is the alignment of the field gradient, cantilever, and crystal. Such alignment is critical, and although difficult, it is not impossible with modern microfabrication technology and feedback control using micro-electro-mechanical actuators [27]. Figure 5 shows a fabricated MEMS device to mount a micro-magnet.

This proposal features two important advantages over schemes using solid-state NMR with impurity dopants [28]. The fabrication difficulty of artificially implanting controlled arrays of spins is avoided in our proposal by using nuclei that are naturally organized into a crystal structure. Also, the use of an ensemble of 10^7 spins avoids the need for the daunting task of reading and initializing single nuclear spins. Indeed, ensemble measurement has given solution NMR a developmental head-start against other existing proposals for quantum

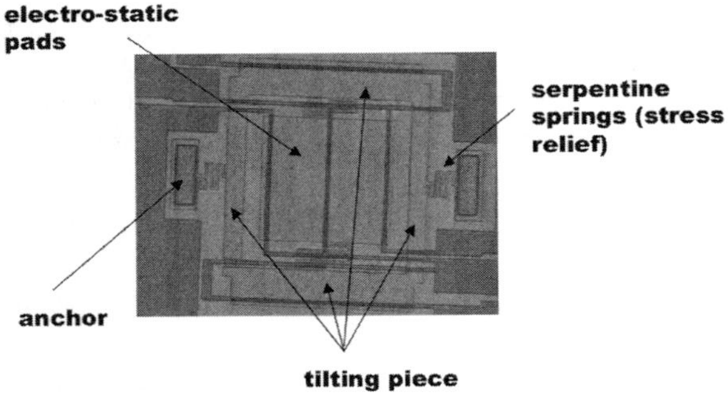

Figure 5. A fabricated Si micro-electro-mechanical structure for mounting a Dy micro-magnet.

computation, since a measurement device does not introduce a devastating decoherence effect in such a weak measurement. It is hoped that the scheme presented here will carry NMR quantum computation onward to the many-qubit regime.

References

1. R. P. Feynman, Int. J. Theor. Phys. **21**, 467 (1982); S. Lloyd, Science **273**, 1073 (1996).
2. P. W. Shor, in Proceedings of the 35th Symposium on the Foundations of Computer Science (IEEE Computer Society Press, Los Alamitos, 1994), p. 124.
3. L. K. Grover, Phys. Rev. Lett. **79** 325 (1997).
4. P. W. Shor, Phys. Rev. A **52**, R2493 (1995).
5. N. A. Gershenfeld and I. Chuang, Science **275**, 350 (1997); D. G. Cory, A. F. Fahmy, and T. F. Havel, Proc. Natl. Acad. Sci. USA **94**, 1634 (1997).
6. R. Schack and C. M. Caves, Phys. Rev. A **60**, 4354 (1999).
7. E. Knill, I. Chuang, and R. Laflamme, Phys. Rev. A **57**, 3348 (1998).
8. S. L. Braunstein *et al*, PRL **83** 1054 (1999).
9. W. S. Warren, Science **277**, 1688 (1997).
10. R. Tycko, Solid State Nuclear Magnetic Resonance **11**, 1 (1998).
11. F. Yamaguchi and Y. Yamamoto, Appl. Phys. A **68**, 1 (1999).
12. T. D. Ladd, J. R. Goldman, F. Yamaguchi, and Y. Yamamoto, Appl. Phys. A **71**, 27 (2000).
13. D. W. Leung, I. L. Chuang, F. Yamaguchi, and Y. Yamamoto, Phys. Rev. A **61**, 042310 (1999).
14. T. D. Ladd, J. R. Goldman, A. Dana, F. Yamaguchi, and Y. Yamamoto, quant-ph/000912 (2000).
15. J. R. Goldman, T. D. Ladd, F. Yamaguchi, and Y. Yamamoto, Appl. Phys. A **71**, 11 (2000).
16. C. P. Slichter, *Principles of Magnetic Resonance* 3rd Ed. (Springer, Berlin, 1990).

17. U. Haeberlin, *High Resolution NMR in Solids: Selective Averaging* (Academic Press, New York, 1976); M. Mehring, *High Resolution NMR in Solids* (Springer-Verlag, Berlin, 1983).
18. S. Lloyd, *Science* **261**, 1569 (1993).
19. W. Van Der Lugt and W. J. Caspers, Physica **30**, 1658 (1964); M. Engelsberg, I. J. Lowe, and J. L. Carolan, Phys. Lett. B **7** 924 (1973); G. Cho and J. P. Yesinowski, J. Phys. Chem. **100**, 15716 (1996).
20. N. Bloembergen, Physica **15**, 386 (1949).
21. J. A. Sidles, Appl. Phys. Lett. **58**, 2854 (1991); D. Rugar, C. S. Yannoni, and J. A. Sidles, Nature **360**, 563 (1992).
22. U. Dürig, H. R. Steinauer, and N. Blanc, J. Appl. Phys. **8**, 3641 (1997).
23. T. B. Gabrielson, IEEE Trans. Electron Devices **40:5**, 903 (1993).
24. A. Abragam, Principles of Nuclear Magnetism (Oxford University Press, 1961).
25. D. Rugar *et al*, Science **264**, 1560 (1994).
26. T. D. Stowe *et al.*, Appl. Phys. Lett. **71**, 288 (1997).
27. S. A. Miller, K. L. Turner, and N. C. MacDonald, Rev. Sci. Instrum. **68**, 4155 (1997).
28. B. E. Kane, Nature **393**, 133 (1998); G. P. Berman, G. D. Doolen, P. C. Hammel, and V. I. Tsifrinovich, Phys. Lett. B **61** 14694 (2000).

QUANTUM COMPUTING IN DIAMOND

M.S. SHAHRIAR, P.R. HEMMER, S. LLOYD, AND A. TURUKHIN

Massachusetts Institute of Technology, Cambridge, MA 02139

Our approach for quantum computing is to use spectral hole-burning (SHB) materials for which a large number of individual atoms can be selectively addressed using lasers. However, the qubits are the spins of individual active atoms, which are known to have long coherence lifetimes in these materials. Qubit initialization and single-qubit manipulations can be performed using laser excited Raman transitions. Because of the large optical Rabi frequencies, we anticipate being able to perform >1000 single-qubit operations within the coherence lifetime. Two-qubit, nonlinear operations will be performed using either cavity QED or optical dipole coupling in conjunction with laser excited Raman transitions. The most promising candidate material is NV-Diamond, for which the optical matrix elements are large, thereby enabling fast qubit manipulation and efficient detection of the final state. Here, we present the details of ours studies of single qubits in this medium. Specifically, Raman excited spin coherences were experimentally observed in N-V diamond color centers via nondegenerate four-wave mixing (NDFWM) and electromagnetically induced transparency (EIT). The maximal absorption suppression was found to be 17%, which corresponds to 70% of what is possible given the random geometric orientation of the N-V center in diamond. In the context of quantum computing, this level of transparency represents the efficient preparation of quantum bits (qubits), as well as ability to perform arbitrary single qubit rotations.

A quantum computer that stores information on two-state systems called quantum bits or qubits must be able to address and manipulate individual qubits, to effect coherent interactions between pairs of qubits, and to read out the value of qubits. Current methods for addressing qubits are divided up into spatial methods, as when a laser beam is focused on an individual qubit or spectral methods, as when a nuclear spin in a molecule is addressed using NMR. The density of qubits addressable spatially is limited by the wavelength of light, and the number of qubits addressable spectrally is limited by spin linewidths. We have developed a method for addressing qubits using a technique that combines spatial and spectral selectivity. The result is a design for quantum computation that provides the potential for a density of quantum information storage and processing many orders of magnitude greater than that afforded by ion traps or NMR. Specifically, this method uses an ensemble of spectrally resolved atoms in a spectral holeburning solid such as NV-diamond.

The basic concept is illustrated by considering a small volume element of a crystal containing a set of impurity atoms. Each atom sees a unique surrounding, so that the resonance frequency for a given transition is different for different atoms. The number of spectrally resolvable bands, N_R, is determined by the ratio of the spectral spread to the width of the individual resonance. We consider a situation where the number of atoms in the selected volume is less than N_R, so that each atom can be addressed individually. For the basic model, we can consider each atom to have two metastable states, representing the spin up and down, for example, coupled to an excited state via optical frequency transitions; this is often called a Λ transition. The energy splitting (ε) between the metastable states establishes the frequency spacing between the sets of atoms that can be coupled to one another via nearest neighbor interactions. Note that in this model two atoms are considered to be spectrally adjacent when the lower frequency optical transition of one is degenerate with the higher frequency optical transition of the other.

The first criterion in making a quantum computer is the ability to identify distinct quantum bits. In order to realize this objective, we must have at most one atom per channel. This can be achieved in NV-diamond by making use of a storage level where all but one atom from each channel can be transferred to for a long time (hours).

The second criterion is the ability to prepare the quantum bits in a pure (0 or 1) state. At the onset, the two metastable states are both occupied. For the case of NMR at room temperature, this implies that the number of spins that are in a pure state is very small. For other systems such as phosphorous in quantum dots, this problem requires that the quantum computer be operated at temperatures in the millikelvin regime. In our model, near-perfect alignment can be produced at much higher temperatures (at least up to 15 Kelvin, accessible by compressor based closed-cycle cryostats), via two-photon induced coherent population trapping, or the dark resonance. The third criterion is the ability perform single bit operations. This can be also achieved by using two-photon induced spin transitions, where the frequency difference is matched to ε, and the optical frequencies are matched to the desired quantum bit. The fourth criterion is that there must be a way to load quantum information in to the system, and a way to retrieve quantum information from the system. In this model, the two laser photons used to excite the qubit can be encoded (in the form of the amplitude and phase of the beat between the two frequencies) with the quantum information during the input process. Similarly, the response of the qubits to a given state of the two photons can be used to read out the qubits. The fifth criterion is that the time needed to perform a single operation should be much less than the dephasing time of the qubits. In NV-diamond, the dephasing time is about 0.1msec, while the time for a single operation is about 100 nsec. The dephasing time can be reduced further by (a) using samples that are free from ^{13}C, and (b) using techniques of dynamic noise suppression developed by us. The sixth criterion is that in order for quantum error correction codes to work properly, the system should be amenable to parallel operations. In diamond it is possible to conduct quantum computing in up to 100 parallel channels at a time, under conditions where the total number of channels is about 300. Finally, the most important criterion is that one must be able to realize a CNOT operation between two nearest-neighbor qubits. We have identified in explicit detail two different methods for achieving this objective in NV-diamond. The first method, applicable to high-density of color centers, uses the direct optical dipole-dipole coupling between two qubits that are very close to each other spatially, and can be turned into spectral neighbors via applying a magnetic field. This method is potentially easier to realize, and can be employed to realize close to 100 coupled qubits for realistic parameters. The second method, applicable to low-density of color centers, uses a high-finesse optical cavity, resonant with a transition common to both bits, to enhance the optical dipole-dipole coupling. This method is potentially more difficult to realize, but can yield more than 300 coupled qubits. Furthermore, this approach is more amenable to parallel operations.

So far, we have demonstrated some of the key features necessary for this scheme to work. Specifically, we have (i) identified NV-diamond as the medium of choice, with necessary energy levels, and a favorable ratio of the time for single operation (100 nsec) and the decoherence time (0.1 msec), enabling up to 10^3 operations (ii) determined explicit models for realizing a CNOT using cavity-induced as well as direct dipole-dipole coupling, (iii) demonstrated near-perfect state preparation of qubits[23], (iv) demonstrated single bit operations on a collection of qubits, and (v) developed theoretical methods for suppressing spin dephasing, which may lead to significant reduction of the decoherence time. Here, we present the details of ours studies of single qubits in NV-diamond. In addition to quantum computing, this work may be useful in other fields such as

The use of optical Raman interactions to excite spin coherences in solid materials has numerous potential applications, ranging from low-power nonlinear optics[1] to high-temperature spectral hole burning memories[2] to solid-state quantum computing[3]. The interest in Raman excitation lies in the fact that the spin coherences can be efficiently excited and manipulated using optical laser fields yet are weakly coupled to the environment and hence have the long coherence lifetimes needed for optical memories and quantum computing. Recently, we performed proof-of-principle experiments in Pr doped Y_2SiO_5 (Pr:YSO) that demonstrated the potential advantage of Raman excited spin coherences for optical storage[4]. However, Pr:YSO has a weak optical oscillator strength ($\sim 10^{-7}$), as do many other spectral hole burning materials. This weak oscillator strength limits many potential applications because the ability to achieve efficient Raman excitation of a spin coherence depends on the product of the oscillator strength and the laser intensity. For example, for high temperature SHB applications, efficient Raman excitation becomes more difficult as the optical homogeneous width increases with temperature. Hence, for a fixed laser power, higher oscillator strength permits higher temperature operation. For quantum computing, the low optical transition rate limits the gate speed and the number of quantum logic operations that can be performed within the spin coherence lifetime[4].

For the current experiment we chose the N-V color centers in diamond because it has a large optical oscillator strength (~0.1), relatively long spin coherence lifetimes (1-100μs range), and has been previously demonstrated to exhibit Raman heterodyne signals[6]. The nitrogen-vacancy (N-V) color center in diamond is created by radiation damage and annealing of Ib diamond, and is described elsewhere[7]. The sample we used is estimated to have about 30 ppm N-V color centers and has an peak optical density about 0.6 (25% transmission) for a 1 W/cm^2 probe intensity at 15 K, on the zero-phonon line, near 637 nm optical wavelength. The optical density is much higher for weaker probe beam. The optical transition has an inhomogeneous broadening of 750 GHz, and a homogeneous width of 50 MHz. Previous work has shown that an applied magnetic field along (111) direction is required to observe Raman heterodyne signals[6], and therefore is needed for Raman excited spin coherences. Figure 1 shows an energy level diagram of the N-V center. The Raman transition frequency (~120 MHz) is determined by the spacing between the S=0 and S=-1 ground state spin sublevels. This spacing is controlled by the magnitude of the applied ~1 kGauss magnetic field. At this field strength, the S=0 and S=-1 ground sublevels for (111) oriented N-V centers are near an anti-crossing (Figure 1a). Under these conditions, a partial mixing of the spin sublevels makes it possible to enhance the Raman transition strength which otherwise would be small due to the small spin-orbit coupling[8].

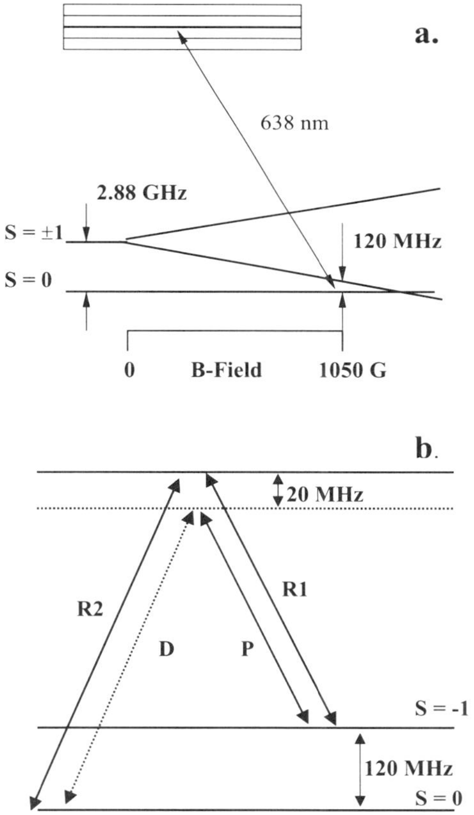

Figure 1. (a) Energy level splitting of N-V diamond as a function of magnetic field strength applyied in the (111) direction. (b) A Λ–shaped three-level system interacting with Raman beams and probe to generate a NDFWM signal at a magnetic field strength of about 1000G.

The laser beams R1 and R2 in Figure 1b act as Raman pump beams to produce a two-photon ground state coherence via coherent population trapping[9]. In some of the experiments, a Raman enhanced non-degenerate four-wave mixing (NDFWM) technique is used to achieve a higher signal to noise ratio, in analogy to experimental techniques used previously to study Pr:YSO[4]. In this case, the laser beam P acts as a probe beam, which is diffracted from a phase grating produced by the two-photon coherence, and generates a diffracted beam D according to phase matching conditions $\mathbf{k}_D=\mathbf{k}_{R2}-\mathbf{k}_{R1}+\mathbf{k}_P$. To further enhance signal to noise, a heterodyne detection scheme is sometimes used to detect beam D. All the laser fields shown are derived from a single dye laser output using acousto-optic frequency shifters. This greatly relaxes dye laser frequency stability requirements since the resonant Raman interaction is insensitive to correlated laser jitter.

To match Figure 1, the Raman beams R1, R2, and probe beam P are downshifted from the original laser frequency 400, 280, and 420 MHz respectively. To generate lineshapes, the frequency of beam R2 was scanned around the 120 MHz Raman transition frequency with the frequencies of beams R1 and P held fixed. Compensating galvos allow changing the frequencies without perturbing the optical alignment. The intersection angle of the Raman beams was about 3.5° in the plane of the optical table. The direction of the probe beam P was optimized for diffraction efficiency but was 3.5° out of the plane of the optical table. An additional beam from the argon laser at 514.5 nm is also directed into the sample to serve as a repump. Without this repump beam, the NV center would exhibit persistent spectral hole-burning due to reorientation of the N-V center in the diamond lattice and no CW signal would be seen after a short time[10]. In contrast to our previous studies on Pr:YSO, the repump beam does not provide spectral selectivity for the four-wave mixing signal, therefore the optical transition is strongly inhomogeneously broadened. All laser beams were linearly polarized and focused into the crystal by a 150 mm focal length lens, producing a spot with a diameter about 100 μm. During the experiment, the sample was maintained at a temperature of 15 K inside the helium flow JANIS cryostat. The magnetic field was applied by a pair of Helmholtz coils that allow fine-tuning of the ground state energy level splitting near the anticrossing point. This magnetic field must be applied within 1° of the crystal (111)-direction and have a strength near 1 kGauss[6].

A representative NDFWM signal is shown in Figure 2 as a function of the frequency difference between Raman beams. The lineshape is taken at intensities substantially below saturation limits (see below). As shown, the Raman linewidth is about 5.5 MHz, which is comparable to the 5 MHz inhomogeneous width of the spin transition[10] and much smaller than both the homogeneous width of the optical transition (~50 MHz) and laser jitter (~100 MHz). This sub-optical transition linewidth is taken as evidence of the Raman process. The maximal diffraction efficiency was found to be 0.5% for a very weak probe. This diffraction efficiency is reasonable given the probe detuning and the absorption due to high optical density of the sample.

To evaluate the relative matrix elements of our Raman system, we investigated the NDFWM signal amplitude as a function of Raman laser beam intensities. The results are shown in Figure 3. The saturation intensities were found to be 36 W/cm^2 and 56 W/cm^2 for the optical transitions R1 and R2 respectively. The fact that the observed saturation intensities for the Raman excited spin coherence are much larger than those predicted for a single isolated color center is due to the large inhomogeneous broadening of the optical transition. For the data of Figure 3, the intensity of the probe beam was 1.6 W/cm^2, which is substantially below its measured saturation intensity (48 W/cm^2). The intensity of the repump beam was around 10 W/cm^2.

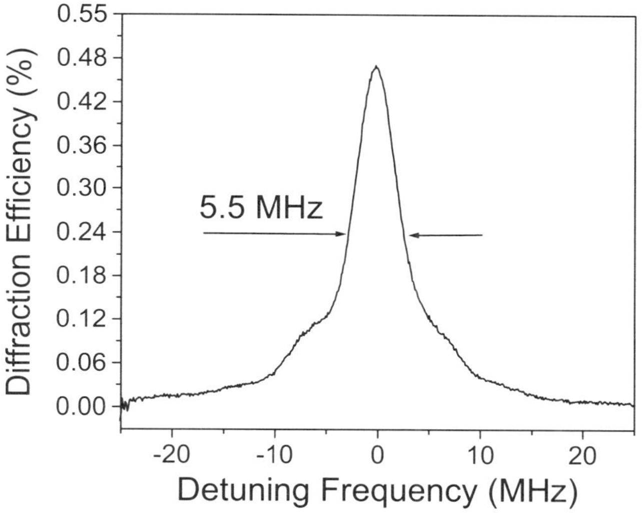

Figure 2. NDFWM signal efficiency at 15 K. Intensities of R1, R2, P, and repump beams were 1.2, 1.6, 5.6 and 10 W/cm^2, respectively. Central difference frequency is 120 MHz.

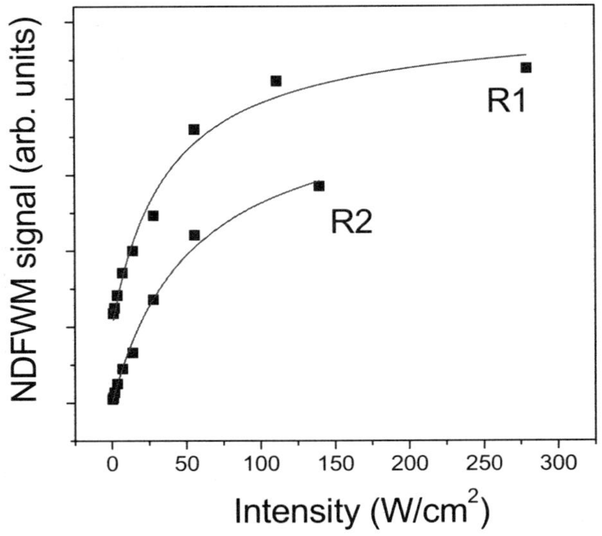

Figure 3. Saturation curves for R1 and R2 beams (plots are vertically shifted for clarity). Intensity of the probe was 1.6 W/cm^2. The intensity of the complimentary Raman beam was about 1.5 W/cm^2 for both curves. The temperature of the sample was 15 K.

Electromagnetically induced transparency[11] was also observed in the N-V diamond color center. To do this, the intensity of Raman beam (R2) was greatly reduced to 1 W/cm^2 and served as a probe, while the other Raman beam (R1) was increased to the maximal available intensity of 280 W/cm^2 and served as the coupling beam. The NDFWM probe beam (P) was blocked during the experiment. The representative experimental trace of the transmission of the probe as a function of Raman detuning are presented in Figure 4. The maximal value of transparency is 17% of the background absorption (optical density 0.3). The background absorption is reduced in the presence of the pump beam due to off-resonant pump-induced persistent spectral hole burning. We would like to point out that the observed value of transparency is large for such a strong inhomogeneous optical broadening. It is in good agreement with a simplified theoretical model which takes into account the inhomogeneous broadening and the fact that only 1 in 4 color centers has the correct orientation relative to the magnetic field. Fitting the experimental EIT spectrum to the model gives an rms Rabi frequency of about 160 MHz. For applications where single atoms will be excited the theoretical model shows that resonant atoms with the correct orientation, driven by such a large Rabi frequency would exhibit close to 100% transparency. Finally, the observed EIT linewidth of 8.5 MHz is substantially smaller than the laser jitter and the optical homogeneous linewidth, and is even smaller than the 24 MHz radiative contribution to the optical linewidth. The fact that width of the EIT peak exceeds the inhomogeneous linewidth of the ground state spin transition is most likely due to power broadening, since a relatively high laser intensity was needed to produce efficient electromagnetically induced transparency in the presence of the large optical inhomogeneous broadening. This work was supported by ARO grant #DAAG55-98-1-0375, AFOSR grants #F49620-98-1-0313 and #F49620-99-1-0224.

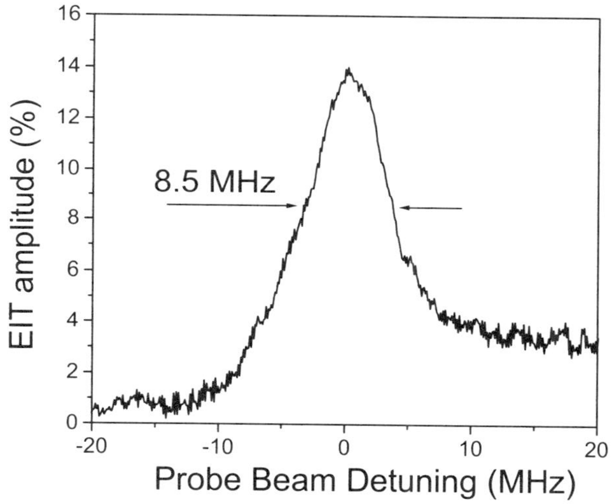

Figure 4. EIT amplitude relative to the peak probe absorption versus probe detuning in NV-diamond at 15 K for a coupling field intensity of 280 W/cm^2 . The intensities of probe and repump were 1 and 10 W/cm^2 respectively. The sloping background is due to the frequency dependent efficiency of the acousto-optic shifter.

References

1. M. D. Lukin, P. R. Hemmer, M. Loffler, and M. O. Scully, Phys. Rev. Lett. **81** (1998) 2675.
2. H. Lin, T. Wang, and T. W. Mossberg, Opt. Lett. **20** (1995) 1658; X. A. Shen, E. Chiang, and R. Kachru, Opt. Lett. **19** (1994) 1246.
3. C. Williams and S. Clearwater, Exploration in Quantum Computing (Springer-Verlag, New York, 1998).
4. B. S. Ham, M. S. Shahriar, M. K. Kim, and P. R. Hemmer, Opt. Lett. **22** (1997) 1849.
5. M. S. Shahriar et al., quant-ph/0007074; M.D. Lukin and P.R. Hemmer, Phys. Rev. Lett. **84** (2000) 2818; T. Pellizzari, S. A. Gardiner, J. I. Cirac, and P. Zoller, Phys. Rev. Lett. **75** (1995) 3788.
6. X.F. He, N.B. Manson, P.T.H. Fisk, Phys. Rev. *B* **47** (1993) 8809; N. B. Manson, X. F. He and P. T. Fisk, J. Lum **53** (1992) 49.
7. A. Lenef, S. W. Brown, D. A. Redman, S. C. Rand, J. Shigley and E. Fritsch, Phys. Rev. B **53** (1996) 13427
8. J. P. O. Martin, J. of Lumin. **81** (1999) 237.
9. H. R. Gray, R. M. Whitley, and C. R. Stroud, Opt. Lett. **3** (1990) 131.
10. N. Manson, private communication.
11. S. E. Harris, Physics Today, July 1997 and references therein.
12. X. A. Chen and R. Kachru, Opt. Lett. **17** (1992) 520.
13. E. vanOort, N.B. Manson, and M. Glasbeek, J. Phys. C **21**(1988) 4385.

QUANTUM COMPUTING WITH ELECTRONS FLOATING ON LIQUID HELIUM

M.I. DYKMAN

Department of Physics & Astronomy, Michigan State University, East Lansing, MI 48824, USA

P.M. PLATZMAN

Bell Laboratories, Lucent Te chnolgies, Murray Hill, NJ 07974, USA

Electrons on a helium surface form a quasi two-dimensional system which displays the highest mobility reached in condensed matter physics. We propose to use this system as a set of interacting quantum bits. We will briefly describe the system and discuss how the qubits can be addressed and manipulated. The working frequency of the proposed quantum computer is ~ 1 GHz. Careful analysis shows that the relaxation rate can be at least 5 orders of magnitude smaller, for low temperatures.

1 Introduction

Finding physical systems that are suitable for quantum computation is a major challenge, since these systems must consist of a large number of quantum objects whose states and interactions can be conveniently manipulated. In addition, the qubits must be sufficiently isolated from other degrees of freedom so that they do not disturb the time dependence of the computational wave function. It should also be possible to read out the final state of the system (the result of the calculation)[1]. Ions and neutral atoms in traps [2], cavity quantum electrodynamic systems [3], bulk NMR systems[4], quantum dots[5], nuclear spins of atoms in doped silicon devices [6], localized electron spins in semiconductor heterostructures[7], and Josephson-junction based systems[8] have been proposed, and in a number of cases proof of the principle has been convincingly demonstrated. Many of these suggestions and some very recent ideas are discussed in these Proceedings. However, for all proposed systems there are high technological and scientific barriers, which must be overcome to achieve a useful quantum computer.

We believe, based on realistic estimates[9], that there already exists a system that is a good candidate for a scalable analog quantum computer (AQC) with an easily manipulated set of qubits. This is a system of electrons floating on the surface of superfluid helium. It is incredibly clean, because the electrons essentially reside in vacuum, and the closest condensed matter is the superfluid helium, which is itself very pure, with a microscopically smooth surface. Electrons on helium display the highest mobility achieved in a condensed matter system so far[10]. They have been extensively studied theoretically and experimentally[11], and are by now well understood. Our estimates indicate [9] that, under realistically obtainable geometry, temperature, and magnetic field, the electrons can function as qubits provided they are confined by micron-size electrodes located below the helium surface.

The system of electrons on helium has several important advantageous features. First of all, different states of a qubit have different electric dipole moments. This makes it relatively simple to address the qubits and to read out their states. A realistic calculation shows that the coherence time of the electron states is $\sim 10^5$ times the reciprocal working frequency of the proposed AQC, which is determined by the Rabi frequency and the energy of qubit-qubit interaction. The typical interqubit distance is $\sim 1\,\mu$m, and therefore fabricating a

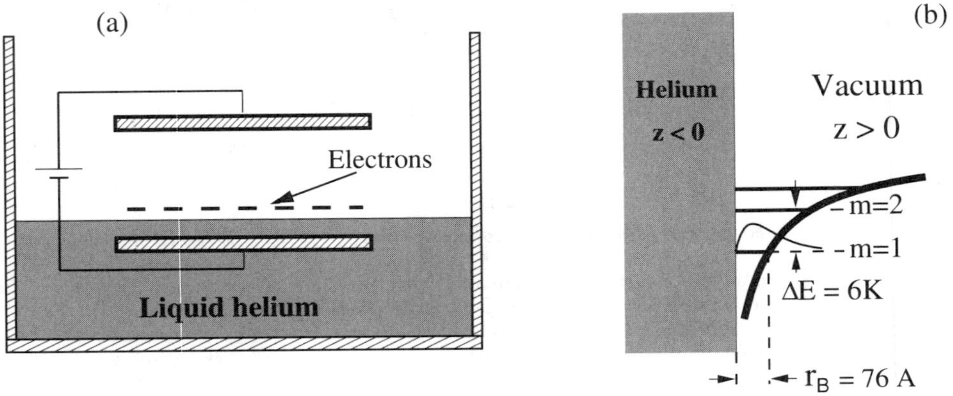

Figure 1. (a) Conventional geometry of experiments on electrons trapped at the helium-vacuum interface. The electrons are attracted to the liquid helium by the image potential and an electrostatic potential of the capacitor, and are free to move in the plane. (b) The hydrogenic energy levels for electron motion in the z direction transverse to the helium surface, along with a typical $m = 1$ ground state wave function are displayed in the image potential for $z > 0$.

multi-qubit system should not require technological breakthroughs.

In what follows we briefly review the physics of electrons on helium, describe the proposed design of the quantum computer [9], and discuss how operations on qubits can be performed.

2 Electrons on helium

The geometry of experiments on electrons trapped above the liquid-helium surface is shown schematically in Fig. 1. This very clean system has been recognized as a perfect tool for investigating many basic concepts of condensed matter physics. Not only the electron energy spectrum and scattering mechanisms are well established[11], but this is one of only a few well-understood strongly correlated electron systems. Correlation effects are strong because, although the system is nondegenerate, the ratio of the Coulomb energy of the electron-electron interaction to the thermal energy $\Gamma = e^2(\pi n_e)^{1/2}/k_B T$ usually exceeds 20 (n_e is the electron density; typically $n_e \sim 10^8$ cm^{-2}). For $\Gamma > 130$, the electrons form a Wigner crystal[12,13], whereas for higher temperatures they form a nondegenerate correlated electron liquid, with unusual transport characteristics [14].

An electron on a bulk helium film is bound in the direction z normal to the helium surface by an image potential of the form $V(z) = -\Lambda e^2/z$ (see Fig. 1),with $\Lambda = (\epsilon - 1)/4(\epsilon + 1)$ ($\epsilon = 1.057$ is the dielectric constant of liquid helium). Because there is a barrier of ≈ 1 eV for penetration into the helium, the electron z-motion is described by a 1D hydrogenic spectrum. The mth state has an energy $E_m = -R/m^2$, with an effective Rydberg energy $R = \Lambda^2 e^4 m_e/2\hbar^2 \approx 8$ K, and effective Bohr radius $r_B = \hbar^2/\Lambda m_e e^2 \approx 76$ Å(Fig. 1).

Electronic transitions between the states in Fig. 1b were observed using microwave spectroscopy[15]. The level spacing could be controlled by a static electric field $\mathcal{E}_\perp$ normal to the helium surface from the capacitor in Fig. 1a. This field Stark-shifted the transition frequencies. The linear Stark shift is ≈ 1 GHz/(V/cm), which will be important for our AQC. The calculated shift was in excellent agreement with the experiment. For $T = 0.4$ K the natural linewidth for the $1 \to 2$ phototransition was estimated from measurements of inhomogeneously-broadened spectra to be below 15 MHz[16].

For $T < 0.7$ K, the only substantial electron coupling to the outside world leading to decoherence effects is to thermally excited height variations $\delta(\mathbf{r}, t)$ of the helium surface, where $\mathbf{r}$ is the electron in-plane coordinate. These variations are described as propagating capillary waves, ripplons. Ripplons are very soft excitations: for the wave vectors $k < 5 \times 10^5$ cm^{-1}, which are of interest for the AQC, the ripplon energies are $\hbar\omega_{\mathrm{r}}(\mathbf{k}) < 4 \times 10^{-3}$ K. Therefore, at 10^{-2} K many such ripplons are present, and the corresponding average mean-square displacement of the surface is determined by thermal fluctuations, with $\delta_T \approx 2 \times 10^{-9}$ cm.

The electron-ripplon coupling energy is

$$H_{\mathrm{e-r}} = e\hat{\mathcal{E}}_{\perp}\delta(\mathbf{r}, t), \tag{1}$$

where $\hat{\mathcal{E}}_{\perp}$ is the effective z-directed electric field on the electron. It is given by the sum of the externally applied field $\mathcal{E}_{\perp}$ and the [nonlocal in $\mathbf{r}$] field from the image potential $\sim 10^2 - 10^3$ V/cm.

In the lowest hydrogenic level it is the weak coupling to ripplons which limits the in-plane mobility of the surface state electrons. It also determines the dephasing rate of the $m \to m'$ photoinduced transitions for unconfined electrons. This dephasing rate is of the order of the intraband momentum relaxation rate τ^{-1}. In the single-electron approximation, $\tau^{-1} = e^2\mathcal{E}_{\mathrm{eff}}^2/4\sigma$, where $\mathcal{E}_{\mathrm{eff}}$ is the known characteristic value of the field pressing the electron to the surface (it is linear in $\mathcal{E}_{\perp}$), and σ is the surface tension of helium. This expression is in good agreement with the measurements [17]. The relaxation rate is record low: from the recent mobility measurements[10], $\tau^{-1} \approx 10^7$ s^{-1}, in agreement with the upper bound on the photoabsortpion linewidth cited before.

Since all measurements refer to a strongly correlated electron system, one may ask if the agreement of the single-electron theory and experiment is not fortuitous. However, in a broad range of T and n_e, the single-electron transport theory is essentially applicable in the absence of a magnetic field [14]. This is no longer true in the presence of a field $B_{\perp}$ normal to the electron layer, which is also of interest for the AQC. It turns out that transport in the correlated electron system can be analyzed in a non-perturbative way[14]. The many-electron conductivity displays a specific dependence on the parameters of the system, and in particular, it is a non-monotonic function of $B_{\perp}$. The theory is in full agreement with the experiment[18], which reassures one that our understanding of the electron system on helium is adequate.

Besides intraband relaxation, the coupling to ripplons gives rise to transitions from the excited $m = 2$ to the ground $m = 1$ hydrogenic band in Fig. 1. A simple estimate for the interaction (1) gives the relaxation rate $T_1^{-1} \sim (R/\hbar)(\delta_T/r_B)^2$. Since $\delta_T/r_B \sim 10^{-3}$, the rate T_1^{-1} is $\sim 10^{-6}$ of the transition frequency ~ 120 GHz. This suggests using the lowest t wo hydrogenic levels of an individual electron as a convenient qubit, whose state can be changed by the application of a resonant microwave field.

For a field amplitude $E_{\mathrm{RF}} = 1$ V/cm, the Rabi frequency $\Omega = |eE_{\mathrm{RF}}\langle 1|z|2\rangle|/\hbar \sim eE_{\mathrm{RF}}r_B/\hbar$ of the oscillations of a resonantly driven electron between the lowest and the first excited state is about 10^9 s^{-1}. This is the working frequency of the quantum qubit we

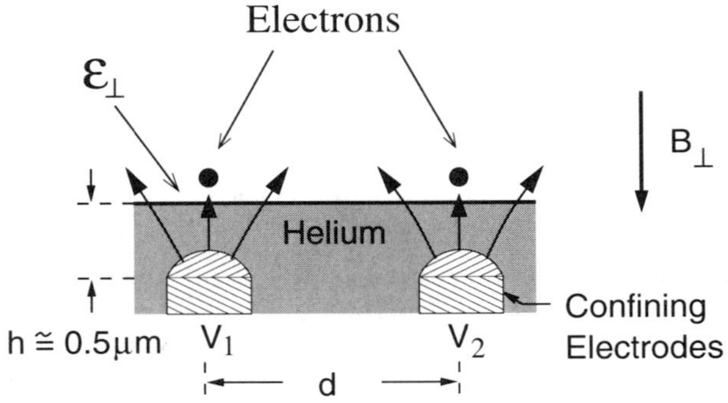

Figure 2. Electrons on helium above a patterned substrate. The rough dimensions ($h \approx d$), the shape of the field lines, and the gate voltages are included. The average distances of the electron from the helium surface in the hydrogenic states $|1\rangle$ and $|2\rangle$ are ≈ 11 nm and ≈ 45 nm, respectively, for small pressing field $\mathcal{E}_\perp$.

are suggesting. Even a laterally unconfined electron has $\Omega T_1 > 10^4$.

3 Qubits using electrons on helium

Recently we suggested[9] that electrons on helium can be used as addressable qubits if they are confined laterally. The geometry is shown schematically in Fig. 2. The lower plate of the capacitor that holds electrons at the helium surface is patterned and forms a set of individually addressable micro electrodes. To make the frequency of resonant energy exchange between neighboring electrons of the order of $\Omega \sim 10^9$ s^{-1}, the distance d between the dots should be $\sim 0.5\,\mu$m. Consequently, the electrodes should be submerged at $h \sim d \sim 0.5\,\mu$m beneath the helium surface. This geometry allows accommodation of $\geq 10^8$ qubits, for a typical area of the helium surface 1 cm^2.

The electrode potentials V_n create both out-of-plane and in-plane fields on the electrons. They can be easily found for a spherical electrode of radius R. If the center of the sphere is at the distance h below the surface, the V_n-dependent part of the pressing field $\mathcal{E}_\perp$ is $\mathcal{E}_n \approx V_n R/h^2$. Then, for the geometry in Fig. 2, the change of the Bohr frequency of the $1 \to 2$ transition by 1 GHz should occur for the increment $\Delta V_n \sim 0.1 - 1$ mV [with account taken of the neutralizing background]. This allows one to make a **one-qubit gate** by tuning the Bohr frequency of the targeted electron to the frequency of the externally applied microwave radiation. Depending on the duration of the radiation pulse $T_{\rm RF}$ with respect to the Rabi frequency Ω, the qubit will be put into an arbitrary superposition of the ground and excited states. In fact, an optimal arrangement is to combine a radiation pulse with the pulse of the control voltage V_n, which tunes the qubit into resonance for a desired amount of time.

Electron confinement gives rise to a further dramatic increase of the lifetime of the excited state T_1, and makes the coherence time T_2 very large. The spectrum of a single pinned electron consists of a ladder of discrete in-plane energy levels connected with the ground and excited hydrogenic states in Fig. 1. The level spacing $\hbar\Omega_\parallel$ can be estimated for spherical electrodes, in which case $m\Omega_\parallel^2 \approx e^2 R(h^2 - R^2)^{-2} + eRV_n h^{-3}$. This spacing

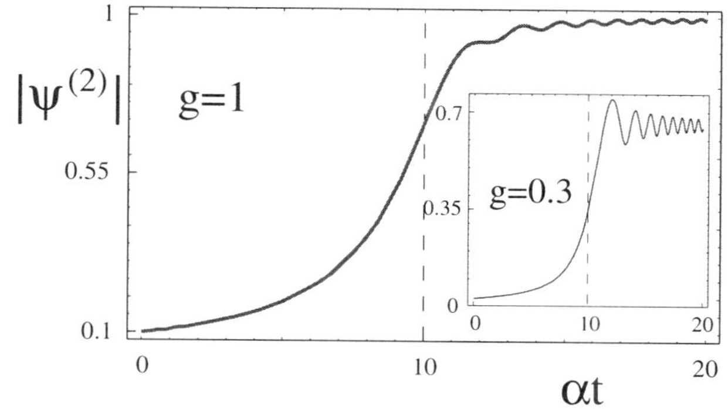

Figure 3. Excitation transfer between the qubits 1 and 2, which are initially ($t < 0$) in the excited and ground states, respectively. The Bohr frequencies of the qubits are linearly changed in time, $\Omega_{1,2}(t) = \Omega_{1,2}(0) \mp \alpha t$ and become equal to each other for $\alpha t = 10$. Solid curves show the amplitude of the excited state of the qubit 2 for different effective coupling parameters $g = \Omega_{\rm sw}/\alpha^{1/2}$.

is not commensurate with $E_2 - E_1$. Therefore a transition from the state $m = 2$ requires emission of a ripplon with energy $\sim \hbar\Omega_{\|}$, which is much higher (~ 0.4 K) than $\hbar\omega_{\rm r}$. We can further suppress one-ripplon decay processes if we apply a magnetic field $B_{\perp}$ normal to the electron layer. It will open up gaps $\hbar eB_{\perp}/mc$ in the in-plane excitation spectrum, which are ≈ 2 K for $B_{\perp} = 1.5$ T.

The dephasing rate T_2^{-1} of a confined electron is determined by a kind of quasi-elastic scattering of thermally excited ripplons, which is similar to the single-electron dephasing rate in quantizing magnetic fields[19]. It contains a factor $(\delta_T/r_B)^4 \sim 10^{-11}$ and is estimated[9] to be $T_2^{-1} < 10^4$ s^{-1} for $T = 10$ mK. The ratio of the working frequency to the relaxation rate is $\Omega T_2 > 10^5$. Other sources of noise and fluctuations can be reduced to the level where they give an even smaller dephasing rate[9].

Two-qubit gates are often discussed in terms of pairing qubits in turn, in isolatation from the rest of the system, and manipulating them in a few fundamental ways. However, in real physical systems interactions are not easily turned on and off. Therefore we would argue that the manipulation of interactions in the whole interacting many qubit system, and the final measurement or collapse of the wave function must be thought of as part of the ultimate computation process. In our computer the qubit-qubit interactions come from the charge on the electrons and are always there. The important ones, as regards the z-motion, are dipolar with energy given by $\sum_{n\neq m} e^2 z_n z_m/2d_{nm}^3$, where d_{nm} is the distance between the nth and mth electrons.

The interelectron coupling allows us to perform both CNOT and swap operations. The latter can be done by sweeping the Bohr frequencies $\Omega_n = [E_2^{(n)} - E_1^{(n)}]/\hbar$ of the electrons past each other, as a result of the voltage change on the underlying electrodes. At resonance, an excitation will be transferred between nearest neighbors over the time $\sim \Omega_{\rm sw}^{-1}$, $\Omega_{\rm sw} = e^2|\langle 1|z|2\rangle|^2/\hbar d^3$. For the parameters in Fig. 2, this time is 10^{-9} s.

Kinetics of excitation transfer between t wo qubits for linear change of the electrode potentials $V_{1,2}$ is illustrated in Fig. 3. The amplitude of the initially occupied excited state of the qubit 1 approaches $\exp(-\pi g^2)$ for large time. Therefore, by adjusting the rate of the potential change, excitation transfer can be made exponentially efficient.

A simple way of measuring the state of a qubit is to apply a pulse of a negative voltage to the underlying electrode. The electron may then tunnel away from the surface. For the geometry in Fig. 1, the tunneling was investigated before[11,20]. The tunneling rate is exponentially larger if the electron is in the excited rather than the ground state. This allows one to detect the electron state[9]. Alternatively, the electron states may be detected by attaching single electron transistors to the control electrodes.

The system of two-dimensional electrons on helium is unique in the context of large AQC systems. It is scalable, easily manipulated, and has perfectly acceptable decoherence times. It has been carefully investigated theoretically and experimentally, and there seems to be no existing technological barriers present for making an AQC using it.

Acknowledgments

MID acknowledges support from the NSF through grant No. ITR-0085922.

References

1. D.P. DiVincenzo, Science **269**, 255 (1995); D.P. DiVincenzo and D. Loss, Superlattices and Microstructures **23**, 419 (1998).
2. J.I. Cirac and P. Zoller, Phys. Rev. Lett. **74**, 4091 (1995); C. Monroe *et al.*, Phys. Rev. Lett. **75**, 4011 (1995); B.E. King *et al.*, Phys. Rev. Lett. **81**, 1525 (1998); Q.A. Turchette *et al.*, Phys. Rev. Lett. **81**, 3631 (1998).
3. Q.A. Turchette *et al.*, Phys. Rev. Lett. **75**, 4710 (1995);
4. N.A. Gershenfeld and I.L. Chuang, Science **275**, 350 (1997); D.G. Cory, M.D. Price, and T.F. Havel, Physica D **120**, 82 (1998); I.L Chuang et al., Phys. Rev. Lett. **80**, 3408 (1998); I.L. Chuang *et al.*, Nature **393**, 143 (1998).
5. D. Loss and D.P. DiVincenzo, Phys. Rev. A **57**, 120 (1998); J.A. Gupta *et al.*, Phys. Rev. A **59**, R10421 (1999); M.S. Sherwin, A. Imamoglu, and Th. Montroy, Phys. Rev. A **60**, 3508 (1999); A. Imamoglu *et al.*, Phys. Rev. Lett. **83**, 4204 (1999).
6. B.E. Kane, Nature **393**, 133 (1998).
7. R. Vrijen *et al.*,Phys. Rev. A **62**, 012306 (2000).
8. V.D. Averin, Solid State Commun. **105**, 659 (1998); Yu. Makhlin, G. Sch¨ on,and A. Shnirman, Nature **398**, 305 (1999); L.B. Ioffe *et al.*, Nature **398**, 679 (1999); Y. Nakamura *et al.* Nature **398**, 786 (1999).
9. P.M. Platzman and M.I. Dykman, Science **284**, 1967, (1999); M.I. Dykman and P.M. Platzman, Fortschr. Phys. **48**, 1095 (2000).
10. K. Shirahama, S. Ito, H. Suto, and K. Kono, "Surface study of liquid-He-3 using surface-state electrons", J. Low Temp. Phys. **101**, pp. 439-44 (1995).
11. "Two dimensional electron systems on helium and other cryogenic substrates", ed. E.Y. Andrei (Kluwer, NY 1997).
12. C.C. Grimes and G. Adams, Phys. Rev. Lett. **42**, 795 (1979).
13. D.S. Fisher, B.I. Halperin, and P.M. Platzman, Phys. Rev. Lett. **42**,798 (1979).
14. M.I. Dykman, M.J. Lea, P. Fozooni, and J. Frost, Phys. Rev. Lett. **70**, 3975 (1993).
15. C.C. Grimes *et al.*, Phys. Rev. B **13**, 140 (1976).
16. A.P. Volodin and V.S. Edelman, Sov. Phys. JETP **54**, 198 (1981).
17. R. Mehrotra *et al.*, Phys. Rev. B **29**, 5239 (1984); A.J. Dahm and W.F. Vinen, Physics

Today **40**, 43 (1987).
18. M.J. Lea *et al.* Phys. Rev. B **55**, 16280 (1997); M.J. Lea and M.I. Dykman, Physica B **251**, 628 (1998); E. Teske *et al.*, Phys. Rev. Lett. **82**, 2772 (1999).
19. M.I. Dykman, Phys. Stat. Sol. (b) **88**, 463 (1978).
20. G.F. Saville, J.M. Goodkind, and P.M. Platzman, Phys. Rev. Lett. **70**, 1517 (1993).

FABRICATION OF THE STRUCTURE FOR QUBITS USING ELECTRONS ON LIQUID HELIUM

JOHN M. GOODKIND AND S. PILLA

University of California, San Diego

In the previous paper[1,2] the system of qubits using electrons on a liquid helium film was described. In this paper we describe the physical realization of the system that we have begun to fabricate. We will not in this brief discussion describe how we intend to operate the system. We will show that we are nano- and micro-fabricating a new type of electronic device that differs from other microelectronic devices in that the final step of the fabrication deposits a layer of helium rather than some other dielectric. The operation of the device will differ in that it manipulates single electrons and it must operate at low temperatures.

1 The qubits; posts and ground plane

For the purposes of this work, we wish to trap and manipulate electrons over selected members of an array of positively charged posts. A schematic drawing of two such posts is shown in Fig 1. In order to accomplish this, we would like the fields from the posts to satisfy the following criteria. 1) The horizontal distance between posts should be ~ 0.5 μ so that the relevant Rabi frequency is ~ 1 GHz. 2) At a horizontal displacement, x, of a fraction of the distance between posts, d, the x component of the field, E_x, from the posts should be at least comparable to the Coulomb field of an electron on its neighbor, ~6.4 kV/m at separation, d~0.5 μ. 3) The vertical component of the field, E_z, over the post should be no greater than a few thousand Volts per meter so that it is a fraction of the field of the image charge in the helium, E_{He} ~ 36 kV/m. 4) The surface of the liquid helium should be a distance, h, above the tops of the posts, sufficiently large that the image field from the posts, for the post diameter employed, is < E_{He}. 5) The ground plane must shield the fields from the traces connected to the posts but the field from the image charge of the electron in the ground plane should also be small relative to E_{He}. This requires that the surface of the helium be at least ~ 0.5 μ above the ground plane. We will show how these requirements can be satisfied for the geometry that we have fabricated.

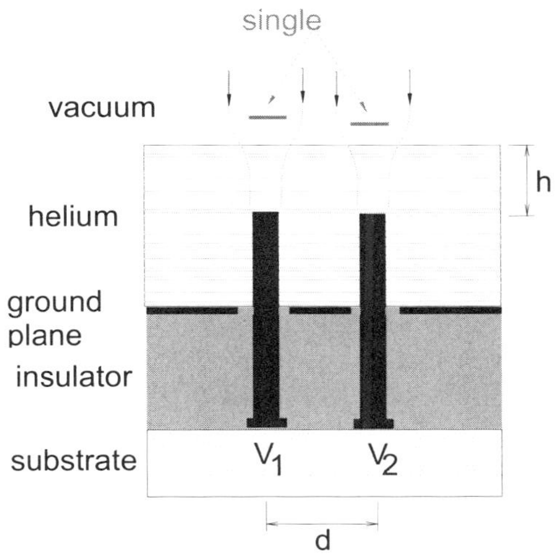

Figure 1. Schematic of the geometry of two posts and ground plane.

The major uncertainty in fabricating this system was whether we could grow posts as small as required with the required length to diameter ratio. Working with Dr. Roberto Panepucci, at the Cornell Nanofabrication Facility, we were able to fabricate the posts shown in the upper frame of Fig 2 with more than adequate aspect ratio.

They were fabricated by writing holes with the VB6, 100 keV ebeam writer into a 1.5 μ thick layer of PMMA. This had been coated onto a gold layer on a silicon wafer. After developing the resist, gold was electroplated into the holes and then the PMMA was dissolved. We then demonstrated that the process could be repeated on top of the traces that supply the Voltages. A preliminary test of the process (on some traces that were damaged before depositing the posts) is shown in the lower frame of Fig 2. We will want the posts to be very uniform in height so that we plan to polish the surface before dissolving the PMMA.

A view of this system on a larger scale looks like a conventional integrated circuit chip with contact pads and traces converging toward the micro-components at the center. The working drawing for the photo- and e-beam lithography is shown in Fig 3. The scale is larger than a conventional chip in order to avoid the need to wire bond 80 leads for each new geometry that we test. The pads are large enough to allow easy alignment with spring contacts permanently installed in the cryostat.

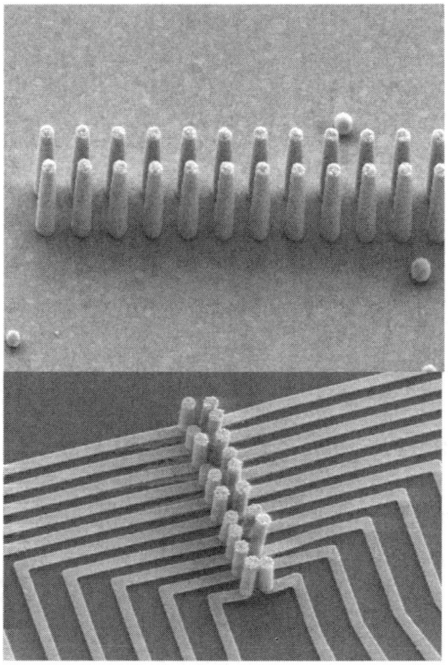

Figure 2. SEM of posts. The upper frame is gold posts deposited on a gold film on a silicon wafer. Center to center separation is 0.5 μ and height is 1.5 μ. The lower frame is posts grown on the trace pattern so that the voltages of the posts may be set individually. The trace pattern in this test was damaged before the posts were deposited so that this is only a test of the plating technique.

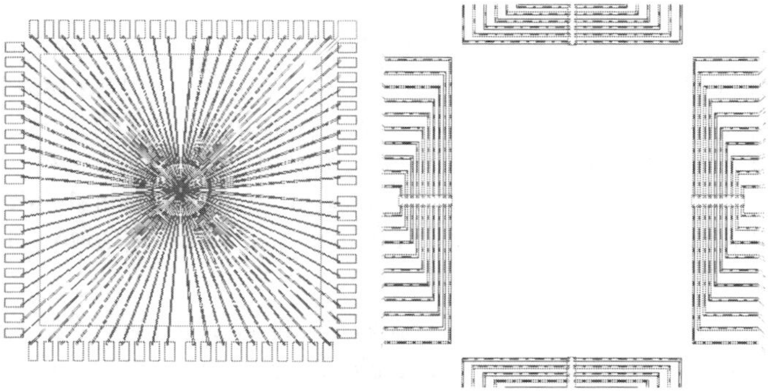

Figure 3. Drawing of leads to posts. The left frame is the full photolithography pattern, 1.5 X1.5 mm. The rectangle connected to a tab on the upper right corner is the ground plane. The right frame is the center section showing the e-beam pattern on which the posts are deposited. The small rectangles are the openings for the posts. 0

In order to optimize the geometry of the posts and ground plane we are computing the electric fields for a variety of geometries. Fig 4 shows computed values of E_x, and E_z over posts separated by 0.6 μ and all at 1V potential relative to the ground plane. The tops of the posts are 0.3 μ above the ground plane and the fields are plotted for three different elevations above the tops of the posts. The posts pass through a rectangular hole in the ground plane which is 1.5 μ by 11 μ. For this geometry, near the center of the array, the peak values of E_x at an elevation of 0.3 μ above the posts is 29 kV/m/V at a distance from the post of 20% of the distance between them. Thus it would equal the Coulomb force with a potential of 0.2 V applied to the posts. For this potential the maximum value of E_z will be ~115 kV/m which is larger than the desired maximum of order 10 kV/m. It would be compensated by applying an appropriate bias between the ground plane and the top plate (see Fig 7). We have not yet computed the fields and equilibrium positions with the electrons in place but it is clear that the equilibrium positions will be increasingly displaced from the positions of the posts as the edge of the array is approached. Therefore it will be desirable to fabricate long arrays and to use only electrons more than one or two positions from the ends.

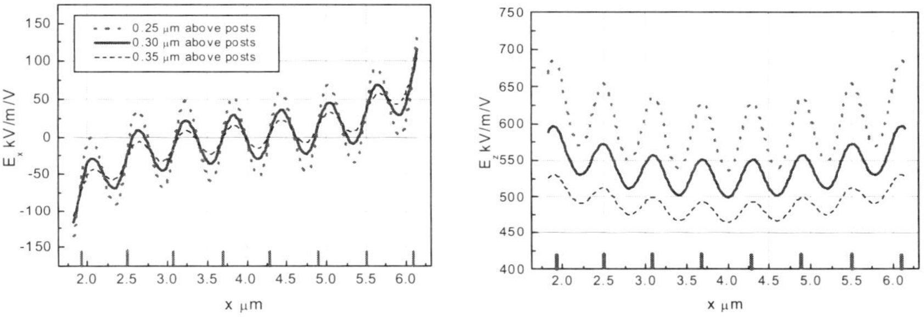

Figure 4. Electric field from two rows of 10 posts each at potential 1V, separated by 0.6 μ, at three different elevations above the tops of the posts. The tops of the posts are 0.1 μ above the ground plane. The fields are computed directly over one of the rows. The positions of the posts are marked by vertical bars along the x axis.

2 Electron source for charging the bits

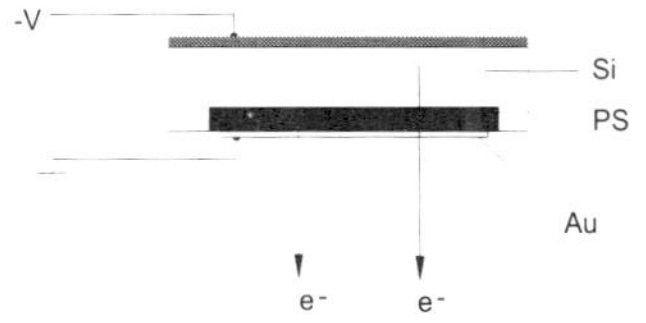

Figure 5. porous silicon diode.

Previous experiments on 2D electrons over liquid helium have used either a hot filament or an electric discharge to spray electrons onto the surface. The former disturbs the system by heating and the latter requires that it be warmed close to 1 K in order to have sufficient gas pressure to sustain the discharge. Both techniques would be too slow for our purposes so that we are developing a tunnel diode source. The most promising approach appeared to us to be diodes fabricated using porous silicon[3]. A thin layer of porous silicon is formed by anodizing a small area of a silicon wafer in a HF solution in bright light. A 10 nm thick layer of gold is then evaporated onto the anodized area. A positive voltage applied to the gold layer relative to the substrate then accelerates electrons out of the porous silicon and through the gold film. These devices have, to our knowledge, been studied only at room temperature. We are attempting to fabricate them so that they will work at low T.

3 Detection of electron states

In the initial stages of this work, we will detect the state of the electrons by applying bias voltages to individual posts that allow the electrons to escape only if they are in the excited state. The escape rates from a pool of electrons on liquid helium have been measured and found to be in agreement with the rates computed by numerical solution of the time dependent Schrödinger equation[4]. The escape rates for electrons in the first excited state and in the ground state differ by orders of magnitude for a given "pulling" field. The computations[4] indicate that for pulse rise times greater than about 0.1 nsec, the tunneling rate will follow the pulse voltage adiabatically. Thus the states of the qubits can be tested in times very short relative to their coherence times. We will detect the escape of individual electrons using micro-fabricated superconducting transition edge detectors.

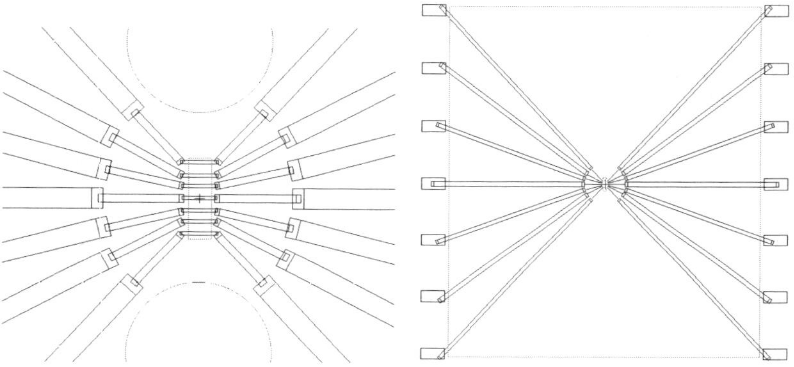

Figure 6 Drawing for the detectors. The left panel is the full circuit. The rectangle is the gold film, top plate that covers the wires. The right panel shows the micro-wires and the opening through the plate over them. The circles are holes for the electrons to pass through from the source (see fig 7).

By making them small their heat capacity can be made sufficiently small so that deposition of 1 eV of energy will heat them above T_C. A positive bias applied to the superconducting element will assure that any electron that escapes will be accelerated to the detector so that all escaping electrons will be detected.

4 The complete system

A schematic diagram of the complete system is shown in Fig 7. The upper and lower plate will be separated by 1 mm. Each will be 15×15 mm. The lower plate is covered with the helium film and the electrons float about 10 nm above the helium surface. The microstructures are exaggerated and of differing scales for illustration.

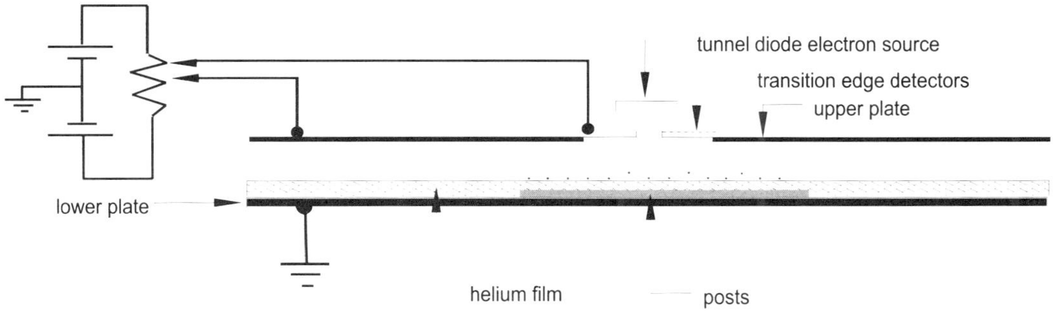

Figure 7. Schematic of the entire system. The upper capacitor plate includes the detectors. The lower plate includes the posts and is covered with the helium film. The electrons float over the posts about 10 nm above the surface of the helium film.

Work supported by the National Science Foundation.

We wish to thank Dr. David Vier for his assistance on the numerical simulations of the fields.

References

1. M. I. Dykman and P. M. Platzman, this conference
2. P. M. Platzman and M.I. Dykman, Science **284**, 1967 (1999)
3. N. Koshida, X. Sheng, T. Komoda, Applied Surface Science **146**, 371 (1999).
4. G. F. Saville and J. M. Goodkind, Phys. Rev. A **5**0, 2059-2067 (1994).

ELECTRON SPIN QUANTUM COMPUTING WITH ENDOHEDRAL FULLERENES

W. HARNEIT, M. WAIBLINGER, K. LIPS, C. MEYER, A. WEIDINGER

Hahn-Meitner-Institut, Glienicker Str. 100, D-14109 Berlin
E-mail: Wolfgang.Harneit@hmi.de

J. TWAMLEY

Department of Mathematical Physics, National Univerity of Ireland
Maynooth, Ireland

We propose to use the electron spin of N@C_{60} (atomic nitrogen inside C_{60}) and similar systems like P@C_{60} as a new type of qubit for quantum computing. The main property of these endohedral fullerenes which makes them suitable for quantum computing application is that the enclosed atoms are almost freely suspended in the inside of the fullerenes and are well shielded from the outside. This shielding leads to extremely long electron spin relaxation times. In addition, fullerenes are fairly easy to manipulate both mechanically and chemically and can be arranged on surfaces, e.g. in linear chains. Our concept for quantum computing using this family of endohedral fullerenes has the following ingredients: (i) the coupling between the qubits is mediated by magnetic dipole-dipole interaction, (ii) gating of the coupling is performed by changing the angle between the external magnetic field and the lines connecting adjacent fullerenes, (iii) addressing specific qubits is realized by charge transfer onto the fullerene cage, (iv) specific arrangements of the qubits are achieved by molecular bonding between fullerenes or by bonding the fullerenes to a substrate.

1 Introduction

Endohedrally trapped nitrogen in C_{60} displays a spectacularly sharp $S = 3/2$ electron paramagnetic resonance (EPR) pattern in both the liquid and the solid state[1], see fig. 1. The fact that the EPR lines remain sharp even in the powder spectrum implies that the encapsulated atom is spherical, sits at the center of the cage and is well shielded from the surroundings. Similarly sharp EPR patterns are observed for N@C_{70} and P@C_{60}[2].

2 Qubit implementation

The quantum information is encoded in the electron spin of a paramagnetic atom. A necessary condition is that the spin lifetime is long compared to the quantum mechanical operation time. The T_1 electron spin relaxation time reaches seconds or more at low temperatures[3] and is thus in a range to provide sufficiently long computation time. The experimental T_2 dephasing time for solid N@C_{60} measured so far is about 20 μs. The final theoretical limit for T_2 is $T_2 \approx T_1$ which may be reached by excluding all unintended spin-spin interactions causing dephasing, in particular those with nuclear spins.

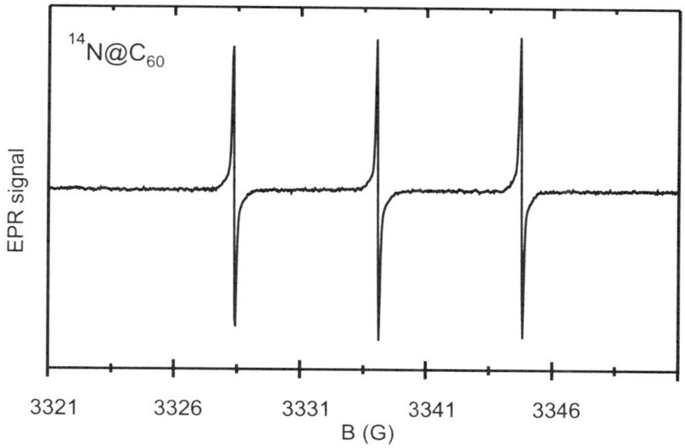

Figure 1. Electron paramagnetic resonance (EPR) pattern of solid $^{14}N@C_{60}$. The triplet splitting is due to the hyperfine interaction with the nuclear $I = 1$ spin.

3 Qubit setting and read-out

The electron spins can be oriented in a strong external magnetic field at low temperatures. The degree of polarization is determined by the ratio of the magnetic Zeeman energy μB and the thermal energy kT. For electrons a high degree of polarization can be reached for $B = 1$T already at pumped liquid Helium temperatures and quasi full polarization at temperatures in the milli-Kelvin range. This is important since essentially pure quantum states can be obtained experimentally and a cold spin reservoir for quantum error correction schemes is available.

The manipulation of the electron spins can be performed by appropriate EPR pulse sequences, in analogy to NMR quantum computing schemes. Besides the high or full polarisation of the spins, an advantage of using EPR is the sensitivity which is orders of magnitudes higher than that of NMR.

Single electron spin read-out is presently under investigation by several groups in connection with quantum computing. Progress is expected in the areas of single electron transistors (SET) which could be very effectively implemented using fullerenes, scanning microscopy using tunneling currents (STM) in which spin-dependent resonance frequencies are detected, or magnetic force resonance microscopy (MRFM).

A more traditional read-out approach uses many identical quantum computers working in parallel and pulsed EPR read-out methods, e.g. applying a $\pi/2$ pulse and observing the phase of the free induction decay. It is noteworthy that using essentially fully polarized spin systems one does not measure average values as in NMR quantum computing.

4 Qubit entanglement via magnetic dipole-dipole coupling

In a high external field, the x and y components of the dipolar field are averaged to zero and the z component is given by

$$B_z = \mu_s(3\cos^2\theta - 1)/r^3 \tag{1}$$

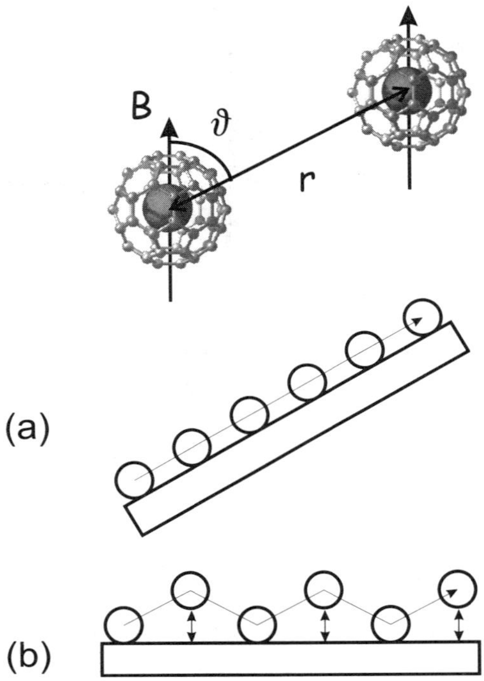

Figure 2. Control of qubit-qubit coupling (J-gate) in a linear chain of endohedral fullerenes via the angle θ of magnetic dipole-dipole interaction. θ is modified by (a) rotating the chain with respect to the external magnetic field or (b) by a displacement of every second fullerene along the direction of the external magnetic field.

where μ_s is the magnetic moment of the generating spin, θ is the angle between the spin direction and the connecting line between the generating spin and the observer at distance r, see fig. 2. This interaction can be used to create the entanglement of adjacent spins.

The maximum coupling strength for two adjacent N@C_{60} molecules at a distance of 1 nm and $\theta = 0$ is $J = 50$ MHz. This number gives a cycle time of $\tau = 1/4J = 5$ ns, which is short compared to the spin relaxation times mentioned above. Thus, the dipolar coupling is in the appropriate range for quantum computing.

5 J- and A-gates

In order to perform calculations of significant length and complexity it is desirable to gain local control of spin resonance conditions for read-in (A-gates) and of dipolar spin-spin interaction (J-gates).

The dipolar interaction can quite simply be controlled (J-gate) via the dipolar angle θ of eq. (1). In a linear chain arrangement, this angle is the same for each pair of spins and can thus be set globally by choosing the angle between chain and external magnetic field. Maximum coupling is achieved at $\theta = 0$, i.e. when all spins are parallel or antiparallel to the chain direction. If the magic angle $\theta = 54.7°$ is chosen, the entire chain is decoupled simultaneously. The angle between two spins in a chain and an external magnetic field can be controlled in two different ways, as shown in fig. 2.

Detuning the resonance frequency of a single spin (A-gate) can be accomplished by electron transfer onto the fullerene cage. In fact, the charge transfer onto N@C_{60} molecules in solution has been studied recently by Jakes[4] who found that the EPR signal (and hence the resonance condition) vanishes when an electron is transferred onto the C_{60} cage. The nitrogen atom seems to be stable under this operation, in accordance with the very weak interaction between nitrogen and C_{60}. A field induced electron transfer onto C_{60} has recently been demonstrated by Schön[5] with a traditional gate structure.

6 Qubit arrangements and upscaling

An important advantage of fullerenes is that they can be simply arranged in specific structures because of their size and molecular properties. A conceptually simple arrangement of endohedral fullerenes for a quantum computer is the formation of linear chains, especially in view of the proposed dipolar coupling. Upscaling the quantum computer to a certain number of qubits requires a corresponding number of distinguishable endohedral fullerenes. The qubits can be distinguished either by using different endohedrals, e.g. ^{14}N@C_{60}, ^{15}N@C_{60}, ^{31}P@C_{60}, or by using an A-gate. A combination of the two is suggested, i.e. different endohedrals are positioned upon an A-gate. Chemical modifications[6] and bonding of endohedral fullerenes are the method of choice to create identical multi-qubit structures.

7 Conclusion

We presented a general concept for electron spin quantum computing using endohedral fullerenes. Many details and realizations have still to be worked out. We think that the electron spin concept of this proposal has advantages compared to NMR quantum computer proposals.

Acknowledgements

This work was supported by the Information Societies Technology programme of the European Union under contract no. IST-1999-11617 (QIPD-DF project, http://planck.thphys.may.ie/QIPDDF/).

References

1. A. Weidinger, M. Waiblinger, B. Pietzak, T. Almeida Murphy, *Appl. Phys.* A **66**, 287 (1998).
2. C. Knapp, N. Weiden, H. Käss, K.P. Dinse, B. Pietzak, M. Waiblinger, A. Weidinger, *Mol. Phys.* **95**, 999 (1998).
3. S. Knorr, A. Grupp, M. Mehring, M. Waiblinger, A. Weidinger, to be published in *Molecular Nanostructures*, eds. H. Kuzmany *et al.*, (World Scientific, Singapore, 2001).
4. P. Jakes, B. Goedde, M. Waiblinger, N. Weiden, K.P. Dinse, A. Weidinger, to be published in *Molecular Nanostructures*, *ibid.*

5. J.H. Schön, Ch. Kloc, B. Batlogg, *Nature* **406**, 704 (2000).
6. E. Dietel, A. Hirsch, B. Pietzak, M. Waiblinger, K. Lips, A. Weidinger, A. Gruss, K.P. Dinse, *J. Am. Chem. Soc.* **121**, 2432 (1999).

DIAMOND QUANTUM DOTS FABRICATED BY ION IMPLANTATION

S.PRAWER, JU-LIN PNEG, JULIUS ORWA, JEFF MCCALLUM, DAVID N JAMIESON, AND LES BURSILL

Spe cialResearch Centre For Quantum Computing Te chnolgy, School of Physics, The Microanalytical Research Center, The University of Melbourne, Parkville, 3010, Victoria, Australia

One possible route to the realization of quantum computing is the use of arrays of quantum dots act as qubits. One method of producing quantum dots is via ion implantation to directly embed foreign atoms into an inert matrix. After annealing, quantum dots of sizes ranging from 1 to 100 nm can be formed. Recently, there has been increased interest in the use of the colour centers in diamond for both single photon sources and to act as the qubits in a quantum computer. Hence there is interest in being able to create arrays of nanocrystalline quantum dots. In this paper we provide the recipe for making nanocrystalline diamond by directly converting the carbon implanted into fused silica into diamond by simple furnace annealing. A scheme is provided by which arrays of single diamond quantum dots could be fabricated.

1 Introduction

One possible route to the realization of quantum computing is the use of quantum dots to acts as the qubits [1]. Many methods have been proposed for the production of quantum dots. One of the most interesting and versatile is the use of ion implantation to embed foreign atoms directly into an inert matrix such as, for example, fused silica. After an appopriate annealing step, quantum dots of sizes ranging from 1-100 nm are formed, usually by a process of precipitation and Ostwald ripening. The size of the clusters is controlled by the original implantation parameters such as dose and ion energy and the annealing parameters such annealing temperature and time and also the annealing ambient. Quantum dots of a wide range of materials from metals to semiconductors have been formed by this method [2]. By using mutiple ion implants quantum dots of compound semiconductors can also be formed, for example ZnS, CdS, and PbS[3], and GaAs [4]

It has long been recognized that the electronic and optical properties of quantum dots depend on their size, with quantum confinement playing an important role [2]. The properties of the dots also depend crucially on surface termination [5] because for dots of the order of 5nm in size or less, a large fraction of the consituent atoms actually lie on the surface. Less obvious is the fact that the crystalline structure of the dots can also be influenced by their size. This is particularly important in the case of metastable phases.

Recently, there has been increased interest in the potential use of colour centers in diamond (specifically the N-V center) to be used either as single photon sources or as qubits in a quantum computer [6]. This provides a strong motivation for the development of methods for the production of quantum dots of diamond. However, at room temperature and pressure, graphite and not diamond is the stable form of carbon.

In this work we report on a successful recipe for the production of diamond quantum dots embedded in a fused silica matrix. The method provides good control over the crystallite size. We show that the stability of the diamond phase depends crucially on the crystallite size, with the diamond phase stable only for dots less than 7nm in diameter. Although the parameter window for the production of diamond quantum dots in this way is small, the simplicity of the method suggests that it may be relatively easy to produce arrays of diamond quantum dots each of which is registered to a gate electrode.

2 Experiment

Optical grade fused silica substrates (sourced from Esco products) were implanted with 1 MeV carbon ions to doses of 5 x 10^{16}, 2 x 10^{17}, and 5 x 10^{17} ions/cm^2. This creates a carbon rich layer 120 nm thick buried at a depth of 1450 nm below the surface with a peak concentration of 2, 8, and 20 x 10^{21} carbon atoms per cm^3, respectively, as measured by secondary ion-mass spectrometry (SIMS). Each of the samples was cut into smaller pieces and annealed at 1100^0C for 1 h in ambients of Ar, O or forming gas (4% hydrogen in argon) at atmospheric pressure.

3 Results

For samples annealed in argon or oxygen, SIMS showed (and high resolution TEM confirmed) that virtually all the carbon diffused out of the substrate. However, when annealing was performed in forming gas, nanocrystals of diamond are formed just below the surface of the fused silica. Figure 1 shows a TEM image of a sample implanted with 5 x 10^{16} C/cm^2 and annealed at 1100^0C in forming gas for 1 h. A layer rich in carbon clusters about 60 nm thick is evident, buried about 140 nm below the surface. SIMS measurements revealed that the end of range, (located much deeper about 1450 nm below the surface) retains only about 2.5% of the original implanted carbon. Hence it is clear that the implanted carbon has diffused towards the surface and formed a layer of nanocrystals close to, but some 140 nm below the surface.

Figure 1 also shows HRTEM images of single isolated diamond nanocrystals, together with a high resolution TEM lattice image in which the interlayer spacing is measured to be 0.21nm which is consistent with the expected spacing from the (111) diamond planes of 0.206 nm.

TEM images diffraction patterns were measured for diamond for carbon nanoclusters of different sizes. The diffraction patterns (not shown here) consist of well defined spots which is as expected from nanocrystalline material. Detailed analysis of the diffraction patterns [7] reveal that only for nanocrystals less than 7 nm in size do the carbon nanocrystals consist of phase pure nanocrystalline diamond. For larger crystallites the diffraction patterns can be indexed either to a metastable form of diamond called n-diamond or to another phase of carbon called i-carbon. Both n-diamond and i-carbon contain a considerable fraction of the carbon atoms in the sp^2 (ie graphite-like) bonding configuration. Electron Energy Loss Spectroscopy confirmed the diffraction results [7]. Hence, two conditions are necessary for the direct conversion of the implanted carbon into diamond, viz., (i) the annealing must be done in an ambient containing hydrogen and (ii) the crystallite size must be less than 7nm in size. The latter is accomplished by limiting the implantation dose and the annealing time. Provided these two conditions are met, fabricating high quality nanodiamond is technologically very straightforward.

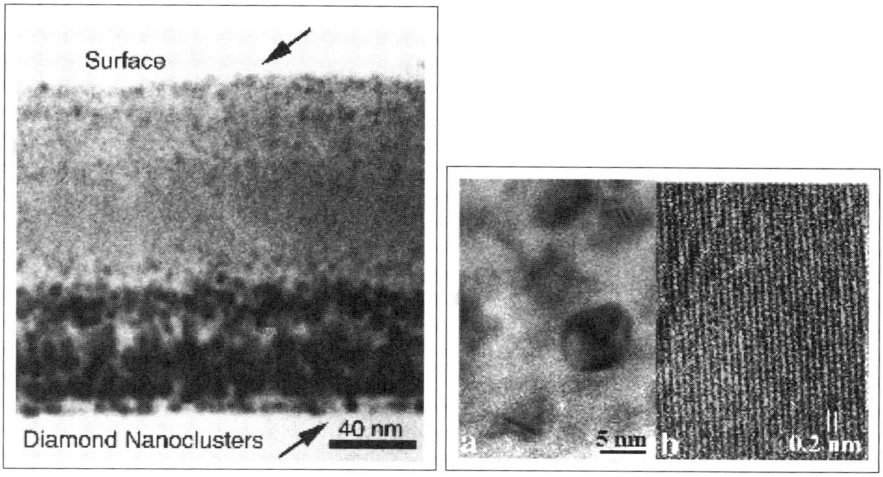

Figure 1. Left panel: Cross sectional TEM Micrograph of fused silica sample implanted with 5 x 10^{16} C/cm^2 and annealed in forming gas at 1100^0C. A layer of nanocrystals has formed about 140 nm below the surface. Right panel: TEM image of a small number of diamond nanocrystals. Note the octahedral habit. The high resolution lattice image shows of the internal structure of a single nanocrystal. The lattice spacing corresponds to the (111) spacing of the lattice planes of diamond.

4 Discussion

We address t wo pertinent issues: Firstly in view of the fact that graphite is the stable form of carbon at room temeprature and pressure, how is it possible for nanodiamond to form from the implanted carbon, rather than some graphitic form of carbon. Secondly how can the above recipe for making nanocrystalline diamond be used potentially for the creation of arrays of nanodiamond quantum dots.

In answering the first question, we note that we only observe the formation of nanodiamond when the annealing takes place in forming gas (ie in an ambient containing hydrogen). We assume that the hydrogen from the forming gas is able to readily diffuse into the fused silica during annealing and is therefore available to stabilize the growth of the diamond nanoclusters. In the presence of hydrogen, small diamond clusters can be more stable than small graphitic clusters. This is because the unterminated diamond surface has more unsatisfied dangling bonds than on a graphitic surface and thus there is significantly more stablization by forming C-H bonds at the growing surfaces. The difference in free energy between bulk diamond and graphite is very small (0.03 eV/atom). For small enough sizes the difference in the free energy associated with the surface termination can more than compensate for the difference in the bulk free energy. Ree et al [8] have calculated that the cross over point at which H terminated diamond nanocrystals are more stable than hydrogen terminated graphite nanocrystals is between 30,000 and 70,000 atoms. The largest clusters which we observed to have a diamond structure were 7 nm in size corresponding to about 32,000 atoms, of which about 20% lie on the surface. Thus our observations are consistent with the prediction from molecular dynamics of the relative stability of hydrogen terminated diamond and graphite surfaces.

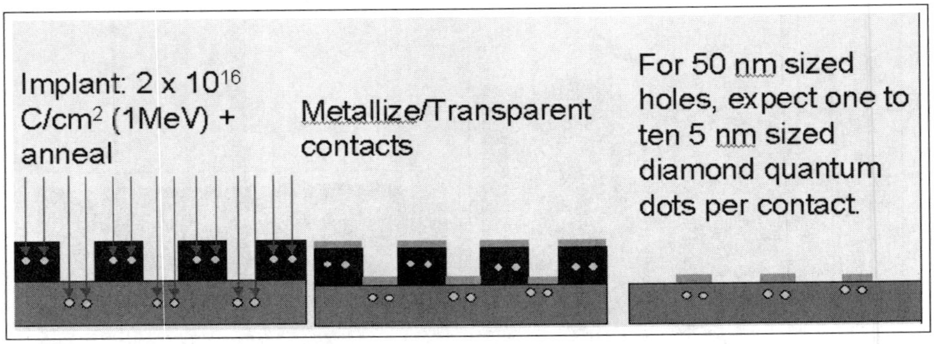

Figure 2. Schematic of a proposal to create arrays of quantum dots using ion implantation

Regarding the second question, the ion implantation route suggests a simple method for the production of arrays of nanocrystals which can be individually addressed. Figure 2 is a schematic representation of one possible realization. The scheme will require very high aspect ratio masks because it is necessary to bury the implanted ions quite a distance below the surface. Hence the required ion energies are quite large (100's to 1000's of keV) and the mask must be thick enough to prevent these ions reaching the surface. Recently, Masuda et al [9] have shown that it is possible to make such masks from Aluminium using an electrochemical technique. Regular arrays of holes about 50 nm in diameter with a separation of about 100 nm can be created by this method. The Al mask can be placed onto the fused silica substrate and exposed to an appropriate dose of carbon ions. Following the ion implantation, the sample is annealed causing the nanocrystals to form near the surface. After this the sample is exposed to a thermal evaporation of electrical contacts which can be chosen to also be transparent (eg ITO). Assuming the use of holes of 50 nm diameter, an ion dose of 2 x 10^{16} C/cm^2 and using the fact that only 2% of the implanted carbon turns into diamond, one would expect between 1 and 10 diamond nanocrystals under each 50 nm sized electrical contact. The conducting tip of a NSOM could then be used to investigate, for example, the shifts of Photoluminescent transitions as a function of applied electric field. Individual addressing and coupling of the dots to an external circuit would require a more sophisticated design for the mask but this would appear to be achievable with present lithographic techniques.

5 Conclusion

We have shown that ion implantation is a viable method for the creation of arrays of quantum dots. Phase pure nanodiamond dots of size less than 7nm have been created by the direct conversion of implanted C into diamond via thermal annealing. Because such a large fraction of atoms lie on the surface in small clusters, surface termination must be taken into account. Indeed, we have shown that size matters not just for electronic structure but also is crucial for the stablization of metastable phases. Finally, we have presented a scheme by which arrays of quantum dots, each addressed by an individual electrode, can be fabricated using contact masks and ion implantation techniques.

References

1. G.D. Sanders, K.W. Kim and W.C. Holton, Phys. Rev B61, 7526, (2000)
2. C.W. White et al., Nucl. Inst. Meth. Phys. Res., B127, 545, (1997)
3. A. Meldrum, E. Sonder, R.A. Zuhr, I.M. Anderson, J.D. Budai, C.W. White, and L.A. Boatner, J. Mater. Res, 14, 4489, (1999)
4. C. W. White, J.D. Budai, J.G. Zhu, and S.P. Wihtrow, Appl. Phys. Lett., 68, 2389, (1996)
5. S.P. Withrow, C.W. White, A. Meldrum, J.D. Budai, D.M. Hembree Jr, and J.C. Barbour, J. App. Phys., 86, 396, (1999)
6. P.R. Hemmer, A.V. Turukhin, M.S. Shahriar, and J.A. Musser, quant-ph/0007114
7. S.Prawer, J.L. Peng, J.O. Orwa, J.C. McCallum, D.N. Jamieson and L.A. Bursill, Phys Rev B62, R , (2000)
8. F.H. Ree, N.W. Winter, J.N. Glosli, and J.A. viecelli, Physica B265, 223, (1999)
9. H. Masuda, M. Watanabe, K. Yasui, D. Tryk, T. Rao, and A. Fujishima, Adv. Mat. 12, 444, (2000)

DOGMA AND HERESY IN QUANTUM COMPUTING

D. P. DIVINCENZO
IBM T. J. Watson Research Center, PO Box 218
Yorktown Heights, NY 10598 USA
e-mail: divince@watson.ibm.com

Important new ideas for the physical implementation of quantum computers are reviewed.

1 Dogma

It may seem today that the physical implementation of quantum computers has only recently been given serious consideration, as a result of the remarkable theoretical developments of the last ten years or so identifying the considerable power of quantum information processing. This would be an incorrect reading of history: many of the earliest workers in the field were very much aware of Rolf Landauer's motto, "information is physical,"[1] and were motivated to explore quantum computing precisely because of the desire to make stronger the linkage between computing and physical reality. What was lacking among the early workers was an the intimate acquaintance with experimental quantum physics required to begin turning these theoretical ideas into reality.

So, while early efforts, like Seth Lloyd's pioneering 1993 paper[2] on quantum computing with spin chains, were on the right track (it was this paper that first got me interested in the subject, despite Rolf Landauer's skepticism), none of the work at that time spoke clearly enough to be comprehensible to any significant part of the experimental community. I have to confess that my own 1994 paper[3], discussing how two-bit gates could be done between spins using an atomic force microscope, falls into this category. Cirac and Zoller[4] were the first workers who were able to put together a compelling scenario for how quantum computing could realistically be attacked in the laboratory; this very quickly converted a noticeable fraction of ion-trap research in the world into quantum computing research.

Upon seeing their beautiful work, I was convinced that a more general approach could be formulated, which would show that almost every area of quantum research could, in a realistic way, approach the problem of building a quantum computer. It was this that led me to put together a general prescription for the physical implementation of quantum computation, codified in five requirements[5], later enlarged to seven[6]. Since I have written about them extensively, I will just quote them *verbatim* from a recent article:

The Requirements

1. *A scalable physical system with well characterized qubits*
2. *The ability to initialize the state of the qubits to a simple fiducial state, such as* $|000...\rangle$
3. *Long relevant decoherence times, much longer than the gate operation time*
4. *A "universal" set of quantum gates*
5. *A qubit-specific measurement capability*
6. *The ability to interconvert stationary and flying qubits*

7. The ability faithfully to transmit flying qubits between specified locations

2 Heresy

It is impressive to me how large an enterprise the building of quantum computers has become. I can cherish the illusion that this is happening as a result of the diligent following of these rules. But, while I think that these rules are a good start, and that they do offer one valid view of how the theory of quantum information processing should be projected onto experiment, I think that we should treat this canon the way physicists usually treat a canon, by criticizing, reinterpreting, and creatively flouting it. Let me give some examples from current work of what I mean:

pre-packaged gates: There is a popular version of requirement 4 which says that "cNOTs and one-bit gates must be doable." At an abstract level, this remains a very serviceable version of this rule (although it has probably always been an inappropriate version to apply to problems in quantum simulation). But the translation of this abstract rule into a physical implementation permits a surprising degree of flexibility. For the cNOT, we often thought that this would require the engineering of very precisely tunable two-qubit interactions (for example, spin-spin interactions), such that the time evolution operator resulting from the engineered time-dependent Hamiltonian $H(t)$ is the cNOT transformation.

This version of the dogma is bunk. One debunker is a paper by Gottesman and Chuang[7], which shows that *no* two-qubit interactions, either of the cNOT type or any other type, need take place after the start of the quantum computation. This requirement is replaced by other potentially difficult, but physically quite different, requirements: first, that the qubits are entangled in particular ways before the start of the quantum computation, and second, that two-qubit measurements, in the Bell basis, be doable in the course of the computation. Creation of entangled states (to be specific, Greenberger-Horne-Zeilinger states are needed in their scheme) itself requires qubit interactions, but creating a specific state is very different from creating a particular unitary transformation: it need not be done by reversible means (for example, relaxation to the ground state can be a good way of getting a particular entangled state), and various corrective strategies like entanglement purification are available for this task.

One might object that nothing much has really been gained, since Bell measurements surely involve interaction between qubits, so it seems that the execution of the quantum program still involves two-qubit interactions, although of a somewhat different form. One might be tempted to postulate that some form of two-qubit interaction is surely necessary during the execution of a quantum computation. But this too is bunk...

cluster-state entanglement: Raussendorf and Briegel[8] have identified a multi-qubit entangled state, the so-called cluster state, with the property that, ideally, quantum computation can be done with it using no subsequent two-qubit operations *at all*, neither unitary operations nor measurements. All steps of quantum computation are completed by one-qubit measurements; the classical communication of these measurement outcomes from step to step carries the computation forward. The cluster state is more intricate than a set of GHZ states, but it is shown[8] that the state can be created by the appropriate application of $\sigma_z\sigma_z$ two-body interactions between nearest neighbors on a square, two-dimensional lattice. It appears that in the non-ideal case (with decoherence present), two-qubit operations are occasionally necessary in order that the state may be teleported from one "cluster" to another. But I think that no one really knows how much simplification along these lines is really possible, so much has been achieved recently that it would be foolish to speculate that we are near the end of the story.

Indeed, this is not the end of the story of gate-scheme simplifications in several respects; I know of at least two other remarkable simplifications that have been discovered in the context of specific physical implementations:

linear optics: Photons flying through a set of intersecting beam paths were nearly the first physical implementation of quantum computing that was envisioned[9,10,11]. But this line of investigation has not gotten very far, largely because the kind of photon-photon interaction required to enact a cNOT interaction directly on a pair of polarization qubits requires monstrously big optical nonlinearities, or absurdly long beam paths (the interaction length required in an ordinary nonlinear crystal is, I think, measured in miles).

You might guess from the above story that more can now be done with this idea; but a direct application of either of the above ideas would require the same difficult nonlinear optics (to create directly GHZ or cluster states). But a recent work[12] has identified a scheme, related to the Gottesman-Chuang approach[7], in which all the necessary work (entangled state creation, gate action by teleportation) can take place with only linear optical components. The only two-photon interactions required are those given by having them simultaneously enter the two input ports of a half-silvered mirror. It was long thought that such a linear interaction could "obviously" not enact a non-trivial gate such as a cNOT; the missing ingredient was use of the full power of photonic measurements, in addition to the necessary preparation of one-photon states. Both one-photon state preparation and detection are non-trivial (especially the latter), but technical progress has been recently reported on both fronts. Interested readers can consult other papers at this conference for details.

We might note that common among all these proposals is an absolute requirement for "on-the-fly" quantum measurements, that is, fast, conclusive measurements, the results of which can immediately effect other physical actions (e.g., the basis in which further measurements are done). We know that this is not an absolute requirement of quantum computation: measurement can be delayed to the end, but at the cost, apparently, of going all the way back to the old "cNOT plus one-bit gate" paradigm for implementation. Well, we have found at least one (perhaps perverse, at first sight) variant on this approach...

exchange-only quantum computing: We have recently[13] offered up an explicit scheme, that, with modest space and time overheads, can do quantum computing without intermediate measurements and without any one-qubit gates. You might ask, what is the point of getting rid of single-bit operations, which are surely the easiest part of the whole game? Well, we believe that there are implementation schemes in which this may not be the case. In our quantum-dot schemes[14], in the popular Kane scheme and its variants[15], and in the surface-acoustic wave/single-electron approach[16], considerable effort will have to be expended, in materials and device complexity, to engineer the capabilities to do implement one-spin Hamiltonians. The technical complexities of fast, accurate single-spin measurements also appear to be daunting in these schemes. On the other hand, there is reason to believe that a particular two-spin interaction, the Heisenberg exchange coupling, will be accurately engineerable, since it corresponds to the natural physics of interaction of the spins of neighboring localized orbitals.

I could go on at greater length about the results above, but there is a lot more dogmabusting out there that I think needs to be mentioned. An even more basic premise of the orthodox discussion is that quantum information is carried by qubits, and that qubits are embodied by quantum-mechanical two-level systems. Well, this is wrong in all sorts of ways:

coding: This is an old part of the story now, and has been largely (but not entirely) accepted into the orthodox viewpoint. At least since the identification of quantum error correcting codes[17], we have understood that we can use the state of multiple two-level systems (three, five, nine,...) to represent a single qubit. There are other embodiments that can also be put under the heading of coding, in which a small part of Hilbert space is singled out in a larger physical system. This has become quite commonplace in the discussions of "macroscopic" quantum coherence in the various superconducting devices that we are interested in now. A new "mesoscopic" system that looks rather promising for quantum computing, using the collective Zeeman coherence of an atomic vapor[18,19], involves a less obvious form of coding. Here the $|0\rangle$ might be represented by all spins down in a large number (perhaps 10^{12}) of atoms, and the $|1\rangle$ by a superposition of states with exactly one spin up. Many promising variants of these sorts of mesoscopic quantum bits are being discussed now[20].

There are, in addition, much more radical departures from the dogma of qubits that are being pursued:

continuous quantum variables: The harmonic oscillator as the fundamental carrier of quantum information is a very different object, apparently, from a qubit. It has an infinite dimensional Hilbert space (these are the "continuous quantum variables"), and it is not a promising undertaking to single out the lowest two energy eigenstates to represent a qubit; since the level spacing is constant, most of the techniques for confining the system to these two levels do not work. Nevertheless, these systems provide a possible alternative starting point for quantum computing, as first pointed out in principle by [21]. Since this first work, there have been demonstrations of teleportation of continuous variables[22], and a series of investigations culminating in the recent work showing how to do fault tolerant quantum computation[23]. Interestingly, the protocol introduced in this last work only involves linear optics as in the above work; the entangled state resources required are quite different. Since Hilbert spaces come in a dazzling variety of shapes and sizes, I wouldn't be surprised to see more of these types of heterodoxies in the future.

Another one of the original tenets was that unitary transformations take place by the precise application of a time-dependent Hamiltonian; both the precision of this operations, and the decoherence that occurs during operations, must be held within stringent limits (a required relative accuracy of 10^{-4} is often quoted). The fact that quantum gates can be done with measurements, which are patently non-Hamiltonian, irreversible acts, already partially demolishes this dogma. But it is also overthrown on other sides as well...

geometric phases: Even when gates are accomplished by Hamiltonian action, it need not be the case that the resultant Unitary evolution depends sensitively, or at all, on the precise time profile of $H(t)$. The sensitive effects are those that depend on the "dynamical phase", that is, roughly speaking, the change of the state that is linked to the change of the energy of the states as a function of time. But we know that even if the dynamical-phase effect is absent (as when the states of the qubit remain eigenstates of the Hamiltonian with constant energy), the state can change due to geometrical-phase effects. The unitary transformations so effected are insensitive, in varying degrees, to the time profile of the Hamiltonian, offering a different form of fault tolerance in quantum computation. The theoretical work in this area has been beautiful[24,25], and, while there has been some work now in the laboratory along these lines[26], I think that most of the impact of these ideas lie in the future.

I would like to mention one last favorite heresy of mine concerning quantum communications. "Flying qubits" will be crucial for many of the exciting applications of quantum information processing; many of us think that these applications will ultimately be much more important than those that can be done with a quantum computer alone. Most have taken on faith that the flying qubit will be a photon, since it travels fast through either fibers or free space, and it doesn't decohere so fast. However, I think that it is at least worth considering the possibility of

electrons as flying qubits. Photons, despite their other virtues, don't interact with other qubits all that well, so perhaps, at least for short distances, electrons could be the flying qubit of choice. Present research has shown that electrons can travel some (modest) distance through a quantum wire without decohering. Some serious theoretical work is proceeding along these lines[27], and I am hopeful for the future of this idea.

3 Outlook

This completes my choice list of current heresies in quantum computing, but, knowing physicists, I am certain that these will not be the last. Seriously, I believe that progress will have to proceed on many fronts, both theoretical and experimental, for us to achieve such a hard thing as quantum information processing. I think that pure theory will continue to produce interesting reasons for why we want to do information processing quantum mechanically. These new algorithms and protocols, along with all the existing ones, will guide us in what kind of hardware we should be aiming for. The job of implementing quantum hardware will take continuous feedback between theory and experiment. Together I think that we will need both a collection of dogmas, around which we can structure our discussions and thinking, and a host of dissents of all kinds, which will permit a reshaping, extending, or perhaps discarding of the dogmas.

I think that we are still thousands of innovations away from having real quantum information processing. If we are lucky, each of our papers has one such innovation (I fear that the ratio is sometimes not so good). So, let's get working!

I am grateful for the support of the National Security Agency and the Advanced Research and Development Activity through Army Research Office contract number DAAG55-98-C-0041.

References

1. R. Landauer, "Information is Physical," Proc. Workshop on Physics and Computation, PhysComp '92 , ed. D. Matzke (IEEE Comp. Sci. Press, Los Alamitos, 1993), pp. 1-4.
2. S. Lloyd, "A potentially realizable quantum computer," Science **261**, 1589 (1993); "Envisioning a quantum supercomputer," *ibid.* **263**, 695 (1994).
3. D. P. DiVincenzo, "Two-bit gates are universal for quantum computation," Phys. Rev. A **51**, 1015 (1995), cond-mat/9407022.
4. J. I. Cirac and P. Zoller, "Quantum computations with cold trapped ions," Phys. Rev. Lett. **74**, 4091 (1995).

5. D. P. DiVincenzo, "Topics in quantum computers," in *Mesoscopic Electron Tr ansport*, Vol. 345 of NATO Advanced Study Institute, Series E, eds. L. Sohn, L. Kouwenhoven, and G. Schoen (Kluwer, 1997), pp. 657-677, cond-mat/9612126.
6. D. P. DiVincenzo, "The physical implementation of quantum computers," Fort. der Physik **48**, 771-784 (2000), quant-ph/0002077.
7. D. Gottesman and I. L. Chuang, "Demonstrating the viability of universal quantum computation using teleportation and single qubit operations," Nature **402**, 390-3 (1999), quant-ph/9908010.
8. R. Raussendorf and H.-J. Briegel, "Quantum computing via measurements only," quant-ph/0010033; H.-J. Briegel and R. Raussendorf, "Persistent entanglement in arrays of interacting particles," quant-ph/0004051.
9. G. J. Milburn, "Quantum optical Fredkin gate," Phys. Rev. Lett. **62**, 2124-2127 (1989).
10. I. L. Chuang and Y. Yamamoto, "Simple quantum computer," Phys. Rev. A **52**, 3489-3496 (1995), quant-ph/9505011.
11. Q. A. Turchette *et al.*, "Measurement of conditional phase shifts for quantum logic," Phys. Rev. Lett. **75**, 4710-3 (1995).
12. E. Knill, R. Laflamme, and G. Milburn, "Efficient linear optics quantum computation," quant-ph/0006088; "Thresholds for linear optics quantum computation," quant-ph/0006120.
13. D. P. DiVincenzo *et al.*, "Universal quantum computation with the exchange interaction," Nature **408**, 339-342 (2000), quant-ph/0005116.
14. D. Loss and D P. DiVincenzo, "Quantum computation with quantum dots," Phys. Rev. A **57**, 120 (1998), cond-mat/9701055.
15. B. Kane, "A silicon based nuclear spin quantum computer," Nature **393**, 133 (1998); R. Vrijen *et al.*, "Electron spin resonance transistors for quantum computing in silicon-germanium heterostructures," Phys. Rev. A **62**, 012306 (2000), quant-ph/9905096.
16. C. H. W. Barnes, J. M. Shilton, and A. M. Robinson, "Quantum computation using electrons trapped by surface acoustic waves," cond-mat/0006037.
17. P. W. Shor, "Scheme for reducing decoherence in quantum computer memory," Phys. Rev. A **52**, R3493-R2496 (1995); A.M. Steane, "Error correcting codes in quantum theory," Phys. Rev. Lett. **77**, 793-797 (1996).
18. M. D. Lukin *et al.*, "Dipole blockade and quantum information processing in mesoscopic atomic ensembles," quant-ph/0011028.
19. D. P. Philips *et al.*, "Storage of light in atomic vapor," quant-ph/0012138.
20. L. M. Duan *et al.*, "Quantum communication between atomic ensembles using coherent light," quant-ph/0003111.
21. S. Lloyd and S. Braunstein, "Quantum computation over continuous variables," Phys. Rev. Lett. **82**, 1784-1787 (1997).
22. A. Furusawa *et al.*, "Unconditional quantum teleportation," Science **282**, 706-709 (1998).
23. D. Gottesman, A. Kitaev, and J. Preskill, "Encoding a qudit in an oscillator," quant-ph/0008040.
24. A. Kitaev, "Fault tolerant quantum computation with anyons," quant-ph/9707021.
25. J. Pachos and P. Zanardi, "Quantum holonomies for quantum computing," quant-ph/0007110.

26. J. A. Jones *et al.*, "Geometric quantum computation using nuclear magnetic resonance," Nature **403**, 869-871 (2000).
27. G. Burkard, D. Loss, and E. Sukhorukov, "Noise of entangled electrons: bunching and antibunching," Phys. Rev. B **61**, 16303-6 (2000), and references therein.

TEMPORAL AND SPATIAL CHARACTERIZATION OF QUANTUM DOTS

Y.HIRAYAMA[a,b,*], K.KANISAWA[a], Y.TOKURA[a], H.YAMAGUCHI[a], and T.FUJISAWA[a]

[a]*NTT Basic Research Labs., 3-1 Morinosato Wakamiya, Atsugi-Shi, Kanagawa 243-0198, Japan,*

[b]*CREST-JST, 4-1-8 Honmachi, Kawaguchi-Shi, Saitama 332-0012, Japan,*

We summarise recent developments in temporal and spatial characterizations of semiconductor quantum dots, which are plausible structures for the realisation of semiconductor quantum computing systems. The pulse technique reveals an extremely long electron relaxation time in a semiconductor quantum dot when the excited state has a different spin configuration from that of the ground state. Single electron emission and capture are detected with a time resolution better than 1μsec by using rf-SET technique. Spatial measurements using low-temperature scanning-tunnelling-microscope in combination with an electron accumulation layer at the InAs surface clearly demonstrate the detection of the electron wave distribution in a semiconductor quantum dot for each individual zero-dimensional mode.

1 Introduction

There is significant interest in quantum computers because they will have vastly enhanced capabilities due to quantum parallelism. In quantum computers, quantum bits (qubits) formed from a coherent two-state system are used in stead of classical bits. Experimental demonstrations of quantum computation have been carried out by using nuclear magnetic resonance (NMR) spectroscopy and atomic systems [1-3]. However, solid-state qubits may be desirable from the viewpoint of scalability. Although a qubit was recently demonstrated by using superconducting systems [4], semiconductor qubits are attracting great interest because of the availability of cutting edge semiconductor technologies, and there have been many qubit proposals based on semiconductors. Among them, qubits based on semiconductor quantum dots are the most plausible; the charge, spin, zero-dimensional mode, and exciton characteristics of quantum dots have a potential to realize qubits [5-15]. For example, the coupled dots schematically shown in Fig. 1 operate as a quantum two-state system when there is a coherent quantum coupling between them. This coupling results in the energy gap opening at the point where the left- and right-dot energy levels cross. The gap formation has already been confirmed experimentally by microwave absorption measurements [16,17]. Therefore, whether electron locates in the left or right dot behaves as qubit in the coupled dot. However, decoherence becomes important in semiconductor systems [18], and a short time scale characterization becomes necessary in order to demonstrate qubit operation. In this paper, we will discuss the relaxation characteristics of electrons in the quantum dot and the monitor of single electron tunnelling as the short time scale characterizations.

---

* Corresponding author. Tel.: +81-46-240-3440; fax.: +81-46-240-4727
E-mail address: hirayama@will.brl.ntt.co.jp (Y.Hirayama)

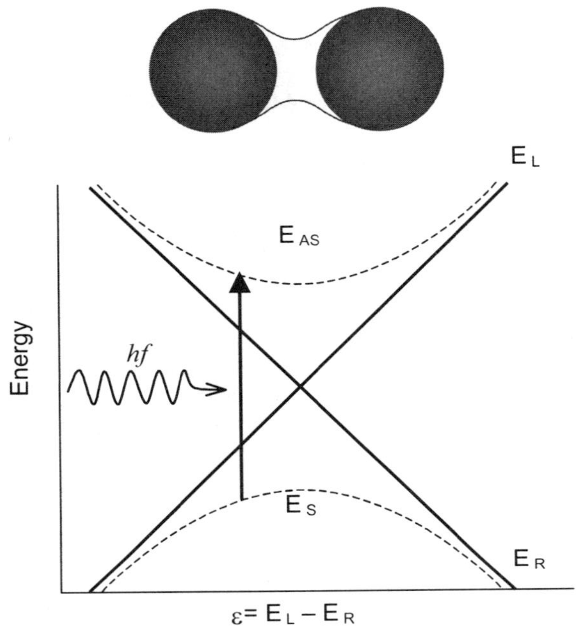

Figure 1. Schematic energy diagram of a coupled dot structure. E_L and E_R are the energy levels of the left and the right dots, respectively. Now we consider the situation where E_L and E_R are controlled by a bias applied to a certain gate. When there is no quantum coupling between the two dots, E_L and E_R simply cross as shown by the solid lines. However, the energy levels separate into symmetric and antisymmetric branches (broken lines) when there is a quantum coupling between the dots. These energy levels operate as a quantum two-level system and can be applied to qubits. The existence of the energy gap between symmetric and antisymmetric levels has been confirmed by microwave absorption experiments.

In semiconductor quantum dots, the zero-dimensional modes have also a possible application to qubits [5,6] and it is important to understand them. In the second half of this paper, we describe the detection of each zero-dimensional mode in a quantum dot with a low-temperature scanning-tunnelling-microscope (LT-STM).

2 Transient current spectroscopy

We recently evaluated carrier relaxation in a semiconductor quantum dot by putting electron only in the excited state using a pulse technique [19,20]. In the actual experiments, we applied a pulse signal to the gate to modulate the electrostatic potential of the quantum dot (Fig. 2(a)). The energy separation between the ground state and excited state was set at 100-800 μeV, which was smaller than the addition energy of a single electron to the dot (about 2 meV for this dot). While the pulse was in the steady state, both the excited and ground states were located above μ_L and μ_R for a period that was long enough to ensure both states were empty. Then a positive pulse with duration of t_p pushed down the potential of the dot such that only the excited state was located in the transport window (see Fig.

2(d)). The window width, i.e. bias voltage μ_L-μ_R, was kept at a small value, around 0.1 mV. The addition energy is larger than the energy separation between the excited and the ground states; thus, both states cannot be simultaneously occupied. This means non-equilibrium transient current flows through the excited state until the ground state is occupied. This transient current is observed in Fig.3 (a).

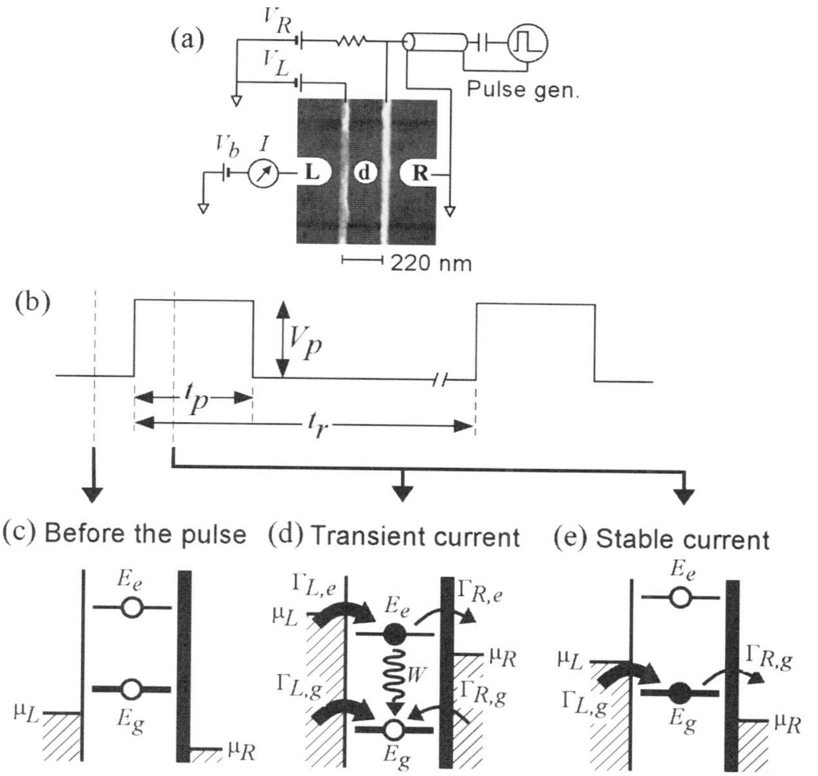

Figure 2. (a) Schematic of the setup for the pulse measurements. All measurements were performed at 150 mK. (b) Schematic of the pulse train used in the measurements. (c) Schematic energy diagram before the pulse. Both the excited state of energy E_e, and the ground state of energy E_g become empty. (d) and (e) Schematic energy diagrams during the pulse. (d) The transient current through the excited state continues until the ground state is occupied. (e) The stable current through the ground state persists during the pulse.

The decay time of the non-equilibrium transport should reflect the rate of energy relaxation, W, from the excited to the ground state. The average number of tunnelling electrons through the excited state $\langle n_e \rangle$ is plotted as a function of the pulse length t_p in Fig. 3 (b). The $\langle n_e \rangle$ is obtained by $I_e t_r / e$, where I_e and t_r are the current through the excited state and the repetition time of the pulse measurement, respectively. The non-equilibrium transport $\langle n_e \rangle$ shows saturating behaviour, which can be fitted well with a single exponential function with a time constant of 2.1 μsec. On the other hand, transport through the ground state $<n_g>$ has a linear dependence on the pulse length. The detailed discussions have revealed that the time constant of 2.1 μsec obtained in Fig. 3 (b) is not determined by W but by the rate of direct tunnelling from the resovours to the ground state ($\Gamma_{L,g}$ and $\Gamma_{R,g}$ in Fig. 2 (d)). This means relaxation time is longer than 2.1 μsec.

When we measure the characteristics in a magnetic field, the energy levels in the quantum dot become complicated due to level crossing and anticrossing among different orbits. Therefore, many combinations of excited and ground states are possible, and we found that they can be classified into two groups. In one group, transient current is clearly seen, and the relaxation time is at shortest longer than a few μsec. In the other, transient current is below our current sensitivity, such as shown in the upper trace in Fig. 3 (a), and the relaxation time should be shorter than a few nsec. This significant difference in relaxation time can be attributed to the spin. If the excited state and the ground state have different spin configurations, the transition should involve the spin-flip process and the rate is expected to be extremely small [21]. Recently, we confirmed that the relaxation time is longer than 100 μsec for some combination of excited and ground states. Although the long relaxation time does not necessarily mean a long phase coherence, it is attractive for spin-related applications of quantum dots.

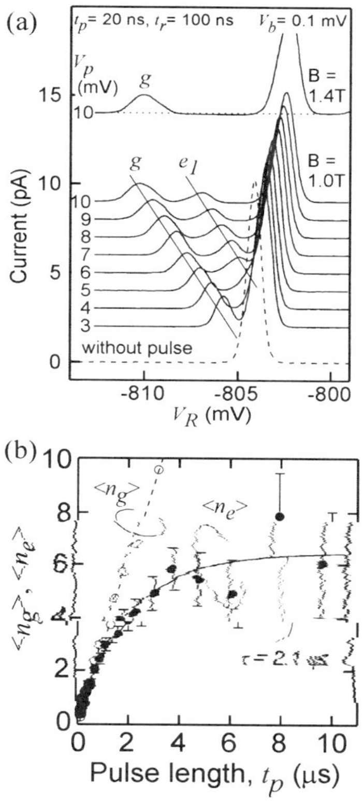

Figure 3 (a) Pulse-excited current for various pulse heights, V_p. Each line is shifted for clarity. Measurements were carried out in a perpendicular magnetic field of B=1 T. The stable current g through the ground state (see Fig. 2(e)) and the transient current e1 through the first excited state (see Fig. 2(d)) are clearly seen. The large peaks connecting to the peak observed without pulse (broken line) come from the stable current through the ground state for the steady situation between pulses. The upper trace measured at B=1.4 T shows no measurable transient current. (b) The average number of tunnelling electrons per pulse, $<n_g>$, for a stable current through the ground state (open circles) and $<n_e>$ for a transient current through the excited state (filled circles). The dashed and the solid lines are fitted to the data with a linear and an exponential function, respectively. Measurements were carried out at B=1.35T.

3 Monitor of single electron tunnelling

Conventional pulse techniques accumulate the signal by repetition of the pulse sequence. Therefore, the average number of tunnelling electrons was measured in Fig. 3. However, it is desirable for semiconductor qubits in the future that the characteristics are detected by the single event. We should develop time-domain measurements of single electron motion in a short time scale in order to follow the fast response of the charge. As a promising technology, the radio-frequency (rf) operated single-electron-transistor (SET) technique has recently been proposed and demonstrated by Schoelkopf et al.

[22]. In this technique, a SET is placed in an *LC* resonator that includes the capacitance of the cables. When the rf carrier signal is supplied at the resonant frequency f_r to the resonator, the reflected (or transmitted) signal is sensitive to the resistance of the SET or the charge on the quantum dot.

We used a transmission-type rf SET in our study. This rf SET has some advantages compared with the reflection-type rf SET discussed in ref.23 and has been applied to measure single electron tunnelling. We used a lateral SET fabricated on an AlGaAs/GaAs structure. This SET shows typical Coulomb blockade oscillations both for dc current and rf transmission when it is measured in a dilution refrigerator with base temperature of 30 mK. The best charge sensitivity is obtained at the largest slope of Coulomb blockade oscillations, and $\delta q = 3.6 \times 10^{-5} e / \sqrt{\text{Hz}}$ for frequencies higher than 1 kHz [24]. For certain gate conditions, we find a random telegram signal (RTS) in our device. This RTS originates from the capture and emission of a single electron at a trap located close to the dot. The specific trap we study has a relatively fast capture (emission) rate and well demonstrates the fast response of the rf SET, as shown in Fig. 4. Signal V_t is high if the trap is occupied by an electron and low if the trap is empty. The wide distribution of the durations that the trap is empty or occupied can be explained by Poisson statistics of each emission and capture processes. As shown in the inset of Fig. 4, we observe a duration as short as a few μsec and a rise/fall time shorter than 1 μsec [24]. If the trap is replaced with another quantum dot or coupled dots, electron motion in these systems can be measured with time resolution shorter than 1 μsec. Although the time resolution of the measurements has to be improved, the rf-SET technique appears to be a powerful tool toward the realisation of semiconductor qubits operation.

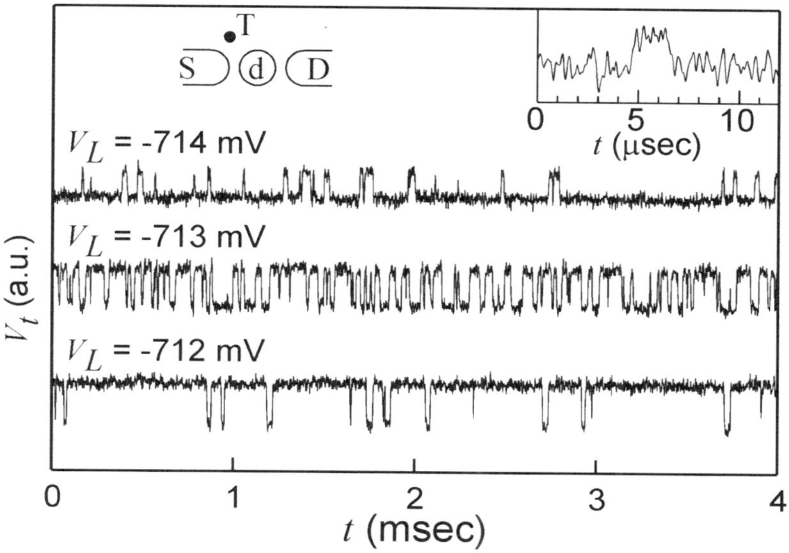

Figure 4 Random telegrah signal (RTS) measured at different gate voltages V_L applied to the narrow gate between the source (S) and dot (D). Signal V_t is high if the trap is occupied by an electron and low if the trap is empty. The right inset shows the shortest duration of the occupied state. The left inset is a schematic diagram of the dot structure, including the location of the trap T.

4 Spatial characterization of quantum dots

The mode in the quantum dot could also possibly be applied to qubits. The zero-dimensional modes have been detected by using transport and/or optical measurements. However, the zero-dimensional modes may become more useful if they are directly detected in semiconductor dots. For artificially manipulated metal systems, electron standing waves and confined states have been observed by LT-STM [25]. However, except for the observation of donor-bound states at a cleaved edge of GaAs and InAs [26], there has been no such report for semiconductor layers and nano- structures. Recently, we observed clear electron standing waves for the two-dimensional electron gas (2DEG) accumulated at the surface of an InAs (111)A thin film as shown in Fig. 5 [27]. We used a constant-current (about 0.2 nA) LT-STM measurement in a fixed tip-sample bias voltage. In this measurement, simultaneously obtained *dI/dV* signal reflects a local density of states (LDOS) of electrons. The modulation amplitude of the bias to detect *dI/dV* was about 10 mV peak-to-peak.

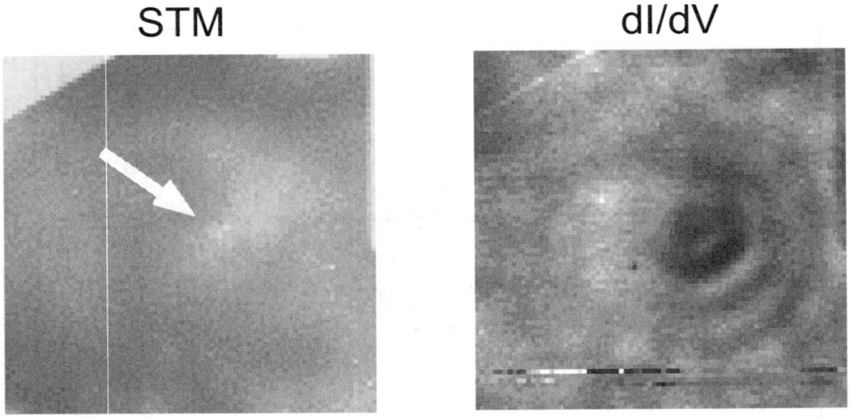

Figure 5. Constant current 67 nm x 67 nm STM image of the InAs (111)A surface and a simultaneously mapped *dI/dV* image taken at 5.3 K. The bias voltage is 0.14 V and this corresponds to the electron tunnelling from the tip to the unoccupied electron state located 0.14 V above the Fermi energy. In the *dI/dV* images, circular LDOS standing waves appear around a point scatterer (shown by an arrow) due to Friedel oscillation. The brighter regions are due to a higher LDOS.

Equilateral triangular defects were often observed at the surface, as shown in the STM images in Fig. 6. Such equilateral triangular regions have a 1/3 or 2/3 monolayer height difference and are the surfaces of stacking fault tetrahedra. The boundaries operate as a potential barrier for conduction electrons. [28] Therefore, the surface-accumulated 2DEG is confined in these equilateral triangular regions. In other words, these defects operate as triangular-shaped quantum dots. The *dI/dV*, i.e. LDOS, results obtained for triangular dots having L of 48.7 and 33.9 nm are shown in Figs. 6 and 7, respectively, where L is a side length of the triangular dots. The three-fold symmetric patterns are clearly observed within the dot. As the bias voltage is decreased, the features retain their three-fold symmetry and appear to migrate towards the center. This tendency agrees very well with the increase in electron wavelength with decreasing energy. In other words, the mode number corresponding to the observed probability distribution decreases with energy. However, in the larger dot shown in Fig. 6, more than two modes are included in the energy separation of about 10 meV, i.e., the modulation amplitude of the *dI/dV* measurement. Therefore, the observed probability distribution patterns represent the average of a few modes in Fig. 6 [27,28].

Figure 6. Constant current STM image of the triangular quantum dot with a side length of 48.7 nm. Measurements were carried out at 5.3 K. Simultaneously mapped *dI/dV* images are shown for bias voltages at 0.14 V (left), 0.10 V (center) and 0.06 V (right).

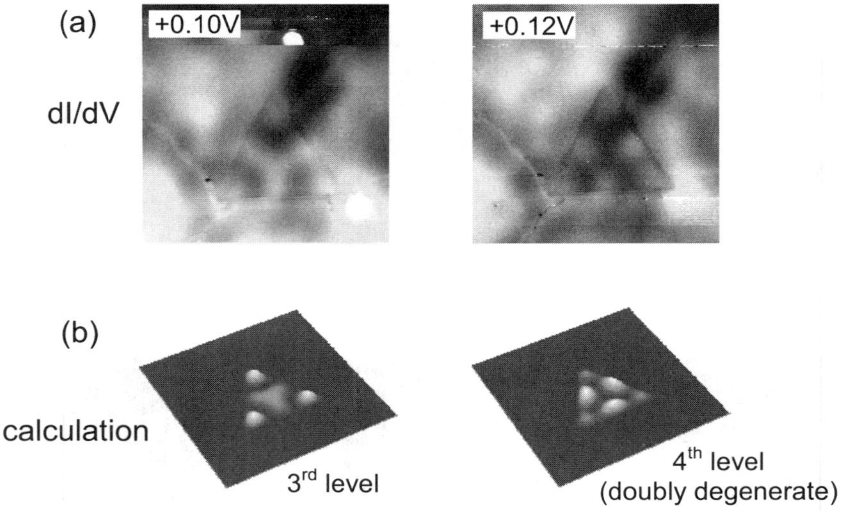

Figure 7. dI/dV images and calculated probability distributions for a triangular quantum dot with a side length of 33.9 nm. The images taken at 0.10 and 0.12 V are third and fourth discrete quantized zero-dimensional modes, respectively.

This situation changes for the small dot shown in Fig. 7. Figure 7(b) shows the calculated probability distribution for the 3rd and 4th modes of the triangular quantum dot. A hard-wall triangular potential with a weak attractive potential near the walls was assumed in the calculation. The calculated probability distribution agrees very well with the experimentally measured distribution in Fig. 7. Furthermore, the estimated energy separation between the 3rd and 4th modes is 140 meV and two different distribution patterns were observed for the bias difference of 0.2 V in Fig. 7(a). Thus, we conclude that each pattern corresponds to a discrete zero-dimensional mode within the triangular quantum dot [28]. These experiments suggest that we can detect the zero-dimensiona mode in a quantum dot by using nanoprobing techniques□although such measurements are still very time consuming.

5 Conclusions

The semiconductor quantum dot system is a promising candidate for realising qubits. Although further development is needed, there have been steady advances in technologies for measuring the characteristics of quantum dots. They have contributed to our understanding of the physics of semiconductor quantum dots. The transmission-type rf-SET technique has demonstrated single electron capture and emission with a time resolution better than 1 μsec. The nonequilibrium current through quantum dot using the pulse technique enables us to estimate the rate of electron relaxation from the excited to the ground states. It is surprising that the relaxation rate is very small for a transition including spin flip and that an electron can stay in the excited states longer than a few ten μsec. Such pulse techniques will be useful to demonstrate the semiconductor qubit operation in the future.

In addition to the temporal evaluations, spatial characterizations have been proceeded for semiconductor quantum systems. The electron wave features are visibly shown by LT-STM measurements for semiconductor two-dimensional electron systems by using electron accumulation at the InAs (111)A surface. The confined zero-dimensional modes have also been measured for the equilateral triangular shaped quantum dot formed in the InAs thin films. We succeeded in detecting an individual zero-dimensional mode for a small-size quantum dot. Although successful measurement is limited to special systems at present, our results suggest the possibility of detecting the mode number of an electron in a quantum dot by using LT-STM. The LT-STM technique may become important as an input and output for qubits

Acknowledgement

We would like to thank M.J.Butcher, D.G.Austing and T.Mukai for their helpful discussions and supports. Part of this work was supported by the NEDO NTDP-98 program.

References

1. Q.A.Turchette, C.J.Hood, W.Lange, H.Mabuchi, and H.J.Kimble, Phys. Rev. Lett. **75**, 4710 (1995).
2. C.Monroe, D.M.Meekhof, B.E.King, W.M.Itano, and D.J.Wineland, Phys. Rev. Lett. **75**, 4714 (1995).
3. N.A.Gershenfeld and I.L.Chuang, Science **275**, 350 (1997).
4. Y.Nakamura, Yu.A.Pashkin, and J.S.Tsai, Nature **398**, 786 (1999).
5. A.Barenco, D.Deutsch, and A.Ekert, Phys. Rev. Lett., **74**, 4083 (1995).
6. N.H.Bonadeo, J.Erland, D.Gammon, D.Park, D.S.Katzer, and D.G.Steel, Science, **282**, 1473 (1998).
7. D.Loss and D.P.DiVincenzo, Phys. Rev. **A57**, 120 (1998).
8. L.H.Openov, Phys. Rev. **B60**, 8798 (1999).
9. A.Imamoglu, D.D.Awschalom, G.Burkard, D.P.DiVincenzo, D.Loss, M.Sherwin, and A.Small, Phys. Rev. Lett., **83**, 4204 (1999).
10. G.D.Sanders, K.W.Kim, and W.C.Holton, Phys. Rev. **B61**, 7526 (2000).
11. S.Bandyopadhyay, Phys. Rev. **B61**, 13813(2000).
12. Tanamoto, Phys. Rev. **A61**, 022305 (2000).
13. X.Hu and S.Das Sarma, Phys. Rev. **A61**, 062301 (2000).
14. J.H.Oh, D.Ahn, and S.W.Hwang, Phys. Rev. **A62**, 052306 (2000).
15. T.Ohshima, Phys. Rev. **A62**, 062316 (2000).
16. T.H.Oosterkamp, T.Fujisawa, W.G. van der Wiel, K.Ishibashi, R.V.Hijman, S.Tarucha, and L.P.Kouwenhoven, Nature **395**, 873 (1998).
17. T.Fujisawa, W.G. van der Wiel, and L.P.Kouwenhoven, Physica **E7**, 413 (2000).
18. T.Fujisawa, T.H.Oosterkamp, W.G. van der Wiel, B.W.Broer, R.Aguado, S.Tarucha, and L.P.Kouwenhoven, Science **282**, 932 (1998).
19. T.Fujisawa, Y.Tokura and Y.Hirayama, ICPS2000(F15), to be published in Phys. Rev. B (RC).
20. T.Fujisawa, Y.Tokura, and Y.Hirayama, to be published in Physica B.
21. A.V.Kaetskii and Yu.V.Nazarov, Phys. Rev. **B61**, 12639 (2000), cond-mat/0003513.
22. R.J.Schoelkopf, P.Wahlgren, A.A.Kozhevnikov, P.Delsing, and D.E.Prober, Science, **280**, 1238 (1998).
23. T.Fujisawa and Y.Hirayama, Jpn. J. Appl. Phys. **39**, 2338 (2000).
24. T.Fujisawa and Y.Hirayama, Appl. Phys. Lett. **77**, 543 (2000).
25. M.F.Crommie, C.P.Lutz, and D.M.Eigler, Science **262**, 218 (1993).
26. M.C.M.M.vanderWielen, A.J.A. van Roji, and H. van Kempen, Phys. Rev. Lett. **76**, 1075 (1996).
27. K.Kanisawa, M.J.Butcher, H.Yamaguchi, and Y.Hirayama, ICPS2000(L23) (unpublished)
28. K.Kanisawa, M.J.Butcher, Y.Tokura, H.Yamaguchi, and Y.Hirayama (unpublished).

THEORETICAL ISSUES IN SPIN-BASED QUANTUM DOT QUANTUM COMPUTATION

XUEDONG HU AND S. DAS SARMA
Department of Physics, University of Maryland, College Park, MD 20742-4111
E-mail: xdhu@physics.umd.edu, dassarma@physics.umd.edu

We review our recent work addressing various theoretical issues in spin-based quantum dot quantum computation and quantum information processing. In particular, we summarize our calculation of electron exchange interaction in two-electron double quantum dots and multi-electron double dots, and discuss the physical implication of our results. We also discuss possible errors and how they can be corrected in spin-based quantum dot quantum computation. We critically assess the constraints and conditions required for building spin-based solid state quantum dot quantum computers.

1 Introduction

It has long been pointed out that quantum mechanics may provide great advantages over classical physics in physical computation.[1,2] However, the expeditious growth of research on quantum computation[3] started after the advent of Shor's factorization algorithm[4] and quantum error correction schemes.[5,6] Among the many hardwares that were proposed are the ones based on electron spins.[7,8,9,10,11] Obviously, electron spin provides a perfect candidate for quantum bits (qubits) as the electron spin sub-Hilbert-space is generally well-defined and its decoherence relatively slow. In addition, electron spin can be manipulated by external magnetic fields and by interacting with other spins through exchange or Heisenberg coupling.

Our study of electron-spin-based quantum computers (QC) has focused on the scheme using single electrons trapped in quantum dots.[8,12,13] Here the qubit is the spin degrees of freedom of a single electron trapped in a gated semiconductor (GaAs) horizontal quantum dot (QD). Single-qubit operations can be achieved by applying an external magnetic field, while two-qubit operations can be built upon exchange interaction between electrons in neighboring quantum dots when the inter-dot potential barrier is lowered and electron orbital wavefunction is allowed to mix. Final measurement in this scheme requires single spin detection, for which various proposals such as spin filters[14] have been suggested.

A working quantum computer needs to satisfy many stringent conditions. The main requirements include initialization, well-defined Hilbert space, weak decoherence, precise coherent manipulation, unitary evolution, and efficient quantum measurement.[15] In the case of spin-based quantum dot quantum computer, electron spin can be initialized by lowering the sample temperature, and applying an external magnetic field to differentiate the spin up and down directions. Because of the three-dimensional confinement and the fact that GaAs conduction band is mainly formed from S atomic orbitals, the trapped electron has very weak spin-orbit coupling and therefore its spin has a small decoherence rate. The spin Hilbert space of a single quantum dot with one electron is well-defined as long as the Zeeman splitting between the two spin levels is much smaller than the single electron excitation energy associated with the quantum dot confinement potential, which requires a sufficiently small quantum dot. The Hilbert space structure is not as clear cut when two quantum dots are brought together for two-qubit operations. The exploration of this part of the spin Hilbert space is one major focus of our recent study, and we will review some of our results in the following section. Coherent control of electrons in mesoscopic nanostructures has been under intensive study by many groups for sometime.[16,17,18,19,20,21,22,23,24] To reliably manipulate individual electrons and their spins, more precise controlling techniques have to be developed, and various noise effects such as those from offset charge oscillations and other volatile sources have to be eliminated or suppressed. Single spin detection is another difficult proposition for experimentalists. Currently available SQUID (superconducting quantum interference device) can detect a single Bohr magneton but requires a long detection time (in the order of minutes), while proposed spin valve techniques have not yet succeeded in detecting the spin of a single electron. Converting the spin signal to a charge signal and then using single electron transistor based charge measurement techniques to measure the electron spin seems one promising experimental method, but whether this technique (or any other spin measurement technique) can be implemented in microsecond type time scales (the typical low temperature spin coherence time is expected to be microseconds in a practical QC) remains to be seen.

In Section 2 we summarize our work on the theoretical exploration of double dot electron spin Hilbert space, including both two electrons and six electrons in a double dot. In Section 3 we discuss possible errors in a quantum dot quantum computer.

2 Electron Spin-based Quantum Computation

We have studied a double quantum dot (each has one trapped electron) as the basic elementary gate for a quantum computer.[13] This can be considered an artificial double dot quantum molecule with two electrons, analogous to the H_2 molecule with one electron on each dot. In the first order approximation, a Heisenberg exchange Hamiltonian can be written to describe the dynamics of the two electrons trapped in the double dot. However, to establish the operability of such an exchange-based quantum computer, we need to clearly delineate the relation between the two-spin Hilbert space and the complete two-electron Hilbert space. For this purpose, we use a molecular orbital calculation to determine the excitation spectrum of two electrons in two horizontally coupled GaAs quantum dots, and study its dependence on an external magnetic field. In order to provide a more accurate description of the low lying excited states, we include the first excited single particle orbital states (P orbitals of the Fock-Darwin states,[25] which are the eigenstate sequence of a two-dimensional electron in an external perpendicular magnetic field). The resulting spectrum (with an example shown in Fig. 1) confirms the belief that the exchange spin Hamiltonian is sufficient to describe the low energy dynamics of the two electrons.

As shown in Fig. 1, the lowest energy states (one singlet and one triplet) are well-separated from the rest of the excited states. Indeed, the higher energy excited states are separated from the lower energy subspace by the smaller of the single particle excitation energy and the on-site Coulomb repulsion energy (the so-called Coulomb blockade energy) in a single dot. This energy separation is at least 5 meV in our cases, much larger than the GaAs electron Zeeman splitting ($\approx 2.55 \times 10^{-2}$ meV/Tesla) and the energy separation between the ground singlet and triplet states (in the order of 0.1 meV). If the adiabatic condition is satisfied when the exchange gate is turned on, the ground states are essentially isolated from the higher energy excited states. The orbital composition of the ground singlet

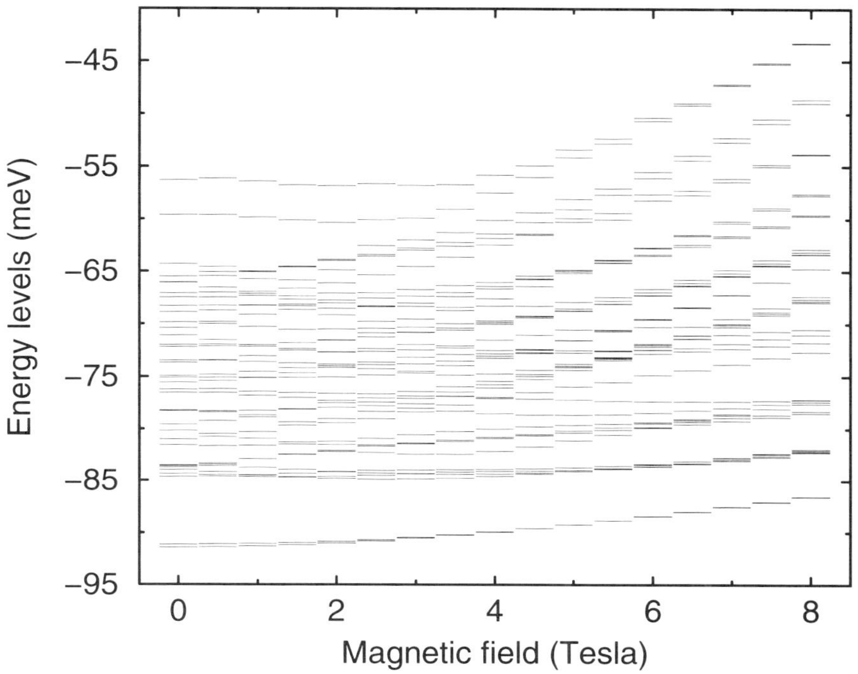

Figure 1. A sample energy spectra of a two-electron double quantum dot in GaAs[13]. Here we show the magnetic field dependence of the two-electron energy spectra obtained in a molecular orbital calculation where both S and P single-electron Fock-Darwin orbitals are used. The dot radius is 20 nm, the inter-dot distance is 30 nm, and the central barrier V_b is 30 meV, corresponding to an actual barrier height of 9.61 meV.

and triplet states are basically ψ_{L0} and ψ_{R0}, the lowest Fock-Darwin orbitals, which are also the orbital states of the electrons when they are in separate quantum dots. Therefore, throughout the duration of the exchange gate (or the lowering of the inter-dot potential barrier), the electron orbital degrees of freedom are frozen, and a spin Hamiltonian such as the Heisenberg exchange Hamiltonian in this case can describe the complete two-electron adiabatic dynamics.

The low-energy spin subspace of the complete two-electron Hilbert space consists of a singlet and three triplet states (whose degeneracy can be lifted by an external magnetic field). The splitting between the singlet and the unpolarized triplet state is the key parameter that determines the speed of two-qubit operations such as an exact swap of the spin states and controlled NOT (CNOT). In essence, the exchange interaction generates a phase difference between the singlet and the unpolarized triplet states (but not the relative weight) in a general two-spin state, thus changing the density matrix of each individual spin. Our numerical results show the singlet-triplet splitting J (also called exchange constant in this context) to be in the range of 0.01 to 1 meV as we vary the inter-dot distance and the size of the individual dots. Such a magnitude of J corresponds to reasonably short operating times (in the order of 0.1 to 10 ns after the adiabatic condition is taken into account) compared to the estimated electron spin coherence time (possibly of the order of microseconds at 50-100 mK temperature) in semiconductor heterostructures.[26,27] Note that low temperature (T) QD-QC operation is essential in this context: first, T must be low to obtain a long spin coherence time; and second, $k_B T$ must be low to satisfy $\Delta \gg k_B T$ condition, where Δ is the separation between the ground and the higher excited orbital levels in the single quantum dot system ($\Delta > 1$ meV in small dots), to maintain the integrity of the spin sub-Hilbert space necessary for quantum computation. Thus, the operation (or the manipulation) time for various gates and magnetic field pulses in QD-QC architecture is constrained from below by the adiabaticity requirement (i.e. it cannot be too fast so that adiabaticity is maintained) and from above by the spin coherence requirement (i.e. the operation must be much faster than the spin coherence time). One must realize that in addition to these stringent operational constraints there is also the severe (and as yet experimentally unsolved) constraints of measuring the final spin state ("read out") which cannot be too slow for successful QD-QC operations.

Our molecular orbital calculation is based on the effective mass approximation in a single envelope-function approach. Due to the sizable band gap in GaAs and the relatively small inter-band coupling, our single envelope function calculation is valid with quite high accuracy. Indeed, we estimated[13] that the effect of the neglected bands and inter-conduction band coupling amounts to about 1% correction to the single particle wavefunctions. To clearly quantify this correction one would have to employ a more sophisticated multi-band calculation or drop the envelope function approach entirely and adopt an atomistic approach, which would automatically include all valence band effects. However, it is unclear, in principle, whether such a first principle t ypeband theory approach would be a meaningful calculational tool for assessing QD-QC operations since the important energy scales (e.g. J) in the problem are 1 meV or less, and first principle band calculations cannot achieve 1 meV type accuracies.

Experimentally, it is very difficult to trap one and only one electron in a gated t wo-dimensional GaAs quantum dot. Thus it has been speculated that perhaps the requirement of single trapped electron per quantum dot as the fundamental qubit may be relaxed in a QD-QC, and that an odd number of electrons in a QD can provide an effective spin-1/2 system as a qubit. To explore this possibility, we have recently done a multi-electron calculation where we have three electrons in one quantum dot, thus six electrons in a double dot.[28] We first carry out a Hartree Fock mean field calculation to produce an orthonormal set of single-particle basis states, then employ a restricted configuration interaction (CI) approach to calculate the energy spectrum. Our results show that the low energy dynamics in the multi-electron case is generally much more complicated than the t wo-electron case because of the single-particle energy level degeneracies in the higher excited Fock Darwin states. This complexity can be alleviated by applying external magnetic fields or deforming the quantum dots, which lift the level degeneracies. Thus we show in Ref. [28] that subject to some constraints spin-based QD-QC architectures are feasible, at least in principle, in multielectron quantum dot arrays.

3 Error correction in quantum dot quantum computer

To evaluate the feasibility of a quantum computer architecture, it is only one of the many steps to study the operating Hilbert space structure. Having a well defined sub-Hilbert space, as in Fig. 1, is a necessary but by no means a sufficient condition for QC operation. Another important step is to assess the importance of all the possible errors (and how to do the relevant error correction) in this architecture. For spin-based quantum dot quantum computer model, we have explored several kinds of operating error (aside from spin decoherence and errors in single spin measurement). For example, we originally speculated that double occupation in each of the double dot during a two-qubit operation might result in a significant error, the rationale being that one needs to keep tags on all the electrons as each represents a distinct qubit. We now believe that adiabatic operations of these two-qubit gates should effectively erase this problem since adiabaticity will limit the two-electron dynamics within the low-energy two-spin subspace as we discussed in the previous section.[29,30] We have done simulations to study the adiabatic condition and the results will be published elsewhere. We no longer believe virtual double occupancy to be a problem.

As the exchange coupling J (the singlet-triplet splitting) is tuned by changing external gate voltage in a QD-QC, thermal fluctuations (or any other types of fluctuations) in the gate voltage will lead to fluctuations in J, thus causing phase errors in the exchange-based swap gate which is crucial for two-qubit operations. We have estimated this error by assuming a simple thermal (white) noise.[13] More specifically, we assume $J = f(V)$ where V is the gate voltage that controls the value of J. Around any particular value V_0, J can be expressed as $J(V) = J(V_0) + f'(V)|_{V_0} (V - V_0)$. During a swap gate between two quantum dots, the phase of the electronic spin wavefunction evolves as $\phi = \int_0^t J(\tau) d\tau/\hbar$. Thus the fluctuation in the phase ϕ is

$$\begin{aligned} \langle \delta\phi^2 \rangle = \langle \phi^2 \rangle - \langle \phi \rangle^2 &= \frac{1}{\hbar^2} \int_0^t \int_0^t \langle \delta J(\tau_1)\, \delta J(\tau_2) \rangle d\tau_1 d\tau_2 \\ &\sim \int_0^t \int_0^t \frac{[f'(\bar{V})]^2}{\hbar^2} \langle \delta V(\tau_1)\, \delta V(\tau_2) \rangle d\tau_1 d\tau_2 \,. \end{aligned} \quad (1)$$

Using Nyquist theorem $\langle \delta V(\tau_1)\, \delta V(\tau_2) \rangle = 4Rk_BT\delta(\tau_1 - \tau_2)$, we obtain the approximate expression for the phase fluctuation:

$$\langle \delta\phi^2 \rangle \sim 4Rk_BT\alpha^2 t/\hbar^2 \,, \quad (2)$$

where α is the upper bound of $|f'(\bar{V})|$. Assuming the swap gate is performed at 1 K (since J is in the order of 0.1 meV $\sim$ 1 K, the experimental temperature can only be lower than 1 K), and the transmission line connecting the gate to the surrounding bath of cryogenic temperature has an impedance of 50 ohm, the rate for phase fluctuation $\langle \delta\phi^2 \rangle / t$ is about 3.2 MHz. The phase error accrued during a swap gate is then about 0.06%. This is quite a small error which is the same order of magnitude as the theoretical tolerance of the currently available quantum error correction codes. To further lower this error rate, one can go to lower experimental temperature and turn up J more gently (which requires longer time but produces smaller α) in the QD-QC operation.

Another possible error in the two-qubit operations of the architecture we studied is caused by inhomogeneous magnetic fields.[31] Such a field may come from magnetic impurities or unwanted currents away from the structure. Magnetic field affects both orbital and spin part of the electron wavefunction. The orbital effect is accounted for by adjusted exchange coupling J, while the spin effect is accounted for through Zeeman coupling terms:

$$H_s = J(\mathbf{B})\mathbf{S}_1 \cdot \mathbf{S}_2 + \gamma_1 S_{1z} + \gamma_2 S_{2z} \,, \tag{3}$$

where $\mathbf{S}_1$ and $\mathbf{S}_2$ refer to the spins of the two electrons; $J(\mathbf{B})$ is the exchange coupling (singlet-triplet splitting); γ_1 and γ_2 are the effective strength of the Zeeman coupling in the two quantum dots. In an inhomogeneous field, $\gamma_1 \neq \gamma_2$, so that the Zeeman terms do not commute with the exchange term in the Hamiltonian (3). We have done a detailed analysis on how to achieve swap with such a Hamiltonian, and found that there is at the minimum an error proportional to the square of field inhomogeneity in the swap. For example, if the initial state of the two electron spin is $|\phi(0)\rangle = |\uparrow\downarrow\rangle$, the density matrix of the first spin after the optimal swap is

$$\rho_1|_{e^{i\theta}=-1} = \frac{1}{1+x^2}|\downarrow\rangle\langle\downarrow| + \frac{x^2}{1+x^2}|\uparrow\rangle\langle\uparrow| \,, \tag{4}$$

where $x = \delta/(2J) = (\gamma_1 - \gamma_2)/2J$. In other words, the first spin can never exactly acquire the state ($|\downarrow\rangle$) of the second spin. Its state will remain mixed and the smallest error from an exact swap is $\frac{x^2}{1+x^2}$, which needs to be corrected. We have estimated[31] that in GaAs a Bohr magneton can lead to an error in the order of 10^{-6}, which is within the capability of currently available quantum error correction schemes.

4 Conclusion

In this paper we have reviewed some of our recent results in our study of electron-spin-based quantum dot quantum computation. We discuss various issues regarding the Hilbert space structure of a double quantum dot artificial molecule. We also describe possible errors in a spin-based QC. Our tentative conclusion is that building a practical GaAs electron-spin-based QD-QC, while being certainly possible in principle, will be extremely difficult in practice (even assuming a working temperature of 100 mK or less) because of severe constraints and limitations arising from the spin coherence time, the adiabatic condition, the magnitude of exchange coupling, the error correction requirement, and the unsolved experimental problem of a reasonably fast (in less than a microsecond) measurement of a single Bohr magneton.

Acknowledgments

We thank financial support from ARDA, LPS, DARPA, and US-ONR.

References

1. D. Deutsch, *Proc. R. Soc.* A **400**, 97 (1985); D. Deutsch and R. Josza, *Proc. R. Soc.* A **439**, 553 (1992).
2. R.P. Feynman, *Int. J. Theor. Phys.* **21**, 467 (1982); *Found. Phys.* **16**, 507 (1986).
3. D.P. DiVincenzo, *Science* **270**, 255 (1995); A. Ekert and R Jozsa, *Rev. Mod. Phys.* **68**, 733 (1996); A. Steane, *Rep. Prog. Phys.* **61**, 117 (1998); C.H. Bennett and D.P. DiVincenzo, *Nature* **404**, 247 (2000).

4. P.W. Shor, in *Proceedings of the 35th Annual Symposium on the Foundations of Computer Science*, ed. by S. Goldwasser (IEEE Computer Society, Los Alamitos, 1994), p. 124.
5. P.W. Shor, *Phys. Rev.* A **52**, 2493 (1995).
6. A.M. Steane, *Phys. Rev. Lett.* **77**, 793 (1996).
7. D.P. DiVincenzo, in *Mesoscopic electron transport* ed. by L.L. Sohn, L.P. Kouwenhoven, and G. Sch¨ on(Kluwer, Dordrecht, 1997).
8. D. Loss and D.P. DiVincenzo, *Phys. Rev.* A **57**, 120 (1998).
9. R. Vrijen et al., *Phys. Rev.* A **62**, 012306 (2000).
10. A. Imamoglu et al., *Phys. Rev. Lett.* **83**, 4204 (1999); A. Imamoglu, *Fortschr. Phys.* **48**, 987 (2000).
11. Special QC issue of *Fortschr. Phys.* **48** (2000).
12. G. Burkard, D. Loss, and D.P. DiVincenzo, Phys. Rev. B **59**, 2070 (1999).
13. X. Hu and S. Das Sarma, *Phys. Rev.* A **61**, 062301 (2000).
14. D.P. DiVincenzo, *J. Appl. Phys.* **81**, 4602 (1997); **85**, 4785 (1999).
15. D.P. DiVincenzo, *Fortschr. Phys.* **48**, 771 (2000).
16. R. C. Ashoori, *Nature* **379**, 413 (1996).
17. F.R. Waugh *et. al.*, *Phys. Rev. Lett.* **75**, 705 (1995).
18. C. Livermore *et. al.*, *Science* **274**, 1332 (1996);
19. R. H. Blick, D. W. van der Weide, R. J. Haug, and K. Eberl, *Phys. Rev. Lett.* **81**, 689 (1998).
20. T.H. Oosterkamp *et. al.*, *Nature* **395**, 873 (1998).
21. R. Kotlyar, C. A. Stafford, and S. Das Sarma, *Phys. Rev. B* **58**, 3989 (1998); R. Kotlyar and S. Das Sarma, *Phys. Rev. B* **56**, 13235 (1997).
22. U. Banin, Y. W. Cao, D. Katz, and O. Millo, *Nature* **400**, 542 (1999).
23. T. Fujisawa *et. al.*, *Science*, **282**, 932, (1998).
24. Y. Ji *et. al.*, *Science* **290**, 779 (2000).
25. L. Jacak, P. Hawrylak, and A. Wójs, *Quantum Dots* (Springer, Berlin, 1997).
26. S. Das Sarma, J. Fabian, X. Hu, and I. Žuti´c,*Superlattice Microst* **27**, 289 (2000) and cond-mat/9912080.
27. J.M. Kikkawa and D.D. Awschalom, *Nature* **397**, 139 (1999); *Phys. Rev. Lett.* **80**, 4313 (1998); J.M. Kikkawa *et. al.*, *Science* **277**, 1284 (1997).
28. X. Hu and S. Das Sarma, LANL preprint cond-mat/0101102.
29. M.B. Ruskai, LANL preprint quant-phys/9906114. *Phys. Rev. Lett.* **85**, 194 (2000).
30. J. Schliemann, D. Loss, and A.H. MacDonald, *Phys. Rev. B* **63**, 085311 (2001).
31. X. Hu, R. De Sousa, and S. Das Sarma, *Phys. Rev. Lett.* **86**, 918 (2001) and cond-mat/0004459.

RESONANT ENERGY TRANSFER IN NOVEL QUANTUM DOT "MOLECULES"

M. J. FERNÉE, J. WARNER, H. RUBINSTEIN-DUNLOP

SRC for Quantum Computing Technology, Centre for Laser Science, Dept. of Physics, University of Queensland, Qld., 4072, Australia.

We have developed a technique for fluorescence activation of colloidal solutions of lead sulfide (PbS) quantum dots using crystalline overgrowth with cadmium sulfide (CdS). This process can proceed in two ways; epitaxial overgrowth or polycrystalline growth. In the first case, strong band-edge fluorescence results and in the second case, strong above-band-edge fluorescence results. We interpret much of the above-band-edge emission as coming from CdS nanocrystallites, which are attached to a PbS surface. Photoluminescence excitation spectra reveal that the above-band-edge fluorescence results from a strong resonant energy transfer between the PbS and CdS nanocrystallites. We attribute this energy transfer to resonant dipole-dipole coupling commonly associated with energy transfer in molecules. The above-band-edge fluorescence is also characteristic of a strongly quantised system in which efficient energy transfer to the band-edge is restricted due to "phonon bottleneck".

1 Introduction

Strong quantum confinement, which gives rise to discreet atomic-like electronic energy levels in semiconductor quantum dots (QD's), was expected to suppress intra-band energy relaxation due to phonon interactions if the level spacings were significantly greater than individual phonon energies[1,2]. However, this "phonon bottleneck" is not readily observed in QD's. This has been attributed to the difference in the degree of quantisation of the electron and hole states and specifically to the difficulty in achieving strong quantum confinement of the holes[3,4].

PbS is one semiconductor where strong quantum confinement of both electrons and holes can be achieved[5]. Band-edge shifts of up to 1.6 eV are easily achieved in nanocrystals of around 2 nm diameter. Such QD colloidal solutions have strong transitions in the visible and UV spectrum[5,6]. In fact PbS QD's provide a greater degree of quantum confinement that almost any other semiconductor.

Unfortunately PbS QD's have a poor photoluminescence efficiency and extremely short exciton lifetimes, which has made them of limited interest. These properties result from the fact that both the quantum-confined electron and hole strongly interact with a large manifold of surface trapping states[7], which then allow rapid energy relaxation. Passivation of these active trapping states has met with limited success at best[8]. Nevertheless, the strong quantum confinement of both electrons and holes suggests that the PbS QD intrinsic states should provide an excellent "artificial atom" and be subject to "phonon bottleneck".

2 Results

Epitaxial overgrowth of a semiconductor with a wider band-gap is a technique that should be suitable for passivating all of the surface trap states in PbS QD's. CdS is a wide band-gap semiconductor, which has a lattice constant of 5.81 Å in its cubic form, producing a lattice mismatch of less than 2% with PbS (which has a lattice constant of 5.91 Å). This suggests CdS would be useful for epitaxial overgrowth.

We grow CdS on PbS QD's in an aqueous solution under a range of low PH conditions unsuitable for individual CdS QD growth[9]. Two different modes of growth are observed; uniform (epitaxial) overgrowth and polycrystalline growth. All colloidal solutions exist somewhere between these two extremes. Epitaxial overgrowth is highly dependent on the quality of the QD surface and the availability of suitable crystal planes, conditions which we presently have only marginal control of.

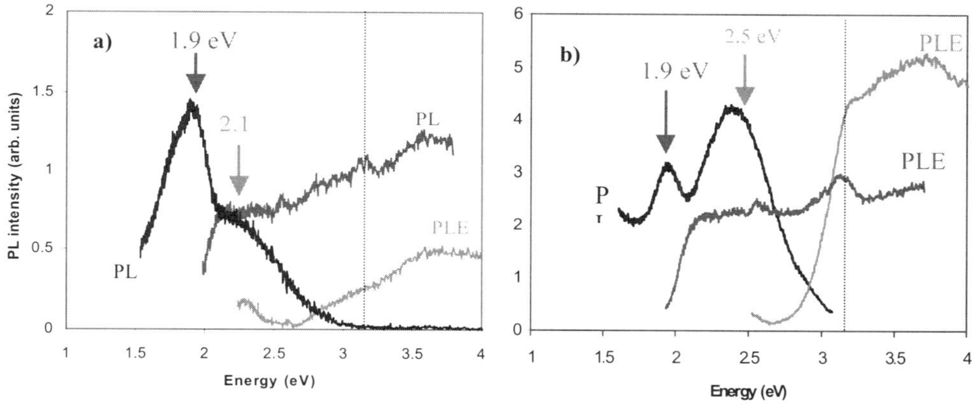

Figure 1. *a) PL and PLE spectra of PbS QD's after surface-state passivation using CdS overgrowth;* b) PL and PLE spectra of QD solution <u>after</u> fluorescence activation of the CdS overgrowth layer.

After the CdS growth phase, the PbS QD colloidal solution photoluminescent intensity increases until it is visible when irradiated with a common UV lamp. In contrast, under these conditions we are unable to observe either uncoated PbS QD fluorescence nor CdS fluorescence. In Figure 1a we present both photoluminescence (PL) and photoluminescent excitation (PLE) spectra of a CdS coated PbS QD solution. We see a prominent PL peak at approximately 1.9 eV while the PLE spectra shows the PbS band-edge at around 2.1 eV. We attribute the 1.9 eV PL peak to direct band-edge recombination. The cause of the 200 meV red shift between the observed PLE band-edge and the PL emission peak is not yet understood. There is also a component of above-band-edge emission in the PL spectra. This component can only be coming from the CdS capped PbS QD's as it is clearly visible under the UV lamp (as orange emission rather than red band-edge emission). It is expected that this emission arises from residual charge trapping states that exist at the PbS/CdS interface. However, it nevertheless suggests that efficient energy transfer to the band-edge does not occur and is therefore direct evidence for phonon bottleneck in PbS QD's.

We next activate fluorescence from CdS by over-coating with a hydroxide layer[10]. In figure 1b we show the effect that this fluorescence activation has on the PL and PLE spectrum of the solution used in figure 1a. While the PL peak at 1.9 eV remains, we now see a strong and broad PL peak near 2.5 eV. Two PLE spectra are taken at the 1.9 eV peak and the 2.5 eV peak. The 1.9 eV PLE curve clearly shows the PbS QD band-edge while the 2.5 eV PLE curve indicates a strong transition near 3.15 eV. This last feature only becomes visible after fluorescence activation of the CdS layer and so is attributed to CdS emission. However, the PL emission peak is located at 2.5 eV, over 600 meV from the PLE feature. As the CdS band-edge is located close to 2.5 eV, most of this emission is attributed to either direct recombination in the CdS layer or trap-state recombination. We attribute much of the extra above-band-edge emission that is observed in figure 1b to emission from incompletely coated PbS/CdS composite particles, which would not fluoresce without the activation step. We also note that the transition in the 2.5 eV PLE curve is directly correlated with a feature in the 1.9 eV PLE curve as well as a distinct feature in the PbS QD absorption spectrum. These correlations are observed in every activated colloidal QD solution that we have prepared. In most cases, the strong transition in the 2.5 eV PLE spectrum corresponds to a sharply reduced coupling to the PbS band-edge emission at 1.9 eV.

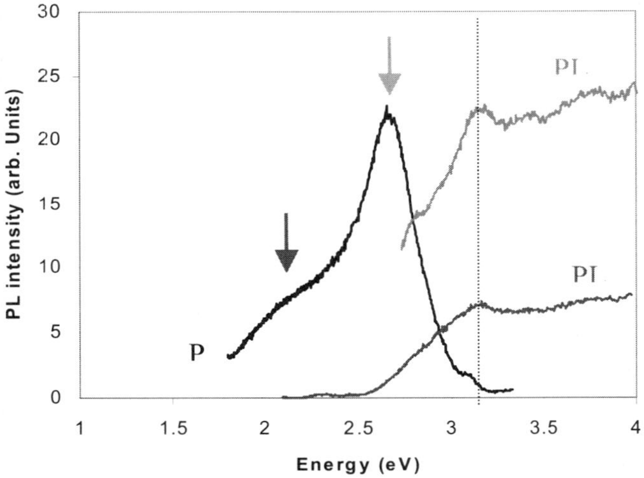

Figure 2. PL and PLE spectra of fluorescence activated CdS QD's grown under similar conditions but in the absence of PbS QD's

In figure 2 we show the PL and PLE spectra of a CdS QD colloidal solution made under conditions unfavourable to CdS QD growth[9] (similar to our CdS overgrowth conditions). The PL spectrum shows a peak at 2.6 eV which we attribute to band-edge recombination, as well as some broad emission below the band-edge which we attribute to trap-state emission. There is no obvious correlation between the CdS QD PL spectrum and the activated CdS coated PbS PL spectrum. The same can be said for the PLE spectra. However, we notice that both PLE spectra show a feature at around 3.15 eV which is far from the band-edge at 2.6 eV but sits surprisingly close to a PbS feature observed in the 1.9 eV PLE spectra of figure 1a and 1b. Furthermore, other CdS QD PLE spectra all seem to show a feature near 3.15 eV, which doesn't vary considerably with QD size.

The above-band-edge fluorescence in activated CdS overcoated PbS QD solutions is most likely a result of resonant energy transfer between polycrystalline CdS crystallites that have grown on PbS QD's. The relative conduction band offset and large mismatch of electron and hole effective masses suggests that direct electron or hole transfer is unlikely. The most likely form of energy transfer between two near resonant excitonic levels in a coupled QD system occurs via dipole-dipole coupling (or the Förster interaction), a mechanism also responsible for energy transfer in molecules[11]. The strong onset of fluorescence shown in the PLE spectra indicates that even QD's with poorly passivated PbS surfaces still show significant energy transfer to the CdS nanocrystallites (which are surface passivated in the activation stage). Studies of PbS energy transfer using femtosecond pump-probe techniques have shown rapid relaxation of the intrinsic excitonic levels on the time scale of picoseconds[6,7]. Therefore the inter-quantum dot coupling strength must be sufficiently large to allow efficient energy transfer to the CdS QD.

The suggested model of inter-QD coupling is shown in figure 3. An above-band-edge excitonic state in PbS is coupled via the Förster process to a nearly energy matched excitoninc level in CdS. Both levels are subject to energy relaxation to either the band-edge or ground state. If the damping rates, λ and γ, are sufficiently small compared to the coupling, F, then the two excitonic levels are coherently coupled. We crudely estimate of the energy transfer rate for a closely coupled system[11] (ie. a QD separation of the order of 1 nm), is of the order 10^{13} s^{-1}, which is at least one to two orders of magnitude faster than the relaxation rates in the system. Under these conditions strong coherent coupling of the two levels is likely.

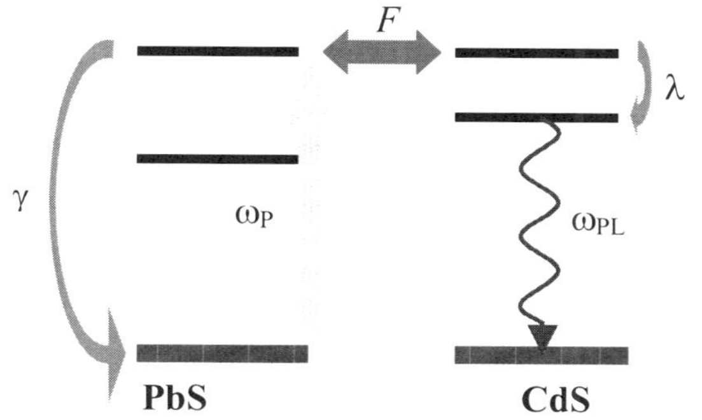

Figure 3. Energy level representation of the coupled quantum dot molecule indicating relevant relaxation processes, dipole-dipole coupling, F, laser pump and photoluminescent emission.

In conclusion, we have developed a technique for fluorescence activation of PbS QD's by overgrowth of CdS (a wide-gap semiconductor). The CdS growth can proceed in two ways; epitaxial growth or polycrystalline growth depending on the quality of the PbS QD surfaces. The expitaxially coated PbS QD's show significant above-band-edge PL which we interpret as evidence of phonon bottleneck in PbS. Further fluorescence activation of the CdS layer reveals a large PL contribution from partially coated PbS QD's that have undergone polycrystalline growth. The PL from the polycrystalline growth QD's arises from rapid resonant energy transfer between PbS and CdS QD's which we attribute to a resonant dipole-dipole coupling similar to that responsible for energy transfer in molecules. Therefore these polycrystalline growth QD's behave like artificial molecules in much the same fashion as single QD's behave like artificial atoms. The strength of the resonant coupling is estimated to be sufficient for observation of coherent effects and the system should be suitable for the preparation and observation of entanglement between the coupled states[12].

References

1. U. Bockelmann, G. Bastard, Phys Rev B **42**, 8947, (1990).
2. H. Benisty, C. M. Sotomayor-Torres, C. Weisbuch, Phys Rev B 44, 10945, (1991).
3. V. Klimov, D. W. McBranch, Phys Rev Lett 80, 4028, (1998).
4. V. Klimov, D. W. McBranch, C. A. Leatherdale, M. G. Bawendi, Phys. Rev. B 60, 13740, (1999).
5. I. Kang, F. W. Wise, J. Opt. Soc. Am. B 14, 1632, (1997).
6. J. L. Machol, F. W. Wise, R. C. Patel, D. B. Tanner, Phys Rev B 48, 2819, (1993).
7. T. Okuno, A. A. Lipovskii, T. Ogawa, I. Amagai, Y. Masumoto, J. Luminesc. 87-89, 491, (2000).
8. L. Guo, K. Ibrahim, et. al., J. Luminesc. 82, 111, (1999).
9. A. Mews, V. Kadavanich, U. Banin, A. P. Alivisatos, Phys Rev B 53, R13242, (1996).
10. L. Spanhel, M. Haase, H. Weller, A. Henglein, J. Am. Chem. Soc. 109, 5649, (1987).
11. W. G. teich, K. Obermayer, G. Mahler, Phys. Rev. B 37, 8111, (1988).
12. L. Quiroga, N. F. Johnson, Phys Rev Lett 83, 2270, (1999).

RECENT RESULTS IN TRAPPED-ION QUANTUM COMPUTING AT NIST

D. KIELPINSKI, A. BEN-KISH, J. BRITTON, V. MEYER, M.A. ROWE, W.M. ITANO, D.J. WINELAND

Time and Fr e quencyDivision, National Institute of Standards and Technology, Mailstop 847, 325 Broadway, Boulder CO 80303, USA
E-mail: davidk@boulder.nist.gov

C.A. SACKETT

Department of Physics, University of Virginia, Charlottesville VA 22904, USA

C. MONROE

Department of Physics, University of Michigan, Ann Arbor MI 48109, USA

We review recent experiments on entanglement, Bell's inequality ,and decoherence-free subspaces in a quantum register of trapped $^9Be^+$ ions. We have demonstrated entanglement of up to four ions [1] using the technique of Mølmer and Sørensen [2]. This method produces the state $(|\uparrow\uparrow\rangle + |\downarrow\downarrow\rangle)/\sqrt{2}$ for t wo ions and the state $(|\downarrow\downarrow\downarrow\downarrow\rangle + |\uparrow\uparrow\uparrow\uparrow\rangle)/\sqrt{2}$ for four ions. We generate the entanglement deterministically in each shot of the experiment. Measurements on the t wo-ion entangled state violates Bell's inequality at the 8σ level [3]. Because of the high detector efficiency of our apparatus, this experiment closes the detector loophole for Bell's inequality measurements for the first time. This measurement is also the first violation of Bell's inequality by massive particles that does not implicitly assume results from quantum mechanics. Finally, we have demonstrated reversible encoding of an arbitrary qubit, originally contained in one ion, into a decoherence-free subspace (DFS) of t wo ions [4]. The DFS-encoded qubit resists applied collective dephasing noise and retains coherence under ambient conditions 3.6 times longer than does an unencoded qubit. The encoding method, which uses single-ion gates and the t wo-ion entangling gate, demonstrates all the elements required for t wo-qubit universal quantum logic.

1 Introduction

Trapped ions are a promising candidate system for implementing quantum computing (QC), with potential applications like efficient factorization of large numbers [5] and efficient searching of large databases [6]. While large-scale QC is still far off, trapped-ion quantum registers have already demonstrated all the essential ingredients of QC, including initialization to a known quantum state [7,8,10], efficient detection of final states [3,10], long qubit coherence times [4,11,12,13], and a universal set of quantum logic gates [1,10,14,15]. Recent work on trapped-ion QC at NIST has used this basic set of tools to produce four-particle entanglement [1], observe violation of Bell's inequality [3], and encode quantum information into a decoherence-free subspace [4].

In the experiments described below, the qubits are electronic states of $^9Be^+$ ions held in a linear RF trap [16] of the t ype described in reference [17]. The relevant electronic energy levels of the $^9Be^+$ ion are shown in Fig. 1. Each ion encodes one qubit in its $^2S_{1/2}|F = 2, m_F = -2\rangle$ and $^2S_{1/2}|F = 1, m_F = -1\rangle$ hyperfine sublevels, denoted by $|\downarrow\rangle$ and $|\uparrow\rangle$, respectively. The ions are readily initialized in $|\downarrow\rangle$ by optical pumping. We detect the number of ions in state $|\downarrow\rangle$ by applying left circularly polarized laser light to excite the $|\downarrow\rangle$

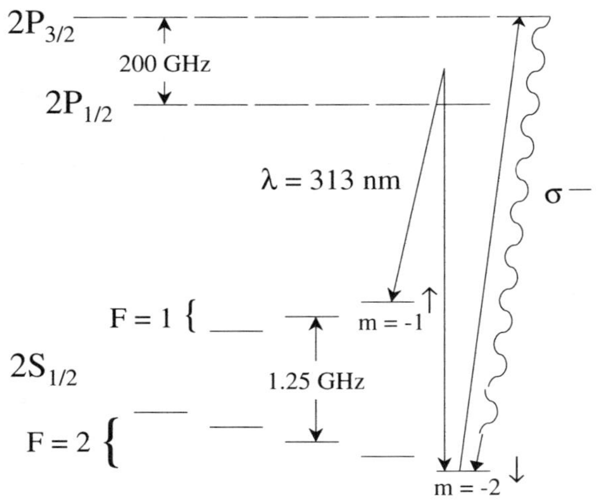

Figure 1. Partial level diagram of the $^9Be^+$ ion, showing the cycling transition for σ_- resonant light and the two-photon off-resonant coupling for Raman transitions. Wavelengths for all transitions are approximately 313 nm.

- $^2P_{3/2}$ transition at about 313 nm. The polarization of the light ensures excitation to the $^2P_{3/2}|F = 3, m_F = -3\rangle$ hyperfine sublevel, which can only decay to the $|\downarrow\rangle$ sublevel by dipole selection rules. Thus an ion in state $|\downarrow\rangle$ can scatter thousands of photons without changing its hyperfine state. On the other hand, state $|\uparrow\rangle$ is spectrally resolved from $|\downarrow\rangle$ by the hyperfine splitting of 1.25 GHz and therefore scatters very little light. The resulting detection accuracy per ion is about 98% in a single shot of the experiment.

We produce controlled rotations of a single qubit by driving stimulated Raman transitions between states $|\uparrow\rangle$ and $|\downarrow\rangle$. The laser beams driving the transitions are tuned about midway between the $^2P_{3/2}$ and $^2P_{1/2}$ states. If the frequency difference $\nu_{\rm diff}$ between the Raman laser beams is equal to the energy splitting ν_0 between states $|\uparrow\rangle$ and $|\downarrow\rangle$, we observe Rabi flopping between $|\uparrow\rangle$ and $|\downarrow\rangle$ corresponding to the evolution

$$
\begin{aligned}
|\downarrow\rangle &\rightarrow \cos\frac{\theta}{2}\,|\downarrow\rangle + e^{i\phi}\sin\frac{\theta}{2}\,|\uparrow\rangle \\
|\uparrow\rangle &\rightarrow \cos\frac{\theta}{2}\quad - e^{-i\phi}\sin\frac{\theta}{2}\,|\downarrow\rangle
\end{aligned}
\tag{1}
$$

Here θ is proportional to the Raman pulse duration and ϕ is the phase difference between the Raman beams at the position of the ion, referred to as the "ion phase". For two ions, we write the ion phases as ϕ_1, ϕ_2. We can effect common-mode changes of ϕ_1 and ϕ_2 by changing the phase of the RF synthesizer that controls the Raman difference frequency. By changing the strength of the trap, we can control the ion spacing precisely, allowing us to make differential changes to ϕ_1 and ϕ_2. These two techniques combined allow us to control ϕ_1 and ϕ_2 independently for two ions, so that we can perform single-qubit rotations on one ion without affecting the other [4].

The potential created by the ion trap is harmonic with axial frequency ν_{CM} of about 5 MHz. Doppler cooling allows us to cool the ions to temperatures $T \sim h\nu_{\text{CM}}/k_B$. The Doppler-cooled ions are strongly coupled by the Coulomb interaction and form a rigid crystal, which in these experiments is a string of ions lying along the trap axis. Because the ions are strongly coupled, their quantized motion is best described in terms of normal modes with frequencies ν_i. For two ions, for instance, the motion of the ions along the axis decomposes into symmetric and antisymmetric normal modes at frequencies ν_{CM} and $\sqrt{3}\nu_{\text{CM}}$. The lowest-lying normal mode for any number of ions is the center-of-mass mode at frequency ν_{CM}. In this mode the ion string moves as a unit.

We can control the motional quantum state of the ions by driving stimulated Raman transitions that couple spin and motion. In our geometry, the wavevector difference between Raman beams lies along the axis of the ion trap, so the Raman beams exert a dipole force only along the axis. Setting ν_{diff} equal to $\nu_0 + m\nu_i$ (m an integer) induces Rabi flopping between states $|\downarrow\rangle|n\rangle$ and $|\uparrow\rangle|n+m\rangle$ with evolution like that of Eq. (1). Here $|n\rangle$ denotes the nth Fock state of the ith normal mode. In particular, we can cool the ions by driving a pulse with $\theta = \pi$ at $\nu_{\text{diff}} = \nu_0 - \nu_i$ (the "red sideband") and optically pumping from $|\uparrow\rangle$ back to $|\downarrow\rangle$. Repeating the cycle several times permits preparation of the ion string in its motional ground state [7,8].

2 Generating Entanglement

We have entangled strings of up to four ions [1] using the technique proposed by Sørensen and Mølmer [2]. In this method, one drives two Raman transitions simultaneously, with the same Rabi frequency on all ions. Fig. 2 shows an energy-level diagram including the relevant transitions. The detuning δ ensures that population transfer between motional states is nonresonant and therefore small. However, the sum of the two Raman transition frequencies is $2\nu_{\uparrow\downarrow}$, so that for any pair of ions the process $|\downarrow\downarrow\rangle \leftrightarrow |\uparrow\uparrow\rangle$ is resonant. For N ions initially in $|\downarrow\rangle$, the resulting evolution produces the state

$$|\psi_e\rangle = \begin{array}{ll} (|\downarrow\downarrow\rangle + |\uparrow\uparrow\rangle)/\sqrt{2} & \text{(two ions)} \\ (|\downarrow\downarrow\downarrow\downarrow\rangle + |\uparrow\uparrow\uparrow\uparrow\rangle)/\sqrt{2} & \text{(four ions)} \end{array} \tag{2}$$

both of which are maximally entangled states.

To implement this technique, we pass one of the Raman beams through an electro-optic modulator (EOM) to generate the two required Raman difference frequencies and uniformly illuminate the ion string with both Raman beams. For two ions, we tune the Raman difference frequencies close to the motional sidebands of the antisymmetric mode; for four ions, we use the mode with frequency $\sqrt{29/5}\ \nu_{\text{CM}}$. In both cases, the mode we choose has equal amplitudes of motion for all ions, so that the Raman Rabi frequencies for all ions are the same under uniform illumination.

In general, the state produced by this method is characterized by some density matrix ρ that approximates the desired state $|\psi_e\rangle$. To show entanglement of the experimental state ρ, it is sufficient to show that the fidelity $F = \langle\psi_e|\rho|\psi_e\rangle$ exceeds 1/2 [1,9]. We have

$$F = (\rho_{\downarrow\downarrow\ldots\downarrow\downarrow} + \rho_{\uparrow\uparrow\ldots\uparrow\uparrow})/2 + |\rho_{\downarrow\downarrow\ldots\uparrow\uparrow}| \tag{3}$$

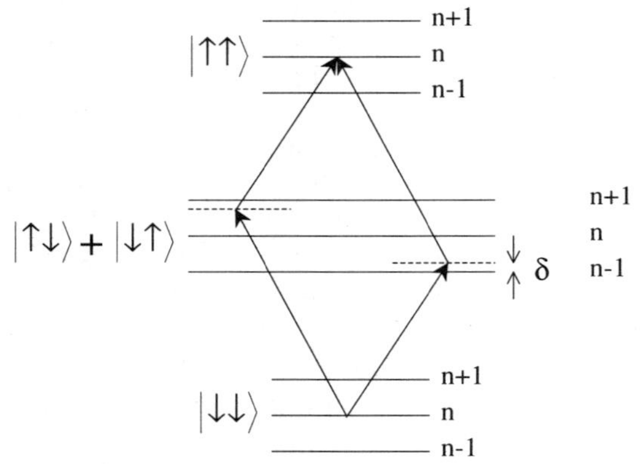

Figure 2. Level diagram for the Sørensen-Mølmer entangling gate in the case of two ions. Here, as opposed to Fig. 1, arrows indicate two-photon Raman transitions. Raman transitions near both red and blue sidebands are applied simultaneously to create the state $(|\downarrow\downarrow\rangle + |\uparrow\uparrow\rangle)/\sqrt{2}$.

We measure the diagonal density matrix elements $\rho_{\downarrow\ldots\downarrow,\downarrow\ldots\downarrow}, \rho_{\uparrow\ldots\uparrow,\uparrow\ldots\uparrow}$ by detecting the number of ions in $|\downarrow\rangle$. To find the far-off-diagonal coherence $|\rho_{\downarrow\ldots\downarrow\uparrow\ldots\uparrow}|$, we rotate all ions together as in Eq. (1) with $\theta = \pi/2$ and with phases $\phi_i = \phi$ relative to the entangling pulse. Then we measure the parity Π of the ion string, defined as $+1$ for an even number of ions in $|\downarrow\rangle$, -1 for an odd number in $|\downarrow\rangle$. It is easy to show that

$$\Pi = 2|\rho_{\downarrow\downarrow\ldots\uparrow\uparrow}|\cos N\phi \tag{4}$$

for N ions, if all other off-diagonal coherences are absent. Sweeping the phase of the RF synthesizer that generates the Raman difference frequency gives a sinusoidal signal of Π vs. ϕ, as shown in Fig. 3. We extract the coherence $|\rho_{\downarrow\downarrow\ldots\uparrow\uparrow}|$ by fitting a sine wave to the data of Fig. 3. The density matrix elements needed to calculate the fidelity are given in Table 1 for two and four ions. For two ions we find $F = 0.83 \pm 0.01$, while for four ions $F = 0.57 \pm 0.02$, demonstrating entanglement in both cases [1]. It is important to note that the data used to calculate F are not post-selected. Since the evolution is deterministic, our method produces an entangled state in each shot of the experiment.

Number of Ions	$\rho_{\downarrow\downarrow...\downarrow\downarrow}$	$\rho_{\uparrow\uparrow...\uparrow\uparrow}$	$\|\rho_{\downarrow\downarrow...\uparrow\uparrow}\|$
2	0.46	0.385	0.43
4	0.35	0.35	0.215

Table 1. Partial density matrix of the state produced by the entangling pulse. From Eq. (3), only these matrix elements are needed to determine whether the experimentally generated state is entangled. Errors are ± 0.01 for two ions and ± 0.02 for four ions.

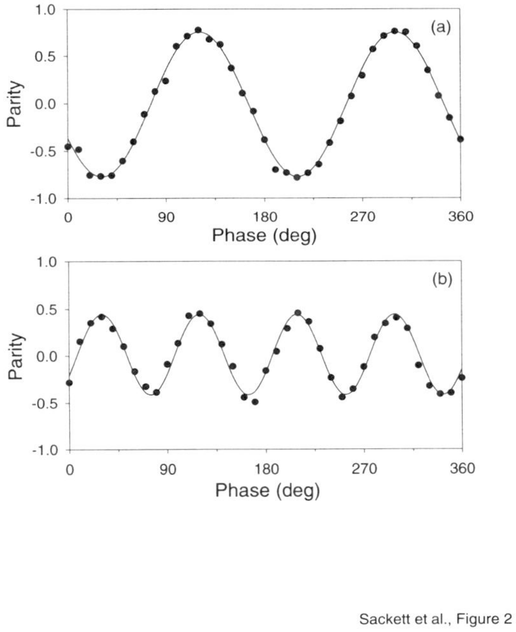

Figure 3. Measurements of the parity Π as a function of RF synthesizer phase ϕ for two and four ions. The data show clear sinusoidal dependence as in Eq. (4), allowing us to extract the far-off-diagonal coherence $|\rho_{\downarrow\downarrow...\uparrow\uparrow}|$.

3 Violation of a Bell Inequality

Classical physics obeys the principle of local realism, which states that objects have definite properties whether or not they are measured and that measurements of an object's properties are not affected by events taking place sufficiently far away. As shown by Einstein, Podolsky, and Rosen [18], quantum mechanics is incomplete if we assume local realism. Bell [19] proved that quantum mechanics is actually inconsistent with local realism. Quantitative conditions for falsifying local realism by experimental measurements were first proposed by Bell [19] and were developed further by Clauser, Horne, Shimony, and Holt (CHSH) [20]. In such an experiment, one prepares a pair of particles in a repeatable way, subjects them to independent classical manipulations parametrized by variables Φ_1, Φ_2, and measures a classical property yielding a value $P_1(\Phi_1), P_2(\Phi_2) = \pm 1$ for each particle. Defining the correlation function $q(\Phi_1, \Phi_2) = P_1(\Phi_1)P_2(\Phi_2)$, one finds that the Bell inequality

$$B(\alpha_1, \delta_1, \beta_2, \gamma_2) = |q(\delta_1, \gamma_2) - q(\alpha_1, \gamma_2)| + |q(\delta_1, \beta_2) + q(\alpha_1, \beta_2)| \leq 2 \tag{5}$$

holds for any values α_1, δ_1 (resp. β_2, γ_2) of Φ_1 (resp. Φ_2) under the assumption of local realism.

In our experiment [3], we apply the entangling operation of Eq. (2) to two ions to prepare our initial state. Then we perform a carrier rotation as in Eq. (1) with $\theta = \pi/2$ and phases ϕ_1, ϕ_2 relative to the entangling pulse phases. Finally, we detect the number of bright ions. The calibration of ϕ_1 and ϕ_2 against the trap strength and the RF synthesizer phase were performed using assumptions from classical physics only [3]. We can therefore identify ϕ_1, ϕ_2 with the classical control parameters Φ_1, Φ_2, so the experiment is of the type proposed by Bell and CHSH. Quantum mechanics predicts that, for an ideal realization of our experiment, $B(\alpha_1, \delta_1, \beta_2, \gamma_2)$ attains a maximum value of $2\sqrt{2}$ for

$$\begin{aligned} \alpha_1 &= -\pi/8 \qquad & \delta_1 &= 3\pi/8 \\ \beta_2 &= -\pi/8 \qquad & \gamma_2 &= 3\pi/8 \end{aligned} \tag{6}$$

contradicting the bound $B(\alpha_1, \delta_1, \beta_2, \gamma_2) \leq 2$ required by local realism.

We repeated the experiment $N_{\text{tot}} = 20\,000$ times for each of the four sets of phases given by Eq. (6). These data are presented as histograms in Fig. 4. We see that the three cases of zero, one, and two ions bright are readily distinguishable. The overlap of histograms corresponding to the three cases gives an inaccuracy of 2% in discriminating between the three cases. This inaccuracy mostly arises from pumping of the $|\downarrow\rangle$ state to the $|\uparrow\rangle$ state because of imperfect polarization of the detection light. Counting the number of events N_0 (N_1, N_2) with zero (one, two) ions bright, we extract the correlation function using

$$q = \frac{(N_0 + N_2) - N_1}{N_{\text{tot}}} \tag{7}$$

and obtain a set of four values of $q(\phi_1, \phi_2)$. We repeated the entire procedure five times to ensure that the measured Bell signal did not vary significantly from run to run. The five sets of correlation functions and the resulting Bell signals are shown in Table 2. The statistical error of each Bell signal was ± 0.01. Averaging the Bell signals together, we find

$$B(-\frac{\pi}{8}, \frac{3\pi}{8}, -\frac{\pi}{8}, \frac{3\pi}{8}) = 2.25 \pm 0.03 \tag{8}$$

an 8σ violation of the Bell inequality Eq. (5). The error in the average Bell signal significantly exceeds the error of each individual measurement of the Bell signal. This discrepancy arises from errors in setting the phases of Eq. (6).

This experiment gives the first violation of the Bell inequality for massive particles obtained using a complete set of correlation measurements. A previous experiment [21]

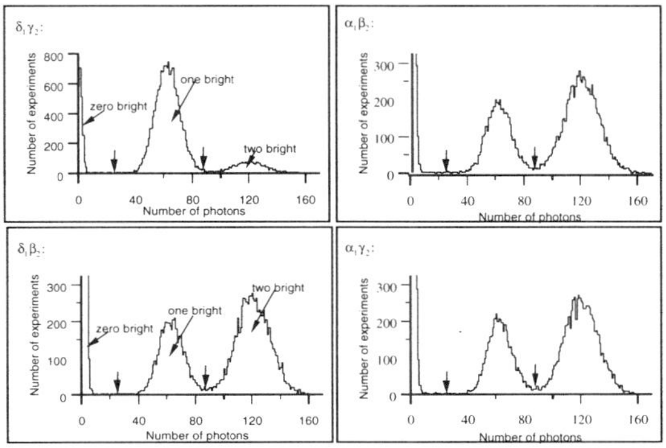

Figure 4. Histograms of fluorescence signals for the four sets of phases given in Eq. (6). The vertical arrows break up the data into cases of zero, one, or two ions bright with 98% accuracy.

Table 2. The correlation values and resulting Bell's signals for five runs of the experiment. The phase angles $\alpha_1, \delta_1, \beta_2, \gamma_2$ are given by Eq. (6). The statistical errors are 0.006 and 0.012 for the values of q and B respectively.

$q(\alpha_1, \beta_2)$	$q(\alpha_1, \gamma_2)$	$q(\delta_1, \beta_2)$	$q(\delta_1, \gamma_2)$	$B(\alpha_1, \delta_1, \beta_2, \gamma_2)$
0.541	0.539	0.569	-0.573	2.222
0.575	0.570	0.530	-0.600	2.275
0.551	0.634	0.590	-0.487	2.262
0.575	0.561	0.559	-0.551	2.246
0.541	0.596	0.537	-0.571	2.245

deduced a violation from an incomplete set of measurements. Furthermore, our data uses the outcome of every shot of the experiment, so the violation is obtained without the use of the fair-sampling hypothesis. In principle, sampling or post-selection of the data allows violation of Eq. (5) even by local hidden-variable theories, the so-called "detector loophole" [22]. Our experiment closes the detector loophole for the first time. However, the "locality loophole" [23] remains open for our data; since the detection events on the two ions occur within each other's lightcones, in principle the detections could influence each other, leading to spurious correlations.

4 Encoding into a Decoherence-Free Subspace

Information stored in a quantum system is liable to be lost through coupling of the system to its environment, a process called decoherence [24,25]. In particular, many proposed quantum computer implementations decohere through a mechanism that couples each qubit to the environment equally strongly [26,27]. Encoding quantum information into a decoherence-free subspace (DFS) [28,29,30] whose states are invariant under coupling to the environment allows information storage even in the presence of this type of decoherence. We have demonstrated the immunity of a DFS of two atoms to environment-induced dephasing and implemented a technique for encoding an arbitrary physical qubit state into the DFS.

The type of decoherence considered here is collective dephasing, in which each qubit undergoes the transformation $|\downarrow\rangle \rightarrow |\downarrow\rangle, |\uparrow\rangle \rightarrow e^{i\zeta}|\uparrow\rangle$ for ζ an unknown phase that is the same from qubit to qubit. One can construct a DFS of two qubits that protects against collective dephasing from the basis states $|\uparrow\downarrow\rangle, |\downarrow\uparrow\rangle$. Any superposition of these states is clearly invariant under collective dephasing, and since there are two states, we can encode one "logical qubit" in this subspace of the two "physical qubits". To demonstrate our encoding technique, we first prepare a state of form

$$|\downarrow\rangle(a|\downarrow\rangle + b|\uparrow\rangle) \qquad |a|^2 + |b|^2 = 1 \qquad a, b \text{ complex} \tag{9}$$

where one of the two physical qubits initially contains the quantum information. In general, we can prepare this state by driving the carrier transition of Eq. (1) on both ions simultaneously, once with $\theta = \beta$, and again with $\theta = \beta$ and ϕ_1 shifted by π. We set $\phi_2 = \alpha$ for both pulses. The final state has $a = \cos 2\beta, b = e^{i\alpha} \sin 2\beta$.

We encode the state of Eq. (9) into the DFS in two steps. First we apply the inverse of the entangling gate, yielding $a(|\downarrow\downarrow\rangle - i|\uparrow\uparrow\rangle) + b(|\downarrow\uparrow\rangle - i|\uparrow\downarrow\rangle)$. Then we drive the carrier with $\theta = \pi/4, \phi_1 = \pi/2, \phi_2 = 0$ to obtain $|\psi_{\rm DFS}\rangle = a(|\downarrow\uparrow\rangle + i|\uparrow\downarrow\rangle) + b(|\downarrow\uparrow\rangle - i|\uparrow\downarrow\rangle)$. The information stored in the physical qubit of ion 2 is now encoded in the logical qubit of the DFS. In the experiment we take $|a| = |b|$, though our method permits preparation and encoding of any state of the more general form.

To read out the encoded information, we reverse the carrier pulse in the encoding and apply the entangling gate to decode $|\psi_{\rm DFS}\rangle$ into $|\downarrow\rangle(a|\downarrow\rangle + b|\uparrow\rangle)$. After decoding, we rotate ion 2 as in preparing the state of Eq. (9) but with the phase on ion 2 changed to α'. We then measure the probability P_2 of finding both ions in $|\downarrow\rangle$. P_2 varies sinusoidally with $\alpha - \alpha'$, and the magnitude of oscillation is equal to the coherence, C, of ion 2. Since we set $|a| = |b| = 1/\sqrt{2}$, ideally $C = 1$. Departures from $C = 1$ measure the effects of both decoherence and imperfect logic. We verified that C is independent of α, thus showing that our encoding method works even if we have no information about the phase of the input state.

To study the effects of decoherence on the DFS-encoded state, we engineered a collective dephasing environment using an off-resonant laser beam with a randomly varying intensity. The beam induces shifts of ν_0 that are common to both ions through the AC Stark effect. The DFS state should resist the dephasing effect of this environment. The coherence of ion 2 in the test state $|\downarrow\rangle(|\downarrow\rangle + e^{i\phi}|\uparrow\rangle)/\sqrt{2}$, however, should be sensitive to collective dephasing. We measure the decay of the test state by simply turning off the encoding and decoding sequences in the procedure used to measure the decay of the DFS-encoded state. We applied the noise beam to the test state and encoded state during a delay time of about 25 μs, as shown in Fig. 5. The coherence without applied noise is about 0.69 for the test state and about 0.43 for the encoded state; they depart from 1 because of imperfect logic gates and detection. For white-noise intensity fluctuations of the Stark-shifting beam, we expect C to decay exponentially for the test state, as shown by the fit line. We also fit the coherence of

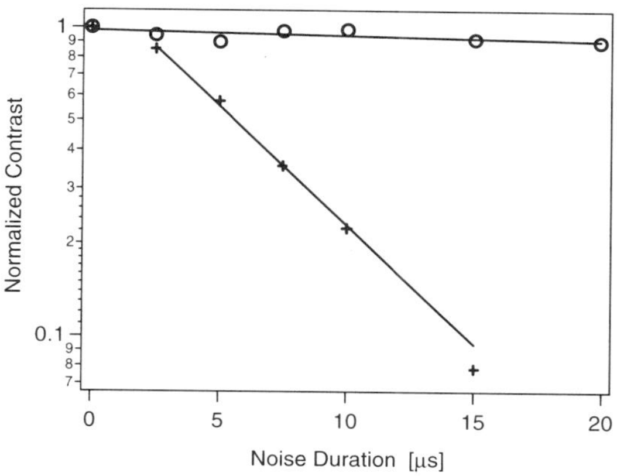

Figure 5. Decay of DFS-encoded state (circles) and test state (crosses) under engineered dephasing noise. The noise is a … decoding. Coherence dat… …tial decay curves. The DF…

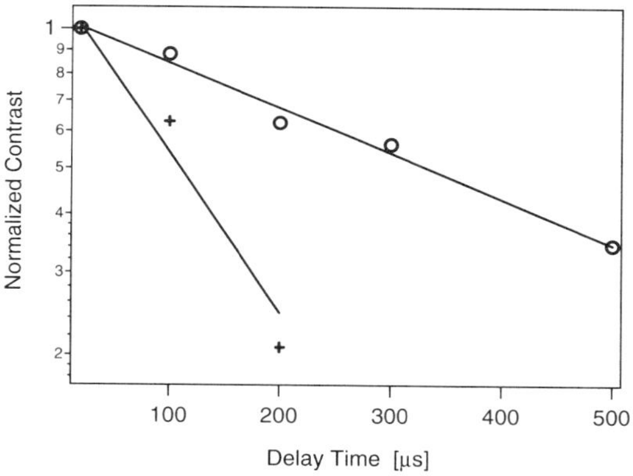

Figure 6. Decay of DFS-encoded state (circles) and test state (crosses) under ambient decoherence. We vary the delay time between encoding and decoding to give the ambient noise a variable time to act. Coherence data are normalized to their values for zero delay time. The fit lines are exponential decay curves for purposes of comparison and are not theoretical predictions.

the DFS state to an exponential decay for comparison. The decay rate of the test state is $0.18(1)\,\mu s^{-1}$, while the decay rate of the DFS state is $0.0035(50)\,\mu s^{-1}$, consistent with zero decay.

We have also measured the storage times of the encoded and test states under ambient conditions in our laboratory. The data are shown in Fig. 6. Here we leave a variable delay

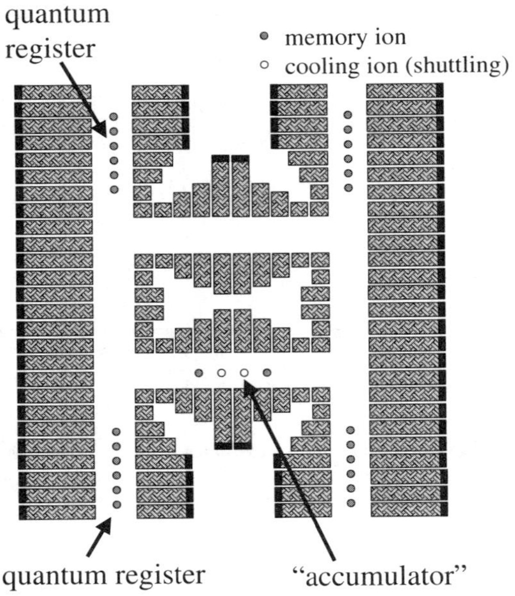

Figure 7. Schematic of the proposed "quantum CCD" architecture for a large-scale ion-trap quantum computer. The "logic ions", which encode the quantum information, are held in a storage region. To perform a logic operation, we shuttle the relevant logic ions into the "accumulator" region, where they interact with lasers. Auxiliary "cooling ions" are provided for sympathetic cooling of the logic ions, which will almost certainly heat up during transfer to the accumulator.

time between encoding and decoding to give the ambient noise time to act. The decoherence of the test state is dominated by ambient fluctuating magnetic fields whose frequencies lie primarily at 60 Hz and its harmonics. These fields randomly shift ω_0 through the Zeeman effect. Since these fields are roughly uniform across the ion string, they induce collective dephasing similar to that created by the engineered environment. We empirically find the decay of both test and encoded states to be roughly exponential, as indicated by the fit lines. The decay rate of the test state is $7.9(1.5) \times 10^{-3}\ \mu s^{-1}$, while the decay rate of the DFS state is $2.2(0.3) \times 10^{-3}\ \mu s^{-1}$. The long lifetime of the DFS state shows that collective dephasing from magnetic field noise is the major ambient source of decoherence for the test state. The loss of coherence of the encoded state is consistent with degradation of the decoding pulses [2,31] due to heating of the ion motional state over the delay time [17].

5 Conclusion and Outlook

We have demonstrated four-particle entanglement, violation of Bell's inequality, and encoding into a decoherence-free subspace using a small quantum register of trapped ions. While these experiments are important steps on the road to quantum computation, the major challenge remains: scaling up the techniques developed for a few ions to build a large-scale quantum computer with thousands or millions of qubits. Fortunately, we have a plausible roadmap to our goal, the "quantum CCD" model of ion-trap quantum computing (Fig. 7). We envision a large ion trap with many independent DC electrodes, which could be used to move ions between storage regions and "accumulator" regions in which quantum logic operations take place. Such a trap could hold thousands of ions, but avoids the slow gate speeds and detection inefficiency associated with large ion crystals by allowing us to work with only a few qubits at a time. Though the ions would most likely heat up while being moved around, the addition of another ion species would enable us to sympathetically cool the ions before performing logic [32]. The DFS encoding demonstrated above would permit us to move qubits around without incurring dephasing from the spatial variation of magnetic fields across the device. In pursuit of the "quantum CCD" goal, we have built a trap supporting a double-well potential with spacing about 1 mm and plan to explore ways of maintaining quantum coherence and entanglement over these large distances.

section*acknowledgmentsThis research was supported by the NSA, ARDA, ARO, and ONR. This paper is a contribution of the National Institute of Standards and Technology and is not subject to U.S. copyright.

References

1. C.A. Sackett, D. Kielpinski, B.E. King, C. Langer, V. Meyer, C.J. Myatt, M. Rowe, Q.A. Turchette, W.M. Itano, D.J. Wineland, and C. Monroe, Nature 404, 256 (2000).
2. A. Sørensen and K. Mølmer, *Phys. Rev. Lett.* 82, 1971 (1999); K. Mølmer and A. Sørensen, *Phys. Rev. Lett.* 82, 1835 (1999); E. Solano, R.L. de Matos Filho, and N. Zagury, *Phys. Rev. A* **59**, 2539 (1999); A. Sørensen and K. Mølmer, *Phys. Rev. A* 62, 022311 (2000).
3. M.A. Rowe, D. Kielpinski, V. Meyer, C.A. Sackett, W.M. Itano, C. Monroe, and D.J. Wineland, *Nature* **409**, 791 (2001).
4. D. Kielpinski, V. Meyer, M.A. Rowe, C.A. Sackett, W.M. Itano, C. Monroe, and D.J. Wineland, *Science* **291**, 1013 (2001).
5. P.W. Shor, in *Proc. 35th Ann. Symp. on Found. Comp. Sci.*, S. Goldwasser, Ed. (IEEE Computer Society, Los Alamitos, CA, 1994), p. 116.
6. L.K. Grover, *Phys. Rev. Lett.* **79**, 325 (1997).
7. F. Diedrich, J.C. Bergquist, W.M. Itano, and D.J. Wineland, *Phys. Rev. Lett.* **62**, 403 (1989).
8. B.E. King, C.S. Wood, C.J. Myatt, Q.A. Turchette, D. Leibfried, W.M. Itano, C. Monroe, and D.J. Wineland, *Phys. Rev. Lett.* **81**, 1525 (1998).
9. This requirement can be relaxed further in some cases, as pointed out by W. Dür and J.I. Cirac (quant-ph/0011025).
10. C. Roos, T. Zeiger, H. Rohde, H.C. Ngerl, J. Eschner, D. Leibfried, F. Schmidt-Kaler, and R. Blatt, *Phys. Rev. Lett.* **83**, 4713 (1999).
11. W.M. Itano, J.C. Bergquist, J.J. Bollinger, J.M. Gilligan, D.J. Heinzen, F.L. Moore, M.G. Raizen, and D.J. Wineland, *Phys. Rev. A* **47**, 3554 (1993).
12. P.T.H. Fisk, *Rep. Prog. Phys.* **60**, 761 (1997).
13. F. Schmidt-Kaler, C. Roos, H.C. Nagerl, H. Rohde, S. Gulde, A. Mundt, M. Lederbauer, G. Thalhammer, T. Zeiger, P. Barton, L. Hornekaer, G. Reymond, D. Leibfried,

J. Eschner, and R. Blatt, *J. Mod. Opt.* **47**, 2573 (2000).
14. C. Monroe, D.M. Meekhof, B.E. King, W.M. Itano, and D.J. Wineland, *Phys. Rev. Lett.* **75**, 4714 (1995).
15. H.C. Nägerl, D. Leibfried, H. Rohde, G. Thalhammer, J. Eschner, F. Schmidt-Kaler, and R. Blatt, *Phys. Rev. A* **60**, 145 (1999).
16. M.G. Raizen, J.M. Gilligan, J.C. Bergquist, W.M. Itano, and D.J. Wineland, *J. Mod. Opt.* **39**, 233 (1992).
17. Q. A. Turchette, D. Kielpinski, B.E. King, D. Leibfried, D.M. Meekhof, C.J. Myatt, M.A. Rowe, C.A. Sackett, C.S. Wood, W.M. Itano, C. Monroe, and D.J. Wineland, *Phys. Rev. A* **61**, 063418 (2000).
18. A. Einstein, B. Podolsky, and N. Rosen, *Phys. Rev.* **47**, 777 (1935).
19. J.S. Bell, *Physics* (N.Y.) **1**, 195 (1965).
20. J.F. Clauser, M.A. Horne, A. Shimony, and R.A. Holt, *Phys. Rev. Lett.* **23**, 880 (1969).
21. M. Lamehi-Rachti and W. Mittig, *Phys. Rev. D* **14**, 2543 (1976).
22. A. Garg and N.D. Mermin, *Phys. Rev. D* **35**, 3831 (1987), and references therein.
23. A. Aspect, *Nature* **398**, 189 (1999).
24. W.H. Zurek, *Phys. Today* **44**, 36 (1991).
25. D. Giulini, E. Joos, C. Kiefer, J. Kupsch, I.-O. Stamatescu, H.D. Zeh, *Decoherence and the Appearance of a Classical World in Quantum Theory* (Springer, Berlin, 1996).
26. *Fortsch. der Physik* **46**, No. 4-8 (1998).
27. *Fortsch. der Physik* **48**, No. 9-11 (2000).
28. D.A. Lidar, I. L. Chuang, and K. B. Whaley, *Phys. Rev. Lett.* **81**, 2594 (1998).
29. P. Zanardi and M. Rasetti, *Phys. Rev. Lett.* **79**, 3306 (1997).
30. L.M. Duan and G.C. Guo, *Phys. Rev. A* **57**, 737 (1998).
31. D.J. Wineland, C. Monroe, W.M. Itano, D. Leibfried, B.E. King, and D.M. Meekhof, *J. Res. NIST* **103**, 259 (1998).
32. D. Kielpinski, B.E. King, C.J. Myatt, C.A. Sackett, Q.A. Turchette, W.M. Itano, C. Monroe, D.J. Wineland, and W.H. Zurek, *Phys. Rev. A* **61**, 032310 (2000).

DEVELOPMENT OF A LINEAR ION TRAP FOR QUANTUM COMPUTING

D. M. LUCAS
Centre for Quantum Computation,
Clarendon Laboratory, Parks Road, Oxford OX1 3PU, U.K.
E-mail: d.lucas@physics.ox.ac.uk

We describe the latest work in our programme to implement a linear ion trap quantum information processor with $^{40}Ca^{+}$. This is an experimental demonstration of read-out for a qubit which is stored in the spin state of the valence electron, in the ground electronic level.

1 Introduction

Since the original suggestion of Cirac and Zoller,[1] the linear ion trap has been widely regarded as one of the most promising systems for the experimental implementation of quantum computing, at least on the scale of ~ 10 qubits.[2,3] The isolated environment of an ion trap provides extremely good protection against decoherence[4] and motional degrees of freedom may be used to implement quantum logic gates.[5]

We have built a linear ion trap for $^{40}Ca^{+}$, with the aim of experiments in quantum information processing. $^{40}Ca^{+}$ is a promising ion species because it offers a moderately high recoil energy, yet the relevant transitions are accessible to solid-state diode lasers. In this paper we report a demonstration of state read-out for a qubit stored in the two Zeeman states of the ground $S_{1/2}$ level of this ion.

2 Quantum computing in $^{40}Ca^{+}$

The low-lying energy levels of $^{40}Ca^{+}$ are shown in figure 1(a). A natural choice for a pair of internal states to be used as a qubit are the $S_{1/2}^{\pm 1/2}$ Zeeman states of the ground level (we use the notation $L_J^{M_J}$ to represent the Zeeman state M_J of the level J in the term L), split by an external magnetic field. Since they are in the ground level these states are extremely long-lived, and coherent control is possible using Raman transitions via an intermediate state far-detuned from one of the S$\leftrightarrow$P transitions. The two beams required can both be derived from one laser, with a precisely-controlled radio-frequency splitting between them applied using acousto- or electro-optical techniques; thus the difficulties of addressing a very narrow transition directly are avoided. The Raman transition would also be used for cooling to the motional ground state, using sideband cooling or the recently-proposed EIT cooling of Morigi *et al.*[6] An alternative choice of qubit in $^{40}Ca^{+}$ is implemented in ref.[7]

3 Read-out for a ground state qubit

The difficulty with using this "ground state" qubit occurs when it is necessary to read out the qubit state (for example, at the end of any applied quantum logic gates, or even to probe the motional state populations after sideband or EIT cooling). A technique which naturally suggests itself for detecting the state of a single ion with high efficiency is that of

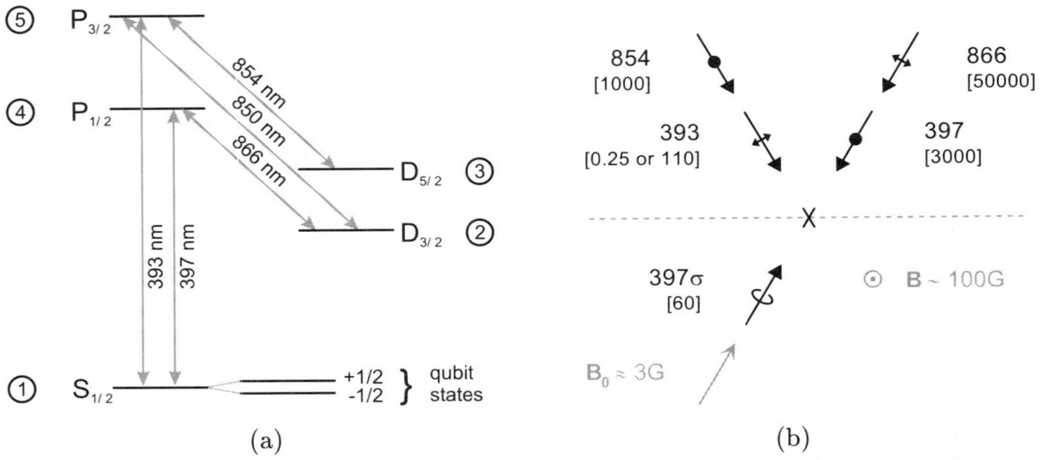

Figure 1. (a) Low-lying energy levels of $^{40}Ca^+$ with allowed transitions (wavelengths to the nearest nm). The proposed qubit states are indicated. The numbering of the levels is used in the text. (b) Directions and polarizations of laser beams in the experiment relative to the applied magnetic fields **B** (out of the page) and $\mathbf{B}_0$ where • represents linear polarization out of the page. Typical beam intensities (W/m^2) are indicated in brackets. The sense of the circular polarization of the 397σ beam is reversed between experimental runs. The position of the ion is indicated by × and the axis of the linear trap by the dashed line.

electron shelving.[8] In the present case this involves transferring the ion selectively from one of the qubit states to the metastable "shelved" $D_{5/2}$ level, where it will remain on average for $\tau(D_{5/2}) = 1168$ ms; if the near-resonant Doppler-cooling radiation on $S_{1/2} \leftrightarrow P_{1/2}$ is then applied (together with $D_{3/2} \leftrightarrow P_{1/2}$ repumping light) the ion will only fluoresce if it was *not* shelved. The presence or absence of fluorescence can be detected with high efficiency in a time short compared with $\tau(D_{5/2})$.

However, the splitting imposed between the $S_{1/2}^{\pm 1/2}$ Zeeman states should be small ($\sim$3 MHz), so as to minimize the effect of fluctuations in the applied magnetic field during coherent qubit manipulations. In this case the qubit states cannot be resolved from each other using the allowed S$\leftrightarrow$P transitions (which have linewidths $\approx$ 23 MHz). Therefore we increase the magnetic field for the read-out phase of the experiment until the Zeeman components of these transitions are spectrally resolved. We may then shelve the $S_{1/2}^{-1/2}$ qubit state by exciting the 393 nm cycling transition $S_{1/2}^{-1/2} \leftrightarrow P_{3/2}^{-3/2}$ with a short pulse of σ^- polarized light, which allows spontaneous decay into the shelved level; due to a favourable $D_{5/2}/D_{3/2}$ branching ratio, the ion is shelved in $D_{5/2}$ with about 89% probability. Provided that the magnetic field is large enough that off-resonant excitation of the $S_{1/2}^{+1/2} \leftrightarrow P_{3/2}^{-1/2}$ transition is negligible, the $S_{1/2}^{+1/2}$ qubit state will be unaffected. Although not exploited in the present experiments, the 11% population 'lost' to $D_{3/2}$ could be recycled without corrupting the qubit state by applying subsequent pulses of 850 nm light on the $D_{3/2}^{-1/2} \leftrightarrow P_{3/2}^{-3/2}$ and $D_{3/2}^{-3/2} \leftrightarrow P_{3/2}^{-3/2}$ transitions; after several complete cycles the probability of $S_{1/2}^{-1/2}$ being shelved in $D_{5/2}$ would then be extremely close to 1.

A simplified rate equation solution for the population transfer during the 393 nm shelving pulse may be found analytically[9] by modelling the system as three levels $[S_{1/2}^{+1/2}, S_{1/2}^{-1/2}, D]$ with populations $[N_+(t), N_-(t), N_D(t)]$ respectively, and two transition rates: (i) the (desired) rate R of transfer from $S_{1/2}^{-1/2}$ to the metastable D states, and (ii) the (undesired) rate R' of transfer from $S_{1/2}^{+1/2}$ to $S_{1/2}^{-1/2}$. Thus

$$\frac{\mathrm{d}}{\mathrm{d}t} N_+ = -R' N_+ \quad ; \quad \frac{\mathrm{d}}{\mathrm{d}t} N_- = R' N_+ - R N_- \quad ; \quad \frac{\mathrm{d}}{\mathrm{d}t} N_D = R N_-$$

where the initial condition $N_-(0) = 1 - N_+(0)$ is given (for example, $N_-(0) = 0.75$ if the ion is prepared in the $S_{1/2}^{-1/2}$ state in 75% of experiments) and $N_D(0) = 0$. The experiment measures the proportion N_3 of the population shelved in $D_{5/2}$, which we find to be

$$N_3(t) = b N_D(t) = \frac{bR}{R - R'} \left\{ \left[N_-(0) - \frac{R'}{R} \right] (1 - \mathrm{e}^{-Rt}) + [1 - N_-(0)] (1 - \mathrm{e}^{-R't}) \right\} \tag{1}$$

where the branching ratios give the numerical factor $b \approx 0.89$ and $t = \tau_{393}$ is the duration of the shelving pulse.

R occurs through resonant excitation of the $S_{1/2}^{-1/2} \leftrightarrow P_{3/2}^{-3/2}$ transition, to which the 393 nm laser is tuned, followed by spontaneous decay to the D states:

$$R = \frac{\Omega_{15}^2}{2\Omega_{15}^2 + \Gamma_5^2} (\Gamma_{25} + \Gamma_{35})$$

where the subscripts refer to levels numbered as in figure 1(a); Γ_{25}, Γ_{35} are the $P_{3/2} \to D_{3/2}, D_{5/2}$ spontaneous decay rates, and $\Gamma_5 = \sum_i \Gamma_{i5}$ is the reciprocal of the $P_{3/2}$ lifetime. The $S_{1/2} \leftrightarrow P_{3/2}$ Rabi frequency Ω_{15} is related to the $\lambda = 393\,\mathrm{nm}$ σ^- laser intensity I_{15} by

$$\Omega_{15}^2 = \frac{3\lambda^3 \Gamma_5}{2\pi h c} I_{15}$$

R' arises from off-resonant excitation of $S_{1/2}^{+1/2} \to P_{3/2}^{-1/2}$ followed by spontaneous decay to $S_{1/2}^{-1/2}$:

$$R' = \frac{\pi}{9} \Omega_{15}^2 \left(\frac{\Gamma_5 / 2\pi}{\delta^2 + \Gamma_5^2 / 4} \right)$$

where the detuning $\delta = \frac{2}{3} \mu_B B / \hbar$ and B is the magnetic field applied during the read-out pulse. These expressions allow the measured $D_{5/2}$ population $N_3(\tau_{393})$ to be related to the basic experimental parameters τ_{393}, B, I_{15} and $N_-(0)$.

Following ref.[9] we define the efficiency of the read-out process to be the difference between the shelved population $N_3(t)$ when $N_-(0) = 1$ and that when $N_-(0) = 0$. However, we allow for imperfect preparation of the initial populations and find that, in two experiments where the initial populations are $N_-^{(1)}(0)$ and $N_-^{(2)}(0)$, the observed efficiency is reduced by a factor $[N_-^{(1)}(0) - N_-^{(2)}(0)]$, to give

$$\epsilon(t) = \left[N_-^{(1)}(0) - N_-^{(2)}(0) \right] \frac{bR}{R - R'} \left(\mathrm{e}^{-R't} - \mathrm{e}^{-Rt} \right) \tag{2}$$

with $N_-^{(1)}(0) > N_-^{(2)}(0)$. For the purposes of testing the read-out scheme, we attempt to ensure that $N_-^{(1)}(0) \approx 1$ and $N_-^{(2)}(0) \approx 0$.

4 Experiment

Much of the apparatus has already been described elsewhere.[10] We give here a brief review and details of the additions necessary for the present experiment.

4.1 Trap

The central vacuum chamber is evacuated to around 2×10^{-11} torr and a cloud of calcium ions is loaded into the linear Paul trap using a calcium oven and electron gun. After loading, the amplitudes of the radio-frequency (6.2 MHz) radial trapping field and the DC axial field are reduced to approximately $100\,\mathrm{V_{PP}}$ and 30 V respectively (giving typical trap secular frequencies of 500 kHz radially and 200 kHz axially). A low-voltage "tickle" oscillation is then applied to one of the end-caps until only a single ion remains in the trap.

4.2 Lasers

Figure 1(b) shows the directions and polarizations of the various laser beams used in the experiment, relative to the low ($B_0 \approx 3\,\mathrm{G}$) and high ($B \sim 100\,\mathrm{G}$) applied magnetic fields. Typical intensities are also shown; beam spot sizes at the trap are $\sim 100 \times 100\,\mu\mathrm{m}$. All five beams may be independently switched, either with mechanical shutters or with acousto-optic modulators. The lasers are of the external cavity grating-stabilized design, and all except the laser at 854 nm are locked to stabilized reference cavities which reduce linewidth and medium-term frequency drift to the 1 MHz level.

The 397 beam for Doppler cooling (to $\sim 1\,\mathrm{mK}$) and fluorescence detection is provided by a frequency-doubled master-slave diode system operating at 794 nm. The same laser system is used for the low-intensity 397σ beam whose circular polarization with respect to $\mathbf{B}_0$ is chosen to prepare the ion in either the $\mathrm{S}_{1/2}^{+1/2}$ or $\mathrm{S}_{1/2}^{-1/2}$ qubit state.

Light from a violet laser diode (393 beam) is used to transfer the ion from the $\mathrm{S}_{1/2}^{-1/2}$ ground state to the shelved $\mathrm{D}_{5/2}$ level, via $\mathrm{P}_{3/2}^{-3/2}$. The intensity at the ion is very low since only ~ 50 transitions are necessary to shelve the ion, and the beam is switched using an AOM in order to give a shelving pulse of well-controlled duration (from $1\,\mu\mathrm{s}$ to 50 ms). The light is linearly-polarized perpendicular to the read-out magnetic field $\mathbf{B}$ and thus half the total beam intensity is available to excite any of the resolved σ components of the $\mathrm{S}_{1/2} \leftrightarrow \mathrm{P}_{3/2}$ transition; here it is tuned to be resonant with the $\mathrm{S}_{1/2}^{-1/2} \leftrightarrow \mathrm{P}_{3/2}^{-3/2}$ component.

The 866 repumping beam is polarized such that it repumps from all $\mathrm{D}_{3/2}^{M_J}$ states, in both the low and high magnetic fields. The 854 beam is used to de-shelve the ion, after the detection phase, in preparation for the next experimental sequence.

4.3 Magnetic fields

The direction of the low field $\mathbf{B}_0$ is aligned parallel to the 397σ optical pumping beam by minimizing the ionic fluorescence due to the 397σ light. For practical reasons, the high field $\mathbf{B}$ is applied perpendicular to $\mathbf{B}_0$, so that the axis of the $\mathrm{S}_{1/2}^{\pm 1/2}$ qubit states must be adiabatically rotated as $\mathbf{B}$ is applied. This is ensured by the $\approx 5\,\mathrm{ms}$ switch-on time-constant of the high-field coils; numerical simulations showed this to be at least an order of magnitude slower than the rate required to introduce a 1% error in the measured state.

4.4 Detection

Fluorescence from the ion at 397 nm is collected in a direction perpendicular to the plane of propagation of the laser beams by a compound lens which subtends a solid angle of 0.20 sr at the trap, and is imaged via a violet filter onto a photomultiplier connected to a gated photon counter. The net count rate from a single ion is around 30 kHz; a counting bin

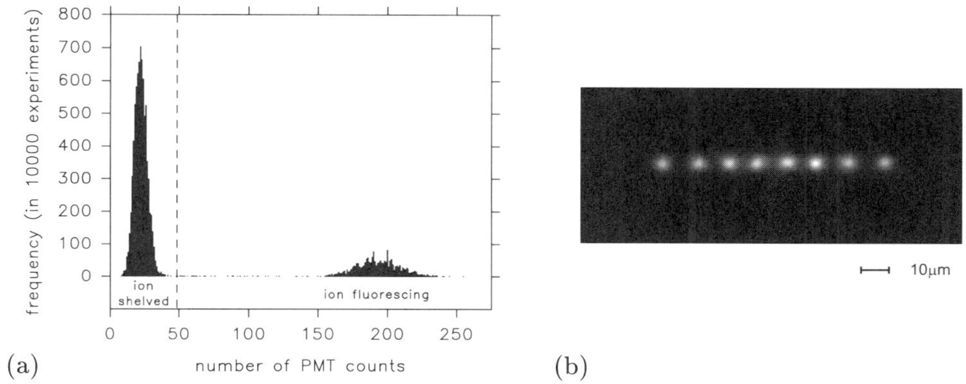

Figure 2. Detection methods. (a) Histogram of photomultiplier counts recorded during the 8 ms detection phase of 10,000 experimental sequences (section 4.5) in which a single ion was first optically pumped towards the $S_{1/2}^{-1/2}$ state, then shelved in the $D_{5/2}$ level as described in the text. The counts when the ion is shelved are due to scattered light from the 397 nm detection beam. A threshold (dashed line) is set to determine the number of experiments in which the ion was shelved; in this data set, the fluorescence is below the threshold in 80% of the experiments ($N_3 = 0.80$). (b) Linear crystal of eight ions, imaged on the intensified CCD camera. The camera provides spatial information, but this particular device does not possess sufficient quantum efficiency at 397 nm to detect whether individual ions are fluorescing or not as rapidly as the PMT.

of 8 ms gives ample discrimination between the "ion fluorescing" and "ion shelved" states, figure 2(a). An intensified CCD camera may alternatively be used for detection, for imaging ion crystals such as that in figure 2(b); in the experiments described here the camera is used only to check for the presence of non-fluorescing (*i.e.* non-$^{40}Ca^+$) ions.

4.5 Experimental sequence

To make a single read-out measurement, the following sequence of events occurs (where • indicates 'on'):

Action	*Duration (approx.)*	*Laser beams* 397	397σ	393	854	866	*Magnetic fields* **B**	$\mathbf{B}_0$
Doppler cooling	15 ms	•				•		•
Optical pumping	10 ms		•			•		•
Adiabatic rotation	30 ms					•	ramping	•
Shelving	1 μs–50 ms			•		•	•	•
Detection	8 ms	•				•		•
De-shelving	2 ms	•			•	•		•

This sequence is repeated many (typically 1000) times to obtain a measurement of the mean $D_{5/2}$ population. The read-out efficiency is deduced by comparing the $D_{5/2}$ population measured when the ion is prepared in the $S_{1/2}^{-1/2}$ state with that when it is prepared in $S_{1/2}^{+1/2}$.

The length τ_{393} of the shelving pulse depends on the intensity of the 393 nm laser beam; two different intensity regimes were investigated, with $\tau_{393} \sim 10$ ms at the lower intensity and $\tau_{393} \sim 10\,\mu$s at the higher one. The total length of the sequence is approximately (90 ms + τ_{393}), when the opening and closing delays of mechanical shutters are included.

The de-shelving phase ensures the ion is returned immediately to the cooling cycle, to avoid having to wait $\sim 1\,\mathrm{s}$ for it to decay spontaneously. The photon counter is gated on for 8 ms during the detection phase, to determine if the ion is shelved or not; and again during the Doppler cooling phase, to check that the ion returned correctly to the cooling cycle after the previous sequence.

5 Results

Results for the shelving efficiency as a function of shelving pulse duration τ_{393} are shown in figure 3. Figure 3(a) shows the measured $\mathrm{D}_{5/2}$ population, where each point represents 1000 experimental sequences. For each value of τ_{393} two runs were carried out, with the sense of the circular polarization of the 397σ beam reversed between them so as to prepare the initial population $N_{-}(0)$ predominantly in either $\mathrm{S}_{1/2}^{-1/2}$ or $\mathrm{S}_{1/2}^{+1/2}$. The data are fitted by equation 1 where the parameters I_{15} (393 nm σ^{-} intensity), $N_{-}^{(1)}(0)$ (for $397\sigma^{-}$ runs) and $N_{-}^{(2)}(0)$ (for $397\sigma^{+}$ runs) are simultaneously floated. The experiment differs slightly from the situation treated in section 3, in that the 866 nm repumping laser is left on during the shelving pulse which prevents any population ending up in the $\mathrm{D}_{3/2}$ level; to a first approximation this recycles about half the lost population back into $\mathrm{D}_{5/2}$ via the qubit state that is being read out ($\mathrm{S}_{1/2}^{-1/2}$) and we use a modified branching ratio $b = 0.94$ in equations 1 and 2.

The observed efficiency ϵ is the difference between the two data sets and is shown in figure 3(b), together with the fitted curve. At the optimum shelving pulse length of around 6 ms the efficiency reaches 60%. Also shown is the ideal efficiency that would be obtained if the initial state preparation were perfect. In a second experiment a higher intensity 393 nm shelving pulse was used, giving a peak observed read-out efficiency of 50% at $\tau_{393} \approx 15\,\mu\mathrm{s}$; the reduction in efficiency appeared to be due to poorer initial state preparation.

Figure 4 shows the shelving efficiency as a function of the magnetic field B applied for read-out, where τ_{393} was fixed at the optimum value of 6 ms. The data are again fitted by equation 1, with independent variable B and the same parameters floated as previously. The field strength was calibrated from the Zeeman shift of the $\mathrm{S}_{1/2}^{-1/2} \leftrightarrow \mathrm{P}_{3/2}^{-3/2}$ component.

Given the simplifications in the limited rate-equation treatment, the fits are satisfactory, and indicate 10–15% imperfection in the state preparation and a 393 nm σ^{-} intensity at the ion of 40–60% of the measured beam intensity in this polarization (implying the beam was not perfectly centred on the ion). We do not account for residual Doppler broadening or laser linewidth, which are small ($\sim$1 MHz) compared to the $\mathrm{S}_{1/2} \leftrightarrow \mathrm{P}_{3/2}$ natural width.

6 Discussion and Prospects

As discussed by Stevens *et al.*,[9] read-out efficiencies significantly less than 100% do not necessarily render quantum computations inefficient and the main cost is the need to re-run an algorithm several times; from figures 3(a) and 4(a) the two different initial states can clearly be discriminated (even when $\epsilon < 50\%$), and could be with significantly less repetitions than the 1000 used here, with optimized parameters.

The observed efficiency of this read-out method appears to be limited mainly by imperfect optical pumping used to prepare the initial states; we have no reason to doubt that with correctly-prepared initial populations the read-out efficiency would approach its theoretical optimum at large magnetic field, or that it could be made close to 100% by repumping

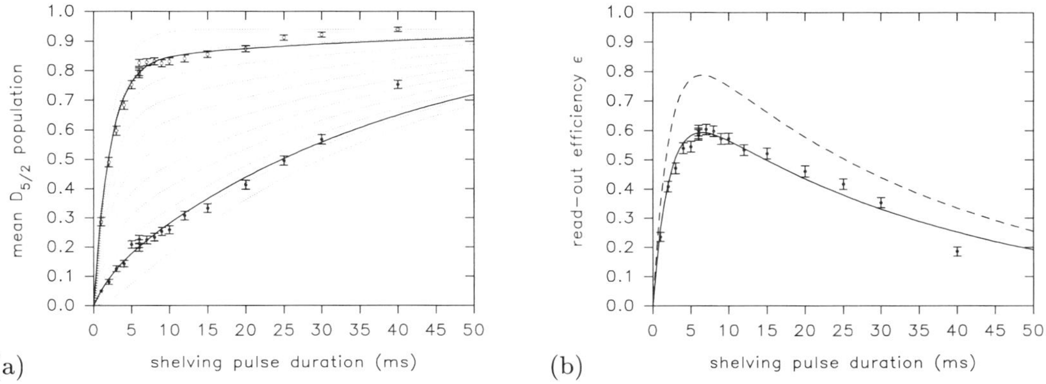

Figure 3. Read-out efficiency as a function of the duration τ_{393} of the 393 nm $S_{1/2}^{-1/2} \rightarrow P_{3/2}^{-3/2}$ shelving pulse, for fixed magnetic field $B = 98\,\mathrm{G}$. Error bars are statistical. (a) Mean $D_{5/2}$ population at the end of the shelving pulse, for a single ion previously pumped with 397 nm σ^- light (○) or 397 nm σ^+ light (●). Each data point represents 1000 experimental sequences. The fit from equation 1 (solid lines) gives $N_-^{(1)}(0) = 0.89(1)$, $N_-^{(2)}(0) = 0.13(1)$, $I_{15} = 0.054(3)\,\mathrm{W/m^2}$. The measured 393 nm σ^- intensity was estimated to be $0.12(4)\,\mathrm{W/m^2}$ at the centre of the beam. The dotted lines show the predictions of the theory for $N_-(0) = 0 \ldots 1$ in ten steps. (b) Observed read-out efficiency ϵ for the same data. The solid line is equation 2 with the above parameters. The dashed line shows the expected efficiency if the initial populations had been perfectly prepared, that is, if $N_-^{(1)}(0) - N_-^{(2)}(0) = 1$.

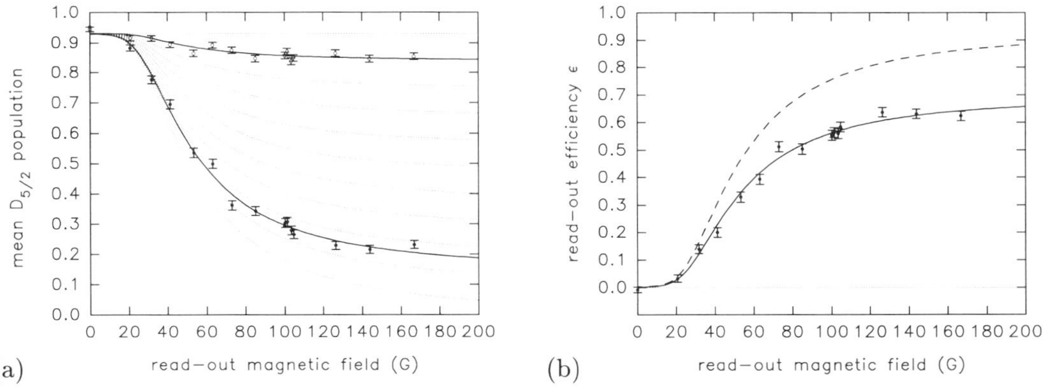

Figure 4. Shelving efficiency as a function of the read-out magnetic field B, for fixed shelving pulse duration $\tau_{393} = 6\,\mathrm{ms}$. Error bars are statistical. (a) Mean $D_{5/2}$ population at the end of the shelving pulse, for a single ion previously pumped with 397 nm σ^- light (○) or 397 nm σ^+ light (●). Each data point represents 1000 experimental sequences. The fit from equation 1 (solid lines) gives $N_-^{(1)}(0) = 0.90(1)$, $N_-^{(2)}(0) = 0.16(1)$, $I_{15} = 0.091(3)\,\mathrm{W/m^2}$. The measured 393 nm σ^- intensity was estimated to be $0.15(5)\,\mathrm{W/m^2}$ at the centre of the beam. The dotted lines show the predictions of the theory for $N_-(0) = 0 \ldots 1$ in ten steps. (b) Observed read-out efficiency ϵ for the same data. The solid line is equation 2 with the above parameters. The dashed line shows the expected efficiency if the initial populations had been perfectly prepared, that is, if $N_-^{(1)}(0) - N_-^{(2)}(0) = 1$.

from $D_{3/2}$ with 850 nm light as mentioned in section 3. The limited state preparation efficiency is probably due to the switching of the high magnetic field necessary in this scheme: at constant low field we obtain very efficient optical pumping (a reduction of the 397 nm fluorescence from an intense 397σ beam by a factor > 50), but when the high magnetic field is switched at ~ 10 Hz as in the read-out experiments the optical pumping is much less efficient (the fluorescence observed during the low-field optical pumping pulse is reduced by a factor ~ 4 only). By varying the direction of the low magnetic field with respect to the optical pumping beam during sequences of read-out experiments we were unable to improve the shelving efficiency, implying that the problem is due to fluctuating transient fields; the fluorescence observed indicates these are at the ~ 1 G level.

Although it might be possible to reduce these magnetic field transients, they will be a more serious problem when it comes to manipulation of the qubit states themselves by driving Raman transitions between them. Then the energy separation of the qubit states, which is first-order dependent on the magnetic field, must be kept constant, leading to a requirement on the magnetic field stability at the mG level. Hence we are currently investigating the possibility of alternative laser read-out methods which do not require the magnetic field to be increased, but still avoid the need for sophisticated laser stabilization.

One such method is to excite Raman transitions $S_{1/2}^{+1/2} \leftrightarrow (P_{1/2}^{-1/2}) \leftrightarrow D_{3/2}^{+1/2}$ followed by spontaneous scattering on the $D_{3/2}^{+1/2} \leftrightarrow P_{3/2}^{+3/2}$ transition to shelve the $S_{1/2}^{+1/2}$ qubit state in $D_{5/2}$; here the selection of Zeeman states is achieved by polarization rather than large magnetic fields. A second possibility is to shelve using the $S_{1/2} \leftrightarrow P_{3/2}$ transition again, but to use electromagnetically-induced transparency to suppress the $S_{1/2}^{-1/2} \leftrightarrow P_{3/2}^{+1/2}$ transition whilst allowing the $S_{1/2}^{+1/2} \leftrightarrow P_{3/2}^{+3/2}$ transition, for example; here a second, relatively intense, pump laser operating close to the $D_{3/2} \leftrightarrow P_{3/2}$ transition is used to provide the EIT. Preliminary calculations suggest that this second scheme can produce a read-out efficiency close to 90% with lasers stabilized to the 1 MHz level and with 1% polarization impurity.

Acknowledgments

None of the work described would have been possible without present and past members of the Oxford ion-trapping group, Paul Barton, Charles Donald, Matt McDonnell, Derek Stacey, J.-P. Stacey, Andrew Steane and David Stevens, or the technical assistance of Graham Quelch. The research is supported by the EPSRC and Christ Church, Oxford.

References

1. J. I. Cirac and P. Zoller, Phys. Rev. Lett. **74**, 4091 (1995).
2. A. M. Steane, Appl. Phys. B **64**, 632 (1996).
3. C. Monroe *et al.*, Phys. Rev. A **55**, R2489 (1997).
4. R. J. Hughes *et al.*, Phys. Rev. Lett. **77**, 3240 (1996).
5. C. Monroe *et al.*, Phys. Rev. Lett. **75**, 4714 (1995).
6. G. Morigi, J. Eschner and Ch. Keitel, Phys. Rev. Lett. **85**, 4458 (2000).
7. H. C. Nägerl *et al.*, Phys. Rev. A **61**, 023405 (2000).
8. W. Nagourney, J. Sandberg and H. Dehmelt, Phys. Rev. Lett. **56**, 2797 (1986).
9. D. Stevens, J. Brochard and A. Steane, Phys. Rev. A **58**, 2750 (1998).
10. P. A. Barton *et al.*, Phys. Rev. A **62**, 032503 (2000).

PROGRESS TOWARDS SIDEBAND-COOLING TRAPPED CALCIUM IONS

D. G. ENZER, M. H. HOLZSCHEITER, D. F. V. JAMES, P. G. KWIAT, S. K. LAMOREAUX , V. SANDBERG, M. M. SCHAUER, J. R. TORGERSON, D. TUPA, R. J. HUGHES
Los Alamos National Laboratory, P.O. Box 1663 MS-H803
Los Alamos, NM 87545, USA

Trapped, laser-cooled ions are being investigated as one possible implementation of a quantum computer. Crystallized ions in a linear trap serve as the individual quantum bits, or qubits. Manipulating both the crystal's motional state and the ions' electronic state generates quantum logic operations. At Los Alamos, we have built a linear rf quadrupole ion trap and have used it to create strings of cooled Ca^+ ions. We have used these crystals to study the phase transition from a linear string into a zigzag configuration as trap parameters are varied; the onset of this phase transition limits the size of the linear quantum register.

Trapped ions are Doppler cooled by irradiating them with a laser beam tuned slightly red of the $S_{1/2}$ to $P_{1/2}$ transition frequency. However, to further cool the ion crystal to its motional ground state, we plan to use sideband cooling on the narrow $S_{1/2}$ to $D_{5/2}$ transition. This calls for a laser much narrower in frequency than the 200 - 500 kHz spacing of motional sidebands in our trap. For this, we have locked a Ti::Sapphire laser to a stable, high finesse cavity . Progress on sideband cooling will be reported.

We consider a method to estimate the maximum number of state manipulations possible for a given number of ions and various ion trap parameters.

Funded by the National Security Agency and ARDA under contract number M0D708600.

Keywords: quantum computation, trapping, cooling, ions

1. Introduction

The overall objective of our experiment[1] is to implement quantum computation (QC) with laser-cooled ions confined in a linear radio-frequency (rf) quadrupole trap. In this scheme[2], a qubit is composed of two long-lived electronic energy levels of a trapped ion.

This system is well suited for quantum computation for several reasons. All essential ingredients of a quantum computer have been demonstrated experimentally, including coherent quantum state manipulation, quantum logic operations and readout. In order to achieve a computational advantage over a classical computer, a quantum computational device must provide the ability to create and maintain a deterministic entanglement between all the qubits; an ion trap is one of only two implementations in which this has been achieved. In contrast to other implementations, the relevant decoherence mechanisms are known and either well-understood (i.e. spontaneous emission)[3,4] or amenable to experimental quantification (ion heating)[5,6], allowing optimized traps to be developed. The ion trap QC architecture should allow experimental studies with tens of qubits, so that ion traps could take experimental quantum computing beyond the limits of classical simulation.

In our experiment, a qubit will be composed of a ground state $S_{1/2}$ sublevel and a metastable $D_{5/2}$ sublevel of a trapped Ca^+ ion. The relevant energy levels of Ca^+ are shown in Fig. 1. The ions will be crystallized into a string along the trap axis after

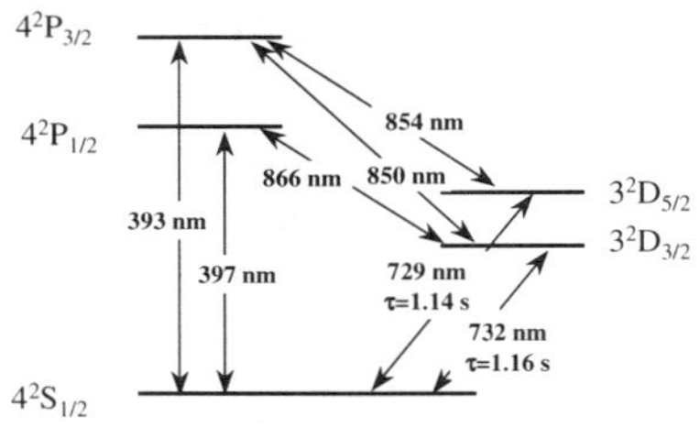

Fig. 1. Relevant energy levels, lifetimes, and transitions of the Ca^+ ion.

Doppler cooling with 397-nm and 866-nm lasers. Although localized, the ions undergo small amplitude oscillations in the trap potential. Their lowest frequency mode is the axial center of mass (CM) motion in which all the ions oscillate in phase along the trap axis. This mode will provide a computational bus. Initially, the ions must be prepared in the quantum ground state of the CM mode. The $S_{1/2} \rightarrow D_{5/2}$ electric quadrupole transition in calcium ions has such a narrow width that upper and lower motional sidebands, separated from the transition frequency by the CM frequency, can be resolved. With 729-nm laser light of suitably narrow linewidth tuned to the first lower sideband, sideband cooling[7] can be used to prepare the bus CM mode in its lowest quantum state. Quantum logic gate operations will be effected by tuning the 729-nm laser to the transition frequency or to a specific sideband, and focusing the beam on individual ions for prescribed times. The laser-ion interaction will coherently change the state of the qubit and the CM mode as the ion undergoes Rabi oscillations between the $S_{1/2}$ and $D_{5/2}$ states. Upon completion of quantum logic operations, the final state of each qubit ($S_{1/2}$ or $D_{5/2}$) can be read out by observing which ions fluoresce and which are dark when driven with 397-nm light.

2. Experimental Apparatus

We have described the Los Alamos quantum computation experiment in detail in an earlier publication[1]. Calcium ions are trapped in a linear rf quadrupole ion trap, with an endcap separation of $2z_0 = 10$ mm and a radial dimension of $r_0 = 1.4$ mm. Typical trap secular frequencies are $\nu_z \sim 100-400$ kHz axially, and $\nu_r \sim 1$ MHz radially. Compensation rods outside and parallel to the trap rods carry DC potentials to center the ions radially in the rf pseudopotential and thereby minimize the ion micromotion. The axial endcaps can carry different DC potentials to center the ions axially.

Laser beams are directed into the trapping volume at a 45 angle to the z-direction. A diode laser beam at 794 nm drives a tapered amplifier and the resulting light is frequency-doubled to produce the Doppler-cooling beam at 397 nm. Diode lasers also produce the repump beam at 866 nm and the cleanup beam at 854 nm. All three diode laser beams have short-term stabilities of around 1 MHz. All three diode lasers are locked to a single low-finesse cavity, which is in turn locked to a stable helium-neon laser, providing long-term stability. Individual quartz plates in the reference cavity are mounted in each of the diode laser beam-paths; tipping the plates allows each laser to be tuned and scanned independently. The 729-nm light needed to drive the $S_{1/2} \rightarrow D_{5/2}$ transition is produced by a Ti::Sapphire laser. We have designed and built a stable reference cavity and associated

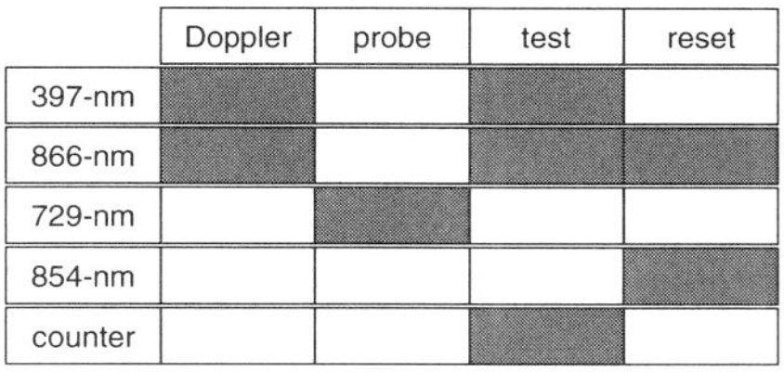

Fig. 2. A sideband-spectroscopy experimental cycle.

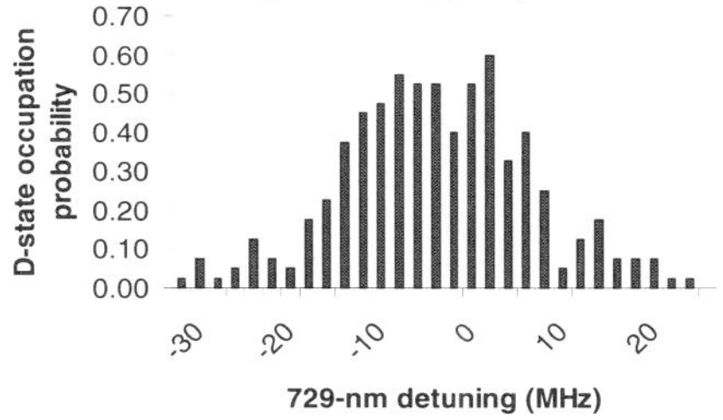

Fig. 3. Sample 729-nm resonance with cooling lasers left on.

optics and electronics to narrow the frequency width of this laser with the Pound-Drever-Hall technique. We have completed this ultra-stable laser system and infer from our electronic signal a short-term linewidth of < 20 kHz.

3. Recent Results

Sideband detection and cooling requires multiple laser beams to be switched on and off for short times (a few microseconds to several milliseconds). Acousto-optic-modulators (AOMs) in each laser's beam-path switch a laser off by steering the beam away from the ion. The timing sequence for sideband detection with a single calcium ion is shown in Fig. 2. First the ion is Doppler cooled. Then, the 729-nm laser is turned on to drive transitions between the $S_{1/2}$ and $D_{5/2}$ states. Then, the cooling lasers are turned on and the ion fluorescence monitored to see if a transition was made. Finally, the system is reset by applying the repump lasers. This sequence is performed repeatedly and the probability for causing transitions to the $D_{5/2}$ state is determined as a function of the 729-nm "prob laser's frequency.

In order to test our system, we went through a similar procedure but kept on the cooling lasers (397-nm, 866-nm) during the 729-nm "probe" phase. This broadened the $S_{1/2} \rightarrow D_{5/2}$ resonance into an easily detectable signal (Fig. 3), much wider than 400 kHz, the expected motional sideband spacing. The test verifies that the basic ingredients for spectroscopy are all working effectively: AOM switching, quantum state readout, and computer control.

4. Optimizing Trap Parameters

Conflicting requirements make it unclear whether to chose a large trap or a small trap for these types of experiments. A small trap has the advantage of higher axial frequencies which give faster ultimate clock speeds yield a lower Doppler cooling limit, and make sideband colling easier. A large trap, however, has less CM heating and produces larger spacing between the ions leading to easier individual addressing. We investigate just how much of a limitation CM heating poses and what ideal clock speed leads to the maximum number of laser operations. We start with five requirements to answer the optimization question "what is the maximum number of operations N_{op} possible for a given number of ions N_{ions}, axial frequency ν_z, trap radius r_0, and applied rf voltage V_{rf}?"

The first requirement is that the ions lie on a line rather than in a zigzag pattern (where they would experience excess micromotion heating). This restricts the ratio of the axial to the radial CM frequencies[8,9], $(\nu_z/\nu_r)^2 < cN_{ions}^{\beta}$ with $c = 2.94$ and $\beta = -1.80$, and gives the maximum trap radius r_0^{max} (Eq. 1) for a given ν_z, N_{ions}, and V_{rf}. The second requirement is that the minimum ion spacing be much larger than the wavelength λ_0 connecting the two qubits states: 729-nm. We insert a safety factor $0 < S_1 < 1$ and require: ($S_1\times$ ion spacing) $> \lambda_0$. Using the minimum ion spacing[10], $(2.018/N_{ions}^{0.559})(e^2/4\pi\epsilon_0 M)^{1/3}(1/(2\pi\nu_z)^{2/3})$, where ϵ_0 is the permittivity of free space and M is the ion mass, we find a maximum axial frequency ν_z^{max} (Eq. 2) for a given number of ions.

The last three requirements limit the maximum number of operations N_{op}^{max} to the product of the clock speed and the time for one of three types of decohering events to occur. The three decohering mechanisms we consider are: spontaneous emission[3,4], CM heating[5,6], and off-resonant transitions to the P states[11]. For the clock speed, we use the axial frequency ν_z times another safety factor $0 < S_2 < 1$. For the effective spontaneous emission time, we use $6\tau_0/N_{ions}$ where τ_0 is the $D_{5/2}$ lifetime and the inclusion of the factor of 6 is described in Ref. 3. For the CM heating time, we use $C M \, {}_z r_0^5/N_{ions}$. The dependence of this CM heating time on M and ν_z coincides with standard analyses such as that of Ref. 6. Dividing the single ion heating time by N_{ions} is justified since on average one of the ions will heat up and destroy the computational coherence in that time. While data from Ref. 6 supports the dependence on r_0 to some power greater than or equal to 4, the exact dependence is still not determined. We choose an r_0^5 dependence[12] but do not expect that our overall conclusions will be very sensitive to the exact exponent. The constant C is within a factor of ten of all the data points reported in Ref. 6 except the two outliers. Finally, for the off-resonant transition time[11], we use $16(\Delta/2\pi)^2(\tau_{ex}^2/\tau_0)(h\cos^2\theta/2M\lambda_0^2)(\lambda_0/\lambda_{ex})^3(1/N_{ions}S_2^2\nu_z^3)$, where Δ is the angular frequency detuning of the (729-nm) laser from the $S_{1/2} \rightarrow P$ transitions, λ_{ex} is the wavelength of the $S_{1/2} \rightarrow P$ transitions, τ_{ex} is the lifetime of the P states, h is Planck's constant and θ is the angle between the laser propagation direction and the trap axis.

Inserting three more safety factors, $0 < S_3, S_4, S_5 < 1$, we get five conditions:

$$r_0^{max} = \sqrt{\frac{eq}{16\pi^2 M}}\frac{\sqrt{V_{rf}}/N_{ions}^{-2/\beta}}{\nu_z} \quad (1)$$

$$\nu_z^{max} = \frac{1}{2\pi}\left(\frac{2.018}{N_{ions}^{0.559}}\right)^{3/2}\left(\frac{e^2}{4\pi\epsilon_0 M}\right)^{1/2}\left(\frac{S_1}{\lambda_0}\right)^{3/2} \quad (2)$$

$$N_{op}^{max,se} = S_3\left[\frac{6\tau_0}{N_{ions}}\right](S_2\nu_z) \quad (3)$$

$$N_{op}^{max,h} = S_4\left[\frac{C M \, {}_z r_0^5}{N_{ions}}\right](S_2\nu_z) \quad (4)$$

$$N_{op}^{max,or} = S_5 \left[16 \left(\frac{\Delta}{2\pi}\right)^2 \left(\frac{\tau_{ex}^2}{\tau_0}\right) \left(\frac{h \cos^2\theta}{2M\lambda_0^2}\right) \left(\frac{\lambda_0}{\lambda_{ex}}\right)^3 \left(\frac{1}{N_{ions} S_2^2 \nu_z^3}\right) \right] (S_2 \nu_z) \quad (5)$$

Here e is the electron charge, q is the pseudo potential parameter[13] taken to be 0.3, and $N_{op}^{max,se}$, $N_{op}^{max,h}$, and $N_{op}^{max,or}$ are N_{op}^{max} as limited by spontaneous emission, heating and off-resonant transitions respectively. Eliminating r_0 from Eq. 4 using Eq. 1, we are left with ν_z^{max} from the ion spacing requirement (Eq. 2) and three expressions for N_{op}^{max}:

$$N_{op}^{max,se} \sim \nu_z \quad (6)$$

$$N_{op}^{max,h} \sim 1/\nu_z^3 \quad (7)$$

$$N_{op}^{max,or} \sim 1/\nu_z^2 \quad (8)$$

¿From these conditions, we can choose an optimal axial frequency. Eq. 6 expresses the need for a large axial frequency so that many operations can be performed before a spontaneous emission event. However, Eqs. 7 and 8 express the need for a smaller axial frequency. In the ion heating case (Eq. 7), this is because larger ν_z's require larger ν_r's (to keep the ions in a line) which require smaller traps which have worse heating rates.

Setting $N_{op}^{max,se}$ equal to $N_{op}^{max,h}$ gives the axial frequency which is the best compromise between spontaneous emission and ion CM heating. Similarly, setting $N_{op}^{max,se}$ equal to $N_{op}^{max,or}$ gives the ideal axial frequency for compromising between spontaneous emission and off-resonant transitions. We compare these two values of ν_z along with ν_z^{max} from Eq. 2 and find the smallest to be the optimal ν_z. For this special value of ν_z, which depends on N_{ions} and V_{rf}, we will be able to perform the largest number of operations within the specified constraints. We note, however, that these criteria do not necessarily lead to a quantum computer that performs calculations at an optimal *speed*. For sheer speed, we would want to maximize ν_z at the expense of N_{op}.

Fig. 4 shows the solution for the maximum value of N_{op} as a function of N_{ions} and V_{rf}, taking all safety factors to be 0.1. The dependence on V_{rf} comes from the ion CM heating condition (Eq. 4) which dominates the ion-spacing and off-resonant transition conditions in the large N_{ions}/small V_{rf} area of the plot. Elsewhere, the ion spacing requirement for ν_z^{max} (Eq. 2) dominates. While it seems promising to be able to achieve 80,000 operations for 1 ion and 270 operations for 20 ions, the axial frequencies needed for these optimal conditions are quite low, as shown in Fig. 5. In fact, if we decide to keep the axial frequency above $\sim$ 400 kHz to achieve adequate Doppler pre-cooling, we are limited to only a few ions in the trap by the ion spacing requirement alone. By relaxing this condition to require a spacing of only ($3.3 \times$ 729-nm) rather than ($10 \times$ 729-nm) (by letting $S_1 = 0.3$), we can get up to 30 ions in the trap with $\nu_z = 400$ kHz. Furthermore, a larger portion of the (N_{ions}, V_{rf}) region shown becomes dominated by either CM heating or off-resonant transitions rather than ion-spaceing. Further details will follow in a published article, but as an example, we could achieve 175 operations for 20 ions by raising V_{rf} to 5000 V. This would require $r_0 \approx 1$ mm, and give rise to a radial electric field of 4×10^6 V/m.

A few conclusions can be drawn from these early studies. First, we emphasize the trade-off between N_{op} and N_{ions}, the number of operations in the computer versus the number of bits. Analytic expressions for this trade-off can easily be found but depend on which process determines the choice of ν_z. Second, parameters *can* be chosen to give large

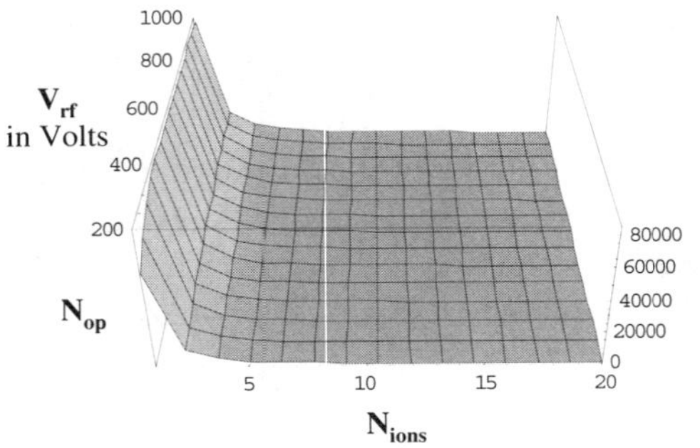

Fig. 4. Maximum number of operations for all safety parameters equal to 0.1.

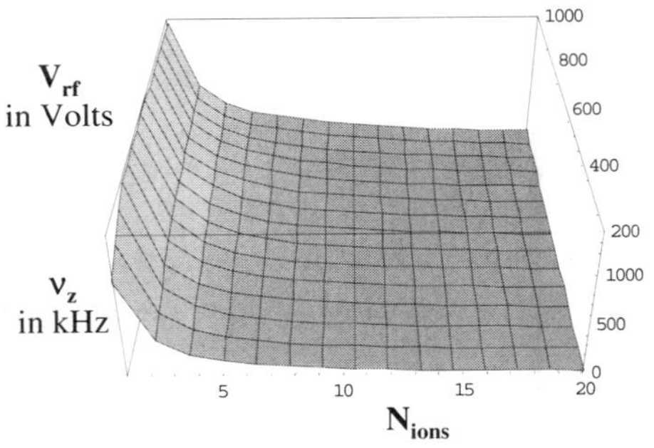

Fig. 5. Optimal axial frequencies for $V_{rf} = 1000$ V, and safety parameters of 0.1.

N_{ions} and N_{op} i.e. 20 and 175 respectively. Third, the ion-spacing requirement seems to be a bigger limitation than expected, however, ion CM heating also plays a large role. Fourth, it seems that small changes to coefficients and/or safety factors can affect which process ultimately limits N_{op}, so these proceses are all important to consider. Finally, other factors to consider are: different ion choices, different logic schemes, Raman-transition schemes, laser linewidths, and off-resonant ion heating transitions.

5. Conclusion

At Los Alamos, we have trapped and imaged calcium ions in a linear rf trap, cooled ions into linear crystals, and observed quantum jumps. We have completed the development of the ultra-stable 729-nm laser required for both sideband cooling and for driving Rabi oscillations between a Ca^+ ion's $S_{1/2}$ and $D_{5/2}$ states. We have built a system to switch multiple lasers to drive the 729-nm $S_{1/2} \rightarrow D_{5/2}$ transition of a trapped ion and measure the resulting quantum state. We have tested our system by performing spectroscopy on the $S_{1/2} \rightarrow D_{5/2}$ transition and are ready to look for the sidebands on the 729-nm line. Finally we propose a way to analyze many of the decoherence, heating, and addressing constraints to arrive at optimal trap parameters and trap frequencies.

1. R. J. Hughes *et al.*, Fortschr. Phys. **46**, 329 (1998).
2. J. I. Cirac and P. Zoller, Phys. Rev. Lett. **74**, 4091 (1995).
3. R. J. Hughes *et al.*, Phys. Rev. Lett. **77**, 3240 (1996).
4. M. B. Plenio and P. L. Knight, Proc. Roy. Soc. **453**, 2017 (1997).
5. D. F. V. James, Phys. Rev. Lett. **81**, 317 (1998).
6. Q. A. Turchette *et al.*, Phys. Rev. A **61**, 063418 (2000).
7. F. Diedrich *et al.*, Phys. Rev. Lett. **62**, 403 (1989).
8. D. G. Enzer *et al.*, Phys. Rev. Lett. **85**, 2466 (2000).
9. J. P. Schiffer, Phys. Rev. Lett. **70**, 818 (1993).
10. D. F. V. James, Appl. Phys. B **66**, 181 (1998).
11. R. J. Hughes and D. F. V. James, Fortschr. Phys. **46**, 759 (1998).
12. D. F. V. James, in *Proc. of the SPIE*, edited by E. Donkor and A. R. Pirich (SPIE, Bellinghn, WA, 2000), Vol. 4047, pp. 60–68.
13. P. H. Dawson, *Quadrupole Mass Spectrometry and Its Applications* (Elsevier Scientific Publishing Company, New York, 1976).

QUBIT UTILIZING CHARGE-NUMBER STATE IN SUPERCONDUCTING NANOSTRUCTURE

J.S. TSAI [a], Y. NAKAMURA [a], and YU. PASHKIN [b]

[a] *NEC Fundamental Research Laboratories, 34 Miyukigaoka, Tsukuba 305-8501, Japan*

[b] *CREST, Japan Science and Technology Corporation, Kawaguchi 332-0012, Japan*

In single-Cooper-pair box, the number of electrons in the box is quantized and they form a single macroscopic quantum charge-number state, corresponding to the number of excess electrons in the box. By making all the electrodes superconducting, we can couple two neighboring charge-number states coherently. In this way one can create an artificial two-level system. Qubit operations were demonstrated in two different control techniques, dc electric-field gate bias and ac field bias. The dc method was unique compared with the commonly used Rabi-oscillation-type operation. Here the system was biased at the degenerate point of the two states so that the dynamical phase does not develop during the operation. This was the first time that the quantum coherent oscillation was observed in a solid-state device whose quantum states involved a macroscopic number of quantum particles. Multiple-pulse experiments were also carried out and phase control was also demonstrated.

Introduction – solid-state electron device

The so called Moore's law that predicting an exponential scalability in the number of the active elements in solid state integrated circuit, doubling the capacity in every 18 months, symbolizes the essence of the immense power associated with the solid state LSI. Based on such observation, it was speculated that to realize a scaled-up version of the quantum computing system consisting of hundreds of qubits, its solid-state electronics version is perhaps indispensable. Among the numerous possible two-level systems, various typical microscopic two-level systems such as polarized photon states, atomic states and nuclear spin states have been utilized to demonstrate the realization of the quantum calculation in a limited scale so far. We have succeeded in demonstrating the first electronic control of 1-qubit achieved in a solid-state device.

Single-Cooper-pair box

We utilized a submicron electron device called a single-Cooper-pair box, to demonstrate the operation of 1-qubit control in solid-state electron device. It is a small superconductive box connected to the outside by a Josephson tunnel junction and a gate capacitor. The Cooper pairs can tunnel into the box through the tunnel junction. However, due to the small capacitance C of the box, the transfer of a Cooper pair would be accompanied by a large increase in charging energy, making the transfer of the Cooper pairs energetically difficult at low bias voltage. When the charging energy E_c associated with the transfer of one electron charge across the tunnel junction, $E_c = e^2/2C$, is smaller than gap energy of the superconductor Δ, the transfer of the quasi-particle can be ignored during the gate operation. The large resistance associated with the tunnel junction suppresses the quantum fluctuation of the charge of the box through the junction, thus the number of Cooper pairs in the box is quantized under a certain gate bias voltage. There are many (typically 10^8 in our experiment) Cooper pairs in the box, but they form a single macroscopic quantum charge-number state, corresponding to the number of excess electrons in the box.

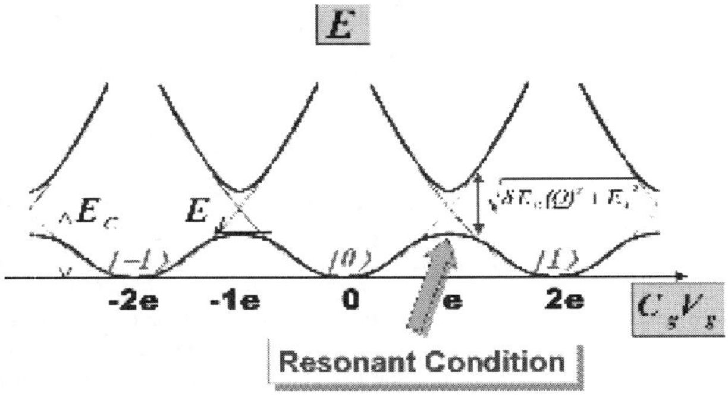

Fig. 1 The band diagram of the single-Cooper-pair box

The Hamiltonian of the system can be written as:

$$H = 4E_C \sum_{n=0,1} (n - \frac{Q}{2e})^2 |n\rangle\langle n| - \frac{E_J}{2}(|0\rangle\langle 1| - |1\rangle\langle 0|)$$

when Josephson energy E_J is small compared with E_C. Figure 1 shows the energy diagram of the system. The energy difference between the upper and lower band is $\Delta E = \sqrt{E_J^2 + \delta E^2}$, where $\delta E = 4E_C(Q/e - 1)$ and $Q = C_g V_g$. The Josephson coupling enables a dissipation-less transfer of an electron pair, thus two neighboring charge-number states, designated as a state $|0\rangle$, and a state with one more Cooper-pair $|1\rangle$, can be coherently coupled. This was achieved by biasing the box at $Q = e$, the resonant condition, by gate, or by using photon-assisted coupling of two states at the off- resonant condition where $Q \neq e$ (Fig. 1). In this way one can create an artificial two-level system.

The result of previous experiments showing that the macroscopic quantum state of superconductivity is extraordinary well isolated from the internal and external dissipation [1] gave us a confidence in such experiment that requires extremely low dissipation. For the detection of the quantum state, we attached an additional tunneling probe to the box, our original method, to study this two-level system. With this probe, we could simply measure the current it carried to monitor the probability of one of the two states that was involved in the coherence.

Device

The typical device picture is shown in Fig. 2 (the photograph and the schematic diagram). The whole structure was made of aluminum thin film (t ~ 20nm). The box has a dimension of 700nm x

50nm, a size quite readily reproducible by an electron-beam lithography. The coherence junction connecting the island and reservoir where quantum coherence was prepared was split into two junctions, forming a SQUID loop. In this way we could control the Josephson energy of the system by an external magnetic field. The tunnel junction was prepared by the angle-controlled evaporations of aluminum through a suspended stencil mask. To couple a wide bandwidth electric signal to the box, a high-speed gate was made employing a coplanar wave-guide design (50Ω). The entire device was made on top of a copper ground plane, with silicon nitride insulating layer in between. The typical Josephson energy of the coherent junction was ~40μeV, the charging energy of the box was typically ~100μeV. The experiments were carried out in an environment with low thermal energy, typically ~30mK or 3μeV, low enough compared with all relevant energies associated with the box.

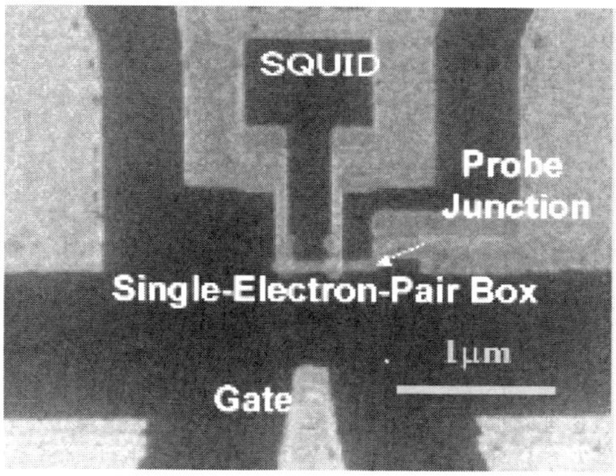

Fig. 2 The device picture.

We attached an additional tunneling probe to the box to study this two-level system. Under the resonant condition, there was a resonant current flowing into the probe junction within the Coulomb blockade region, when the probe junction was appropriately biased. This was the current carried by the so-called Josephson-quasiparticle peak (JQP) resonant cycle [4]. The JQP current signifies the cyclic production of a coherent state between states $|0\rangle$ and $|1\rangle$, followed by decoherence of the coherent state due to transfer of two quasiparticles to the probe electrode.

Time-domain experiment - Qubit

We carried out an experiment designed to observe the dynamics of the coherence, namely, the coherent oscillation. First, as the initial state, we prepared a pure state $|0\rangle$ by biasing the system far from the resonant condition. Next, we applied a sufficiently fast voltage pulse to the gate (a non-adiabatic gate operation) to create a degenerated charge-number state at the resonant condition. The voltage rise time had to be fast enough to initiate a non-adiabatic transition [7]. At the degenerate point, two neighboring charge states $|0\rangle$ and $|1\rangle$ were coherently coupled by the

Josephson energy, so that the two states were forced to undergo a quantum oscillation during the time Δt that corresponds to the gate pulse width. At the end of pulse, the system was brought back to the original bias point. At this moment, the state of the system retained a coherent superposition of states $|0\rangle$ and $|1\rangle$.

At this point, a sampling measurement of state $|1\rangle$ by the attached probe junction was carried out. This was done by measuring the current carried by the probe junction under the repeated pulsing experiments. Under an appropriate voltage bias condition of the probe junction, the $|1\rangle$ state would decay to $|0\rangle$ state by emitting two quasi-particles (electrons) to the probe electrode with a tunneling rate Γ. As a result, the current we monitored reflected the averaged occupation probability of the $|1\rangle$ state after sampling for many events. By sweeping Δt with an increment of 1ps or so, we succeeded in tracing out the time evolution of the quantum oscillation (Fig. 3). The period of the oscillation in this case was expected to be $\hbar/E_J$.

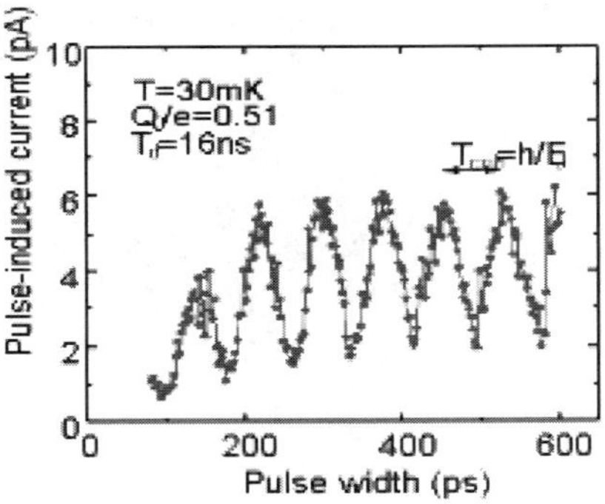

Fig. 3 The quantum oscillation observed

This was the first time that the quantum coherent oscillation was observed in a solid-state electron device whose quantum states involved a macroscopic number of quantum particles. By adjusting Δt of the pulse, as well as the strength of the magnetic field, one can achieve the generation of any desired linear superposition of the two states, or the 1-qubit-control operation.

Since the probe junction attached to the box had a sampling time of typically 6 to 9ns for the most of the experimental devices we used, this probing time restricted the upper limit of the decoherence time in our experiment. The experimentally confirmed coherent oscillation lasted for about 5ns, when biased at the resonant condition. The oscillation period was typically 100ps, so this corresponds to about 50 gate operations. Considering the sampling rate of the probe junction, the observed 5ns oscillation was perhaps not so unreasonable.

Multi-pulse experiment

To study the decoherence under the off-resonant condition, we carried out a multi-pulse experiment [6]. Here, two consecutive voltage pulses were applied to the gate before the probe junction executed the observation (quasiparticle tunneling). The first pulse would bring the system into a coherent state. After this pulse, the system would be biased at the off-resonant condition but still maintaining coherence. Then, during the inter-pulse period t_d, the quantum state will acquire an extra quantum phase dependent factor $exp\left(-i\frac{\Delta E}{\hbar}t_d\right)$. Then, when we applied a second pulse with time delay t_d before the measurement, the probability of the $|1\rangle$ state would be modified by the length of t_d.

A spin representation of the system as shown in Fig. 4 illustrates the above physics geometrically [6]. Here, a surface of a sphere defined by a "spin" vector represents the linearly superposed state of the $|0\rangle$ and the $|1\rangle$ states. The Josephson energy represents the "magnetic field" in x direction, and energy difference $\delta E = 4E_C(Q/e-1)$ represents the "magnetic field" in z direction. When the system is put into the quantum coherence, the spin vector would undergo a precession under the influence of the resultant "magnetic field" $\Delta E = \sqrt{E_J^2 + \delta E^2}$. Thus the axis of rotation can be controlled by selecting the operating point Q that provides a desired $\delta E(Q)$. One can immediately see that the qubit operation biased at the degenerate point is unique compared with the commonly used Rabi-oscillation-type operation. In this case, because of the degeneracy ($\delta E=0$), the unwanted dynamical phase of the system does not develop during the operation.

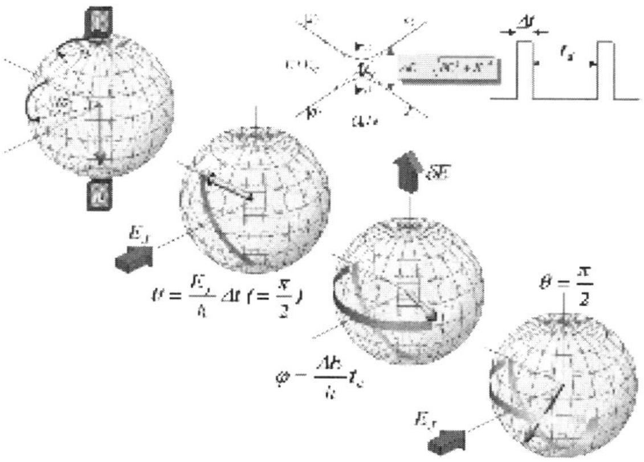

Fig. 4 The spin representation of the device.

The precession around x and z-axes could be traced by sweeping the inter-pulse delay time t_d and sampling the probe current (probability of $|1\rangle$ state). The result of the experiment indeed showed a quantum oscillation with oscillation period of $\Delta E/\hbar$ as expected [6]. The decoherence time at the off-resonant condition was found to be about 1ns, much shorter than the decoherence time found at the resonant condition. Probably the shorter decoherence could be attributed to the fluctuation in Q of the system due to the infamous background charge fluctuation. Such fluctuation would generate a large fluctuation in ΔE at the off-resonant condition, but not so much at the resonant condition. By projecting the observed low-frequency 1/f background charge noise up to the relevant frequency of the experiment, we indeed obtained a roughly similar off-resonant decoherence time estimation [6]. In addition, the control of the phase of the quantum precession (by selecting the time delay t_d) in the two-pulse experiment also corresponded to the operation of the quantum phase gate.

Summary

Utilizing small Josephson junction device having large charging energy, we have performed experiment corresponded to the realization of the 1-qubit operation, the first of such demonstration in a solid-state electron device. Both rotation gate (controlling the amplitude of state $|1\rangle$) and phase gate (controlling of the phase of the state) were demonstrated. The experimentally confirmed decoherence time at the resonant condition was more than 5ns, and was probably limited by the probe measuring time. The decoherence time at the off-resonant condition was about 1ns. The much faster decoherence found at this bias condition could be attributed to the background charge fluctuation in the device. The qubit operation at the resonant point is a unique approach compared with the other more common Rabi-oscillation qubits. It is unique in the sense that during the operation, the system would not acquire quantum phase, thus less susceptible to the fluctuations that affect the phase.

This work has been supported by the CREST project of Japan Science and Technology Corporation (JST).

References

1. J.S. Tsai, A.K. Jain, and J.E. Lukens, Physical Review Letters, **51**, 316, (1983)
2. Y. Nakamura, C.D. Chen, and J.S. Tsai, Physical Review Letters, **79**, 2328, (1997)
3. Y. Nakamura, Yu. A. Pashkin, and J.S. Tsai, Nature, **398**, 786, (1999)
4. T.A. Fulton, P.L. Gammel, D.J. Bishop, L.N. Dunckleberger, G.J. Dolan, Physical Review Letters, **63**, 1307, (1989)
5. J. R. Rubbmark, M. M. Kash, M. G. Littman, and D. Kleppner, Physical Review A, **23**, 3107, (1981)
6. Y. Nakamura, Yu. A. Pashkin, and J.S. Tsai Proceeding to the "International Workshop on Macroscopic Quantum Coherence and Computing (MQC^2)", in press

UNIVERSAL QUANTUM GATES FOR SINGLE COOPER PAIR BOX BASED QUANTUM COMPUTING

P. ECHTERNACH[1], C. P. WILLIAMS[2], S.C. DULTZ[1], S. BRAUNSTEIN[3], AND J. P. DOWLING[2]

[1]*Microdevices Laboratory, Jet Propulsion Laboratory, California Institute of Technology, Pasadena, California*

[2]*Quantum Computing Group, Jet Propulsion Laboratory, California Institute of Technology, Pasadena, California*

[3]*University of Wales, Bangor, United Kingdom*

1 Introduction

Several schemes have been proposed for implementing quantum computer hardware in solid state quantum electronics. These schemes use electric charge[1,2,3], magnetic flux[4,5,6], superconducting phase[7,8,9,10], electron spin[11,12,13,14], or nuclear spin[15,16] as the information bearing degree of freedom. Each scheme has various pros and cons but that based on harnessing quantized charge is especially appealing because the necessary superconducting circuitry for such a qubit can be fabricated using present day e-beam lithography equipment, and quantum coherence, essential for creating superposed and entangled states, has been demonstrated experimentally[17]. Moreover, the fidelity and leakage of such gates is understood[18]. These qualities make the SCB-based qubit a strong contender for the basic element of a quantum computer. Indeed, today's e-beam fabrication technology is sufficiently mature that it would be a simple matter to create a quantum circuit having thousands of quantum gates within a matter of a few hours! Of course, it remains to be seen whether such large-scale quantum circuits could be operated coherently en masse. Nevertheless, the relative ease of fabricating SCB-based quantum gates leads one to consider computer architectural issues related large scale SCB-based quantum circuits.

From an architectural perspective, the existing proposals for SCB-based qubits and quantum gates are sub-optimal. For example, the scheme of Schön et al.[19] uses the time at which a gate operation begins as one of the parameters that determine the unitary operation the gate is to perform. While this is certainly allowed physically, and could even be argued to be ingeniously efficient, it is not a good decision from the perspective of building reliable and scaleable quantum computers. If the starting time is a parameter, a given quantum gate would need different implementations at different times. Moreover, as the computation progressed, timing errors would accumulate leading to worsening gate fidelity. Furthermore, Schön et al. also use the duration of the gate operation as a free parameter that determines the unitary transformation the gate is to perform. Again, this is a poor decision from a computer architecture perspective, as it means that different gates would take different times making it difficult to synchronize parallel quantum gate operations in large circuits. To address both of these problems we have developed an approach to universal quantum computation in SCB-based quantum computing that specifically avoids using time as a free parameter. Instead, our gates operate by varying only voltages of magnetic fluxes in a controlled fashion.

To make a practical design for a quantum computer, one must specify how to decompose any valid quantum computation into a sequence of elementary 1- and 2-qubit quantum gates that can be realized in physical hardware that is feasible to fabricate. The set of these 1- and 2-qubit gates is arbitrary provided it is *universal*, i.e., capable of achieving any valid quantum computation from a quantum circuit comprising only gates from this set. Traditionally the set of universal gates has been taken to be the set of all 1-qubit quantum gates in conjunction with a single 2-qubit gate called controlled-NOT. However, many equally good universal gate sets exist[20] and there might be an advantage in using a non-standard universal gate set if certain gate designs happen to be easier to realize in one hardware context than another[21,22]. Certainly it has been known for some time that the simple 2-qubit exchange interaction (i.e., the SWAP gate) is as powerful as CNOT as far as computational universality is concerned. It makes sense therefore, to see what gates are easy to make and then extend them into a universal set. This is the strategy pursued in this paper. In particular, we show, in the context of SCB-based qubits, that we can implement any 1-qubit operation and a special (new) 2-qubit operation called "the square root of complex SWAP" (or "$\sqrt{iSWAP}$" for short). We then prove that, taken together, $\sqrt{iSWAP}$ and all 1-qubit gates is universal for quantum computation.

2 SCB-based Qubits

A Single Cooper Pair Box is an artificial two-level quantum system comprising a nanoscale superconducting electrode connected to a reservoir of Cooper pair charges via a Josephson junction. The logical states of the device, $|0\rangle$ and $|1\rangle$, are implemented physically as a pair of charge-number states differing by $2e$ (where e is the charge of an electron). Typically, some 10^9 Cooper pairs are involved. Transitions between the logical states are accomplished by tunneling of Cooper pairs through the Josephson junction. Although the two-level system contains a macroscopic number of charges, in the superconducting regime they behave collectively, as a Bose-Einstein condensate, allowing the two logical states to be superposed coherently. This property makes the SCB a candidate for the physical implementation of a qubit.

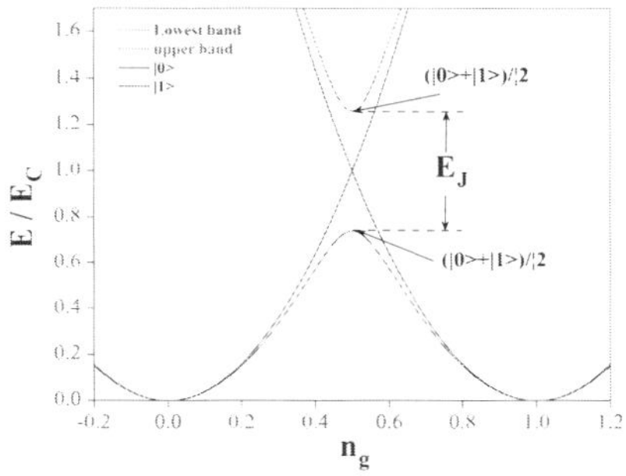

Figure 1. The level diagram for an SCB-based qubit.

The SCB-qubit gained prominence in 1999 when Nakamura et al. demonstrated coherent oscillations between the $|0\rangle$ and $|1\rangle$ states[17]. This was the first time such macroscopic coherent phenomena had been seen experimentally and distinguishes the SCB approach from other solid state schemes in which similar macroscopic coherences have recently been demonstrated[5,6].

Our qubit consists of a *split* tunnel junction as this allows us to control the Josephson: V controls the number of excess Cooper pairs—by varying the externally applied magnetic flux according to:

$$E_J(\Phi_{ext}) = 2E_J^{\text{intrinsic}} \cos\left(\frac{\pi \Phi_{ext}}{\Phi_0}\right) \tag{1}$$

where Φ_0 is the quantum of magnetic flux and $E_J^{\text{intrinsic}}$ is given by the Ambegaokar-Baratoff relation in the low temperature approximation: $E_J^{\text{intrinsic}} = \frac{h\Delta}{8e^2 R_N}$ (in which h, Δ, and R_N are Planck's constant, the superconducting energy gap and the normal tunneling resistance of the junction respectively). Figure 2 shows a schematic diagram of our qubit.

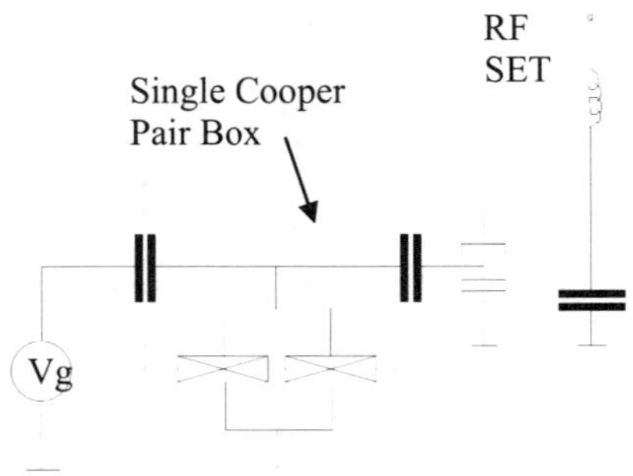

Figure 2. Schematic diagram of a single SCB-based qubit with an adjoining RF SET readout.

The Hamiltonian for the qubit is $H = 4E_C \sum_n (n - n_C)^2 - E_J(\Phi_{ext})\cos(\phi)$, where n is the number of excess Cooper pairs on the island, $n_C = \frac{C_g V}{2e}$, $E_C = \frac{(2e)^2}{2C_\Sigma}$, $C_\Sigma = C_J + C_g$, and ϕ is the difference in phase of the superconducting state across the junction. In the basis of excess Cooper pair number states, $|n\rangle$, restricting the gate charge interval to be $0 \le n_C \le 1$, and choosing the zero of energy to be at $E_o=E_C(1/2-n_c)^2$, the Hamiltonian reduces to:

$$\hat{H}_1 = \begin{pmatrix} -\frac{1}{2}E(V) & -\frac{1}{2}E_J(\Phi_{ext}) \\ -\frac{1}{2}E_J(\Phi_{ext}) & +\frac{1}{2}E(V) \end{pmatrix} \tag{2}$$

where $E(V) = E_C\left(1 - \frac{C_G V}{e}\right)$ and $E_J(\Phi_{ext}) = 2E_J^{\text{intrinsic}}\left|\cos(\frac{\pi \Phi_{ext}}{\Phi_0})\right|$. The two parameters V and Φ_{ext} can be adjusted to achieve different Hamiltonians and hence different 1-qubit quantum gates.

3 One-Qubit Gates

The 1-qubit Hamiltonian, $\hat{H}_1$, acting for a time Δt induces a 1-qubit quantum gate operation given by:

$$\hat{U}_1 = \exp\left(-\frac{i\hat{H}_1 \Delta t}{\hbar}\right) \tag{3}$$

We assume that the Hamiltonian can be switched on and off quickly so that the interval Δt is sharp. The fact that $\hat{H}_1$ has a symmetric structure means that we are only able to implement a limited set of primitive unitary transformations. Nevertheless, it turns out that these primitive transformations can be composed to achieve *arbitrary* 1-qubit gates. The proof is via a factorization of an arbitrary 2×2 unitary matrix into a product of rotation matrices. Specifically, the matrix for an arbitrary 1-qubit gate is described mathematically by[23]

$$\hat{U}(\alpha,\theta,\beta)=\begin{pmatrix} e^{i\left(\frac{\alpha+\beta}{2}\right)}\cos\left(\frac{\theta}{2}\right) & e^{i\left(\frac{\alpha-\beta}{2}\right)}\sin\left(\frac{\theta}{2}\right) \\ -e^{-i\left(\frac{\alpha-\beta}{2}\right)}\sin\left(\frac{\theta}{2}\right) & e^{-i\left(\frac{\alpha+\beta}{2}\right)}\cos\left(\frac{\theta}{2}\right) \end{pmatrix} \tag{4}$$

Such a matrix can be factored into the product of rotations about just the *z*- and *x*-axes.

$$\hat{U}(\alpha,\theta,\beta)=\hat{R}_z\left(\alpha-\frac{\pi}{2}\right)\cdot\hat{R}_x(\theta)\cdot\hat{R}_z\left(\beta+\frac{\pi}{2}\right) \tag{5}$$

where $\hat{R}_z(\xi)=\exp(i\xi\sigma_z/2)$ is a rotation[§] about the *z*-axis through angle ξ, $\hat{R}_x(\xi)=\exp(i\xi\sigma_x/2)$ is a rotation about the *x*-axis through angle ξ, and $\sigma_i : i\in\{x,y,z\}$ are Pauli spin matrices.

It is therefore sufficient to configure the parameters in $\hat{H}_1$ to perform rotations about just the *z*- and *x*-axes to achieve an arbitrary 1-qubit gate. From equations (2) and (3), we find that $\hat{R}_z(\xi)$ can be achieved within time Δt by setting $\Phi_{ext}=\Phi_0/2$, and $V=\dfrac{\xi\hbar e}{C_G E_C \Delta t}-\dfrac{e}{C_G}$. Similarly, $\hat{R}_x(\xi)$ can be achieved within time Δt by setting $\Phi_{ext}=\dfrac{\Phi_0}{\pi}\cos^{-1}\left(\dfrac{\xi\hbar}{2E_J^{\text{intrinsic}}\Delta t}\right)$, and $V=\dfrac{e}{C_G}$. These settings cause, within time Δt, $\hat{U}_1$ to take the form $\hat{R}_z(\xi)=\begin{pmatrix} e^{i\xi/2} & 0 \\ 0 & e^{-i\xi/2}\end{pmatrix}$ or $\hat{R}_x(\xi)=\begin{pmatrix}\cos(\xi/2) & i\sin(\xi/2) \\ i\sin(\xi/2) & \cos(\xi/2)\end{pmatrix}$ respectively.

Thus, by the factorization given in equation (5), an arbitrary 1-qubit gate can be achieved in the SCB-based approach to quantum computing in a time of $3\Delta t$.

Note that the only free parameters used to determine the action of the 1-qubit gate are the external flux Φ_{ext} and the voltage V. The time interval, Δt, over which the Hamiltonian needs to act to bring about an *x*- or *z*-rotation, is fixed by the physics of the particular substrate, e.g., Aluminium or Niobium, used for the qubit. Although we could also have used Δt as an additional control parameter, such a choice would complicate integration of quantum gates into parallel, synchronous, quantum circuits.

[§] The doubling of the angle arises because of the relationship between operations in SO(3) (rigid-body rotations) to operations in SU(2).

4 Qubit Gates

To achieve a 2-qubit gate, it is necessary to couple pairs of qubits. In our scheme, two qubits are coupled using two tunnel junctions connected in parallel. This allows the coupling to be turned on or off as necessary. A schematic for the 2-qubit gate is shown in Figure 3.

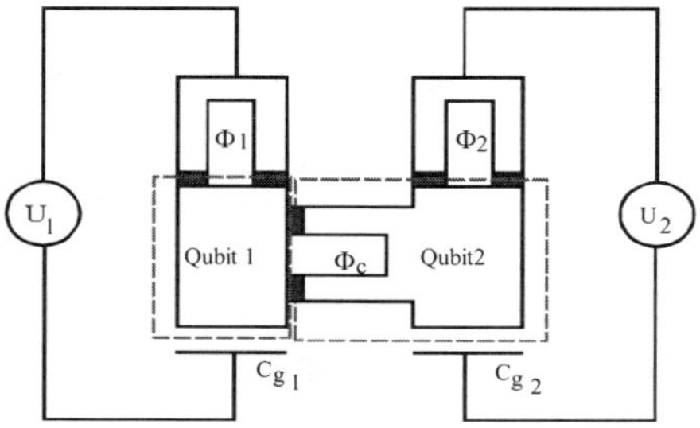

Figure 3. Schematic diagram of a pair of coupled qubits.

The Hamiltonian for the coupled pair of qubits is given by:

$$\hat{H}_2 = E_{C_1}(n_1 - n_{C_1})^2 + E_{C_2}(n_2 - n_{C_2})^2 - E_{J_1}(\Phi_1)\cos(\phi_1) - E_{J_2}(\Phi_2)\cos(\phi_2) - E_{J_C}(\Phi_C)\cos(\phi_1 - \phi_2) \tag{10}$$

where the subscripts 1, 2, and *C*, refer to parameters of qubit 1, qubit 2 and the coupling between them respectively. Assuming again that the zero of energy is at $E_o=E_{c1}(1/2-n_{c1})+E_{c2}(1/2-n_{c2})^2$

$$\hat{H}_2 = \begin{pmatrix} -\frac{E_1}{2}-\frac{E_2}{2} & -\frac{1}{2}E_{J_2}(\Phi_2) & -\frac{1}{2}E_{J_1}(\Phi_1) & 0 \\ -\frac{1}{2}E_{J_2}(\Phi_2) & -\frac{E_1}{2}+\frac{E_2}{2} & -\frac{1}{2}E_{J_C}(\Phi_C) & -\frac{1}{2}E_{J_1}(\Phi_1) \\ -\frac{1}{2}E_{J_1}(\Phi_1) & -\frac{1}{2}E_{J_C}(\Phi_C) & \frac{E_1}{2}-\frac{E_2}{2} & -\frac{1}{2}E_{J_2}(\Phi_2) \\ 0 & -\frac{1}{2}E_{J_1}(\Phi_1) & -\frac{1}{2}E_{J_2}(\Phi_2) & \frac{E_1}{2}+\frac{E_2}{2} \end{pmatrix} \tag{11}$$

where $E_1=E_{c1}(1-2n_{c1})$ and $E_2=E_{c2}(1-2n_{c2})$. The 2-qubit quantum gate induced by this Hamiltonian is

$$\hat{U}_2 = \exp(-\frac{i\hat{H}_2 t}{\hbar}) \tag{12}$$

We can specialize $\hat{U}_2$ to a particular form by setting $n_{C_1} = n_{C_2} = \frac{1}{2}$, $E_{J_1} = E_{J_2} = 0$. These values induce the 2-qubit gate

$$\hat{U}_2 = \begin{pmatrix} 1 & 0 & 0 & 0 \\ 0 & \cos\left(\frac{E_{J_C}\Delta t}{2\hbar}\right) & i\sin\left(\frac{E_{J_C}\Delta t}{2\hbar}\right) & 0 \\ 0 & i\sin\left(\frac{E_{J_C}\Delta t}{2\hbar}\right) & \cos\left(\frac{E_{J_C}\Delta t}{2\hbar}\right) & 0 \\ 0 & 0 & 0 & 1 \end{pmatrix} \tag{13}$$

Specializing further by setting $E_{J_C} = \hbar\pi/(2\Delta t)$ we achieve a 2-qubit gate that we call the "square root of complex SWAP", $\sqrt{iSWAP}$:

$$\sqrt{iSWAP} = \begin{pmatrix} 1 & 0 & 0 & 0 \\ 0 & \frac{1}{\sqrt{2}} & \frac{i}{\sqrt{2}} & 0 \\ 0 & \frac{i}{\sqrt{2}} & \frac{1}{\sqrt{2}} & 0 \\ 0 & 0 & 0 & 1 \end{pmatrix} \tag{14}$$

5 Universal Quantum Computation

The set of all 1-qubit gates together with controlled-NOT is known to be universal for quantum computation. As we have already shown that it is possible to implement any 1-qubit gate in the SCB context, we can prove that all 1-qubit gates and $\sqrt{iSWAP}$ is also a universal set by exhibiting a construction for *CNOT* using only 1-qubit gates and $\sqrt{iSWAP}$. The following gate sequence achieves *CNOT* up to an unimportant overall phase factor of $\exp(i\,3\pi/4)$:

$$\begin{aligned} CNOT \equiv e^{i\frac{3\pi}{4}}\left(I \otimes R_y\left(\frac{\pi}{2}\right)\right)\cdot(NOT \otimes NOT)\cdot\left(R_z\left(\frac{\pi}{2}\right)\otimes R_z\left(\frac{\pi}{2}\right)\right)\cdot\exp\left(i\frac{\pi}{4}(\sigma_z \otimes \sigma_z)\right)\cdot\ldots \\ \ldots\cdot\sqrt{iSWAP}\cdot(R_z(\pi)\otimes I)\cdot\sqrt{iSWAP}\cdot(NOT \otimes NOT)\cdot\left(I \otimes R_y\left(-\frac{\pi}{2}\right)\right) \end{aligned} \tag{12}$$

Thus a controlled-NOT operation can be implemented within the SCB-based approach to quantum computing. If each primitive gate operation, i.e., each 1-qubit rotation or a square root of complex SWAP, takes time Δt then controlled-NOT is implementable in time $9\Delta t$.

6 Experimental program

We have fabricated SCB-based qubits using electron beam lithography and a standard shadow mask technique with double angle evaporation of Aluminum. The resulting junctions were 100nmx50nm in size. To couple the voltage pulses necessary for manipulation of the qubits we use a coplanar wave guide structure designed to have a 50 Ohm impedance and therefore minimize unwanted reflections. The magnetic field to generate the external flux will be created by a wire in close proximity to the SCB. This wire is a short circuit termination of another coplanar waveguide structure.

Figure 4. An SCB-based qubit fabricated in Aluminum using e-beam lithography.

In close proximity to the SCB sits a Single electron transistor which will be operated in the RF mode. The resonant circuit needed for operation of the RF-SET is provided by an inductor and capacitor defined by optical lithography. Testing of the fabricated structures are in the preliminary stages. We have equipped our dilution refrigerator with the necessary microwave equipment to perform the experiments. On a first run, we were able to cool the mixing chamber down to 50 mK. Initial tests are concentrating on observing the resonances of the on-chip resonant circuits. We have observed resonances close to the design frequencies with Q values up to 150. Our next step is to operate the SETs as RF-SETs and characterize its performance. We will then test the SCB and will attempt to measure the coherence times for various operating points.

7 Conclusions

We have designed a realizable set of quantum gates to support universal quantum computation in the context of SCB-based quantum computing. In selecting our universal gate set we paid special attention to two principles of good computer design, namely, that each gate operation should take a fixed and predictable length of time, and that the operations needed to bring about the action of a particular gate should not depend upon the time at which the gate operation begins. Earlier proposals for SCB-based universal quantum computation did not satisfy these criteria. We have fabricated SCB-based qubits using existing state-of-the-art e-beam lithography at the JPL Microdevices Laboratory. There appears to be no impediment to fabricating large-scale quantum circuits that manipulate SCB-based qubits.

References

1. A. Shnirman, G. Schön, and Z. Hermon, Phys. Rev. Lett. **79**, 2371 (1997).

2. D. V. Averin, "Adiabatic Quantum Computation with Cooper Pairs", Solid State Communications, **105**, (1998), pp.659-664.
3. Y. Makhlin, G.Schön, and A. Shnirman, Nature **398,** 305 (1999).
4. M. Bocko, A. Herr and M. Feldman, "Prospects for Quantum Coherent Computation Using Superconducting Electronics," IEEE Trans. Appl. supercond. 7, (1997), pp.3638-3641.
5. J. R. Friedman, V. Patel, W. Chen, et al., "Quantum superposition of distinct macroscopic states," Nature **406**, 43 (2000).
6. C. H. van der Wal, A. C. J. ter Haar, F. K. Wilhelm, *et al.,* "Decoherence of the Superconducting Persistent Current Qubit," Science, **290**, 773 (2000).
7. G. Blatter, V. Geshkenbein, and L. Ioffe, "Engineering Superconducting Phase Qubits," http://xxx.lanl.gov/abs/cond-mat/9912163 (1999).
8. L.B. Ioffe, V.B. Geshkenbein, M.V. Feigelman, A.L. Fauchere, G. Blatter, "Quiet SDS Josephson Junctions for Quantum Computing," Nature **398**, 679 (1999).
9. A. Zagoskin, "A Scalable, Tunable Qubit, Based on a Clean DND or Grain Boundary D-D Junction," http://xxx.lanl.gov/abs/cond-mat/9903170 (1999).
10. A. Blais and A. Zagoskin, "Operation of Universal Gates in a DXD Superconducting Solid State Quantum Computer," http://xxx.lanl.gov/abs/cond-mat/9905043 (1999).
11. D. Loss, D. DiVincenzo, "Quantum Computation with Quantum Dots," Phys. Rev. A 57, 120 (1998).
12. D. Loss, G. Burkard, and E. V. Sukhorukov, "Quantum Computing and Quantum Communication with Electrons in Nanostructures," to be published in the proceedings of the XXXIVth Rencontres de Moriond "Quantum Physics at Mesoscopic Scale", held in Les Arcs, Savoie, France, January 23-30, (1999).
13. D.P. DiVincenzo, G. Burkard, D. Loss, and E. V. Sukhorukov, "Quantum Computation and Spin Electronics," to be published in "Quantum Mesoscopic Phenomena and Mesoscopic Devices in Microelectronics," eds. I. O. Kulik and R. Ellialtioglu, NATO Advanced Study Institute, Turkey, June 13-25, (1999).
14. R. Vrijen, E. Yablonovitch, K. Wang, H. Jiang, A. Balandin, V. Roychowdhury, T. Mor, D. Di Vincenzo, "Electron Spin Resonance Transistors for Quantum Computing in Silicon-Germanium Heterostructures," http://xxx.lanl.gov/abs/quant-ph/9905096 (1999).
15. B. Kane, Nature, 393, 133 (1998).
16. B. Kane, "Silicon-based Quantum Computation," http://xxx.lanl.gov/abs/quant-ph/0003031 (2000). Submitted to Fortschritte der Physik Special Issue on Experimental Proposals for Quantum Computation.
17. Y. Nakamura, Yu. A. Pashkin, and J. S. Tsai, Nature **398**, 786 (1999).
18. R. Fazio, G. Massimo Palma, and J. Siewert, "Fidelity and Leakage of Josephson Qubits," Physical Review Letters, **83**, 25 (1999), pp.5385-5388.
19. G. Schön, A. Shnirman, and Y, Makhlin, "Josephson-Junction Qubits and the Readout Process by Single Electron Transistors," http://xxx.lanl.gov/abs/cond-mat/9811029 (1998). See the paragraph following their equation (7).

20. D. Di Vincenzo, "Two Bit Gates are Universal for Quantum Computation," Physical Review A, **51**, 2, February (1995), pp.1015-1022.

AN INTEGRATED SUPERCONDUCTIVE DEVICE TECHNOLOGY FOR QUBIT CONTROL*

K. K. BERGGREN[a], D. NAKADAb, M. J. O'HARA[a], T. P. ORLANDO[b], E. M. MACEDO[a], R. SLATTERY[a], AND T. WEIR[a]

a) MIT Lincoln Laboratory, 244 Wood St., Lexington, MA, 02420

b) MIT, Dept. of Electrical Engineering and Computer Science, Cambridge, MA 02139

We will outline an effort to develop the building blocks of a large-scale quantum computer based on persistent-current states of a SQUID. Superconductive electronics, which have clock rates above 100 GHz, are an attractive choice for qubit control because they allow many control operations to be executed in the estimated 1 ms before substantial decoherence occurs in the qubit. Critical to this project is the integration of quantum devices with conventional superconductive electronics that will be used to provide classical circuitry for control and readout functions for the quantum computer. Chips with these dissimilar devices will be bonded together using a flip-chip technique to permit inductive coupling between the chips. We have demonstrated bonding of two chips with less than 2 μm of linear misalignment in a configuration robust to repeated thermal cycling between room temperature and 4.2 K. Ultimately, we hope that this effort will help better define the path towards scaling existing small demonstrations of quantum devices into a full-scale, useful, quantum computer.

1 Introduction

Because of strong interactions between pairs of electrons in a superconductor, the phase of the pair wavefunction is continuous within a superconducting loop. Across a Josephson junction, however, discrete steps in the phase can exist in the presence of magnetic fields, resulting in a potential *vs.* phase landscape that can be tuned with classical external fields. Bocko, Herr, and Feldman noted in 1997[1] that coherent control of the wavefunction in this potential landscape could be used to perform quantum computation in principle, but a key question remained: would the coherence time of the wavefunction be long enough in practice to build a large-scale, useful quantum computer? Calculations suggested that the answer was yes—coherence times of ~ 1 ms were predicted to be possible.[2] Two recent experiments, one by Friedman et al.[3] and the other by van der Wal et al.[4] have also helped answer this question by observing superposition of macroscopic quantum states in a superconducting loop. These experiments have paved the way to demonstrations of coherent control of this system.

Both Friedman and van der Wal used superconductive SQUID magnetometers as a transducer for reading out the magnetization of the quantized states of the superconductive loop (the qubit). The success of their experiments was enabled by the fortuitous convergence of two superconductive technologies: a device exhibiting macroscopically observable quantum coherence and a sensor exhibiting nearly ideal sensitivity to variations in magnetic flux. Importantly, there

* This work was supported by the Department of the Air Force under Contract F19628-00-C-0002. Opinions, interpretations, conclusions, and recommendations are those of the authors and are not necessarily endorsed by the United States Air Force.

exists a classical electronics family based on superconductivity, single-flux quantum (SFQ) electronics, that is well suited to integration with analog field sensors, and to the precise synthesis of analog voltages.[5, 6] Superconductivity therefore provides a triumvirate of optimal technologies for quantum computing that can be integrated into a single package: an ultra-sensitive field transducer, a qubit with long coherence time, and an ultra-fast and ultra-precise classical electronics family.

In this work we showed that chips compatible with single-flux quantum (SFQ) circuitry[6] can be bonded to, and exchange classical information with, chips that could carry a qubit. This fabrication technology enables the development of a hybrid quantum computer that uses dissimilar qubit and control technologies. We first measured the coupling between a control line and a DC SQUID on a single chip. We then flipped a second chip, containing only a transformer circuit, onto this chip, and measured the resulting change in the magnetic coupling between the control line and the DC SQUID. We thus demonstrated that current could be transferred onto and off of a flipped chip. In a quantum computing implementation, the DC SQUID and control line would be on a chip with classical control circuitry, while the flipped chip would carry the qubits.

If these technologies are so compatible, why bother with packaging at all? Why not use them to fabricate an entire quantum computer on a single chip? Ultimately, this approach may work well. Whatever technology is ultimately chosen for quantum computation, however, hybridization of dissimilar technologies is likely to be required in order to develop a full-scale quantum computer. Any quantum computation will require extensive classical support circuitry for decoherence mitigation and qubit calibration, and it is unlikely that this support can be executed using an identical fabrication technology to that used for the qubit. Hybridization solves this problem by integrating dissimilar technologies. It further enables flexibility in design choices as the field progresses: either qubit or control technologies can be modified without influencing the other technology.

2 Experiment

Our primary goal in the design of this circuit was to establish the existence of substantial magnetic coupling between the control and qubit chips. We had originally planned to use the transitions of flux into and out of the RF SQUID as a signature of this coupling, but we discovered that this was not necessary: the existence of magnetic coupling could be unambiguously demonstrated using simply the RF SQUID as a component of a transformer on the qubit chip that would interact with the control chip. To show this, we observed a change in the effective mutual inductance between the control current and the DC SQUID with and without the presence of the flipped chip, which could only be explained by the existence of magnetically induced currents in the flipped chip. By effective mutual inductance we mean the sum of all the mutual couplings acting both directly and indirectly between the control current and the DC SQUID.

Fig. 1 shows the schematic diagram of the circuit used in this experiment. The control current was coupled most strongly to a transformer on the flipped chip, in turn coupled to the RF SQUID on the flipped chip, which itself was strongly coupled to a transformer on the control-chip. Finally, the DC SQUID was coupled strongly to the control-chip transformer. Because there were no ground planes in this circuit, all of the circuit elements showed non-negligible coupling to each other, which is not shown in Figure 1, but was accounted for in our theoretical work.

The basic chip fabrication process is a submicron, Nb/AlO_x/Al/Nb trilayer process on 150-mm-diameter wafers, where the layers are defined by UV photolithography and subsequent subtractive processing. Insulating oxide is deposited and planarized using chemical-mechanical polishing.[7] The chips described in this paper were fabricated at a J_c of ~ 700 A/cm^2.

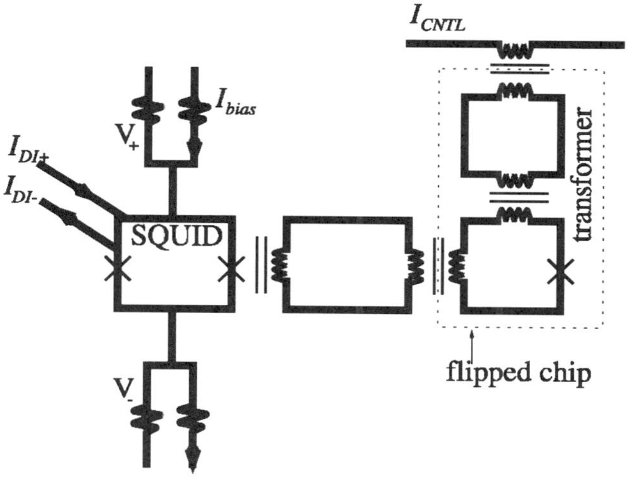

Figure 1: Schematic diagram of test circuit containing a DC SQUID on the "control" chip and an RF SQUID and transformer on the "qubit" chip. Without the transformer, the coupling between the control line and the DC SQUID is very weak. Not shown are two current lines attached to the DC SQUID that can be used to inject flux directly into the DC SQUID.

After fabrication the chips were diced out of the wafers and packaged using a flip-chip aligner-bonder. Our bonding process differed from past superconductive flip-chip bonding work,[8] in that we did not require electrical connections between the circuits, and therefore did not "bump" the chips prior to bonding. Permabond 910, the adhesive used, has very low viscosity, making it easy to dispense, however the short setting time (~ 45 sec) made alignment challenging. In fact the delay between dispense of the adhesive and bonding of the chips was usually several times the nominal setting time of the adhesive (~ 3 minutes). Despite this non-ideality, excellent adhesion was achieved. Figure 2 shows the final cross-section of the packaged device.

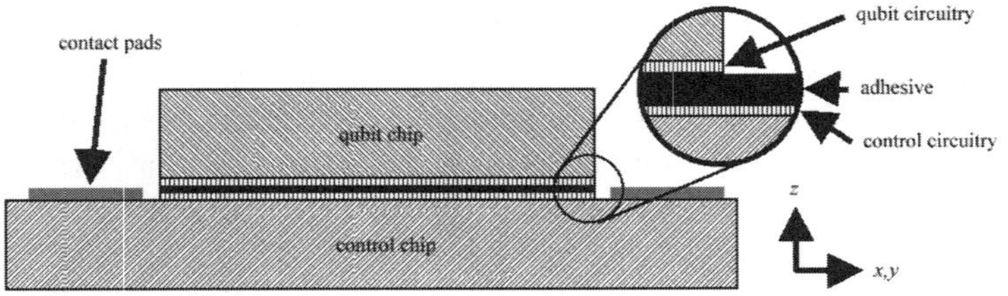

Figure 2: Cross section of the packaged chip including the "qubit" chip, which here contains only a transformer, and the "control chip", which in this case includes a DC SQUID and a current control line, independently coupled to the transformer on the qubit chip

The mutual inductance between the control line and the DC SQUID was determined by measuring the shift in the threshold curve[9] of the DC SQUID as the control current was increased. If the peak threshold is plotted against the control current, the slope of this curve equals $M^* T/\Phi_o$ where M^* is the effective mutual inductance, T is the period of the threshold curve in units of current, and Φ_o is the flux quantum. T can be obtained from the threshold curve by fitting a hyperbola to two adjacent lobes of the threshold curve and taking the difference of the lobe maxima.

3 Presentation and Analysis of Results

Alignment after flip-chipping was assessed using two methods: first through glass slides fabricated with test structures specifically designed for this purpose, and second by polishing the back of the flipped chip after bonding, and imaging through the silicon using infrared microscopy. The misalignment was found by these two methods to be consistently less than ~ 3 μm. Nowhere was the misalignment greater than 4 μm, and on some chips the alignment was better than 1 μm.

The chip-to-chip *z*-direction separation was determined by dicing the chips after testing, polishing the cut surface, and imaging the cross-section in an electron microscope. Using this method, a chip-to-chip separation of ~ 1.6 μm was observed. The observed separation was probably due to the presence of particulates of varying sizes between the chips prior to bonding. It may be necessary in the future to subject the chips to cleaning prior to flip-chip bonding in order to reduce this thickness.

In Figure 3 the shift in SQUID threshold with and without the flipped chip are compared. The effective mutual inductance calculated from the slope of the data shown increased from 31 fH to 72 fH when the flipped chip was included, a change of 41 fH. Each lobe of this data is fit well by a simple hyperbola. The fitting constants that were thus obtained were used to determine Ipeak in the data shown in Figure 3b. Calculations using 3D electromagnetic modeling software showed that the observed change in mutual inductance was reasonable given the uncertainty in our knowledge of chip-to-chip alignment and separation for the precise device that was tested.

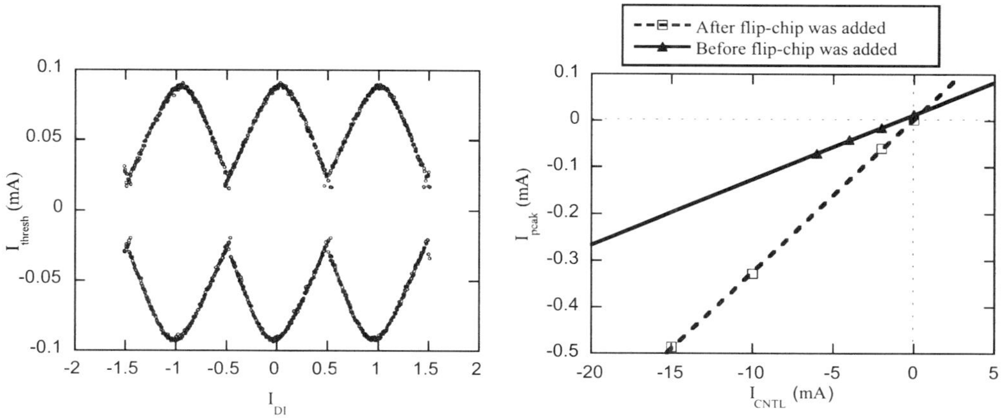

Figure 3 : A) Typical threshold curve of the DC SQUID. B) The observed movement of the SQUID threshold curve is plotted versus the applied control current for the same chip, before and after a flip-chip containing a flux-coupling transformer was added. With the transformer, the flux is coupled more efficiently.

4. Discussion and Conclusions

The main result of this paper is the demonstration of a flip-chip process for superconductive electronics that can be applied to qubit devices. This process was used to couple current magnetically from a carrier chip, onto a flipped chip, and back onto a different area of the carrier chip where the resulting magnetic field was sensed with a DC SQUID. Ultimately the flipped chip will be used for qubit circuitry, and the carrier chip will contain classical control electronics and field transducers for determining the state of the qubit.

The utility of this result is not limited to the field of quantum computing: this process would benefit any technology that uses a low-temperature sensor and could benefit from the application of SFQ electronics. The separation of fabrication processes could be critical to applications in which, for example, a classical SQUID used for magnetic-field detection needs to be coupled at extremely high speeds to control circuitry. In such a case, the parasitic inductance of a bump-bond might be undesirable, or the added cost and complexity of the bump processing might be prohibitive. In addition, monolithic processing of the device might be unfeasible. Large-scale superconductive digital circuits might also benefit from some close variant of the process described here: with the inclusion of a means of transferring DC bias currents between the two chips, this technology would allow increased circuit integration without requiring decreases in speed as signals go off chip.

Acknowledgements

We would like to thank the Lincoln Laboratory microelectronics fabrication team for help in processing these circuits, and George Fitch for assistance in testing, Dan Baker, Skip Hoyt, Peter Daniels, and Deb Landers for assistance in fabrication and packaging, and Gerry Sollner, Lenny Johnson, Alfredo Anderson, Dan Oates, Jay Sage, Hans Mooij, Donald Crankshaw, Marc Feldman, and Mark Bocko for helpful conversations.

References

1. M. F. Bocko, A. M. Herr, and M. F. Feldman, "Prospects for Quantum Coherent Computation Using Superconducting Electronics," IEEE Trans. Appl. Supercond. **7**, 3638 (1997).
2. Lin Tian, L. S. Levitov, Caspar H. van der Wal, J. E. Mooij, T. P. Orlando, S. Lloyd, C. J. P. M. Harmans, and J. J. Mazo, "Decoherence of the Superconducting Persistent Current Qubit," URL http://xxx.lanl.gov (cond-mat/9910062); to be published in Proceedings of the NATO-ASI on Quantum Mesoscopic Phenomena and Mesoscopic Devices in Microelectronics, I. O. Kulik and R. Ellialtioglu, eds.
3. J. R. Friedman, Vijay Patel, W. Chen, S. K. Tolpygo, and J. E. Lukens, "Quantum Superposition of Distinct Macroscopic States," Nature **406**, 43 (2000).
4. Caspar H. van der Wal, A. C. J. ter Haar, F. K. Wilhelm, R. N. Schouten, C. J. P. M. Harmans, T. P. Orlando, Seth Lloyd, and J. E. Mooij, "Quantum Superposition of Macroscopic Persistent-Current States," Science **209**, 773 (2000).
5. S. P. Benz, C. A. Hamilton, C. J. Burroughs, Jr., T. E. Harvey, L. A. Christian, and J. X. Przybysz, "Pulse-Driven Josephson Digital/Analog Converter," IEEE Trans. Appl. Supercond. **8**, 42 (1998).
6. For a review, see K. K. Likharev, "Rapid Single-Flux Quantum Logic," in The New Superconducting Electronics, H. Weinstock and R. Ralston, eds., Kluwer, Norwell, Mass., 1993, p. 423.
7. M. B. Ketchen, D. Pearson, A. W. Kleinsasser, et al., "Sub-μm, Planarized, Nb-AlOx-Nb Josephson Process for 125 mm Wafers Developed in Partnership with Si Technology," Appl. Phys. Lett. 59, 2609 (1991); K. K. Berggren, E. M. Macedo, D. A. Feld, and J. P. Sage, "Low Tc Superconductive Circuits Fabricated on 150-mm-Diameter Wafers Using a Doubly Planarized Nb/AlOx/Nb Process," IEEE Trans. Appl. Supercond. **9**, 3271 (1999).
8. K. E. Yokoyama, G. Akerling, A. D. Smith, and M. Wire, "Robust Superconducting Die Attach Process," IEEE Trans. Appl. Supercond. 7, 2631 (1997); M. Maezawa, H. Yamamori, and A. Shoji, "Chip-to-Chip Communication Using a Single Flux Quantum Pulse," IEEE Trans. Appl. Superconduct. 10, 1603 (2000); R. D. Sandell, G. Akerling, and A. D. Smith, "Multi-chip Packaging for High Speed Superconducting Circuits," IEEE Trans. Appl. Supercond. **5**, 3160 (1995).
9. See, for example, T. Van Duzer and C. W. Turner, Principles of Superconductive Devices and Circuits, 2nd ed., Prentice Hall, Upper Saddle River, N.J., 1999, pp. 260-263.

IMPLEMENTING QUANTUM COMPUTERS THAT "ENERGETICALLY" DEFEND AGAINST DECOHERENCE

ARI MIZEL

Pennsylvania State University, 104 Davey Laboratory, University Park, PA 16802, USA
E-mail: ari@phys.psu.edu

In a recent paper, we proposed a novel ground-state scheme for quantum computer realization. The approach is time-independent and inherently defends against decoherence by possessing an energy barrier to excitation. A proof was given that time-independent qubits and their time-dependent counterparts can execute the same algorithms. A comparison of the implemention requirements of the two types of qubits was presented. Here, ground-state quantum computation is reviewed, including an analysis of its strengths and weaknesses. We then discuss proposals for building a realistic ground-state quantum computer. Utilizing the wealth of creative suggestions that have been made for time-dependent quantum computation, we consider which ideas could be adapted to the ground-state case. Josephson junction implementations are found to be particularly attractive. The design of a ground-state quantum computer using Josephson junction arrays is outlined.

1 The Ground-State Approach to Quantum Computation

Consider a lone qubit participating in a quantum algorithm. As the algorithm proceeds, the qubit's state $|\psi(t)\rangle$ changes in time. Initially, the qubit's state is $|\psi(t=t_1)\rangle$. Since a qubit state is a superposition of two basis vectors $|0\rangle$ and $|1\rangle$, this initial state can be specified with the two complex numbers $\langle 0|\psi(t=t_1)\rangle$ and $\langle 1|\psi(t=t_1)\rangle$. After the first step of the algorithm, the state becomess $|\psi(t=t_2)\rangle$, which can be specified with two more complex numbers. If there are N-1 steps in the algorithm, then 2N complex numbers are sufficient to detail the progress of the qubit from its initial state, through intermediate states, to its final state. One can imagine a different means of instantiating the same 2N numbers. Rather than storing the numbers in a time-dependent state $|\psi(t)\rangle$ defined on a 2 dimensional Hilbert space, one could store them in a time-independent state $|\phi\rangle$ defined on a 2N dimensional Hilbert space. The development of the state $|\psi(t)\rangle$ in time would then be replaced by the development of $|\phi\rangle$ through Hilbert space. This is the essential idea behind ground-state quantum computation. To make this notion more precise, let us establish 2N basis vectors in the 2N dimensional space occupied by $|\phi\rangle$. We label these basis vectors by $|0,i\rangle$ and $|1,i\rangle$, where the index $i = 1, \ldots,$ N. We can say that $|\phi\rangle$ develops through Hilbert space just as $|\psi(t)\rangle$ develops in time if the following mapping holds:

$$\langle 0|\psi(t=t_i)\rangle = \langle 0,i|\phi\rangle \qquad \text{and} \qquad \langle 1|\psi(t=t_i)\rangle = \langle 1,i|\phi\rangle \ . \tag{1}$$

How can we make this relation hold? A physical system with state $|\psi(t)\rangle$ can be forced to develop in time in a computationally meaningful fashion if it is subjected to an appropriate time-dependent Hamiltonian. But how can we force a physical system with state $|\phi\rangle$ to develop through Hilbert space in the computationally meaningful fashion specified by the mapping (1)? This can be achieved by subjecting the system to an appropriate time-independent Hamiltonian. If the time-independent Hamiltonian is chosen so that its ground-state is the desired $|\phi\rangle$, then the system will have state $|\phi\rangle$ as long as it is not excited. We have recently demonstrated how, given any quantum algorithm, to select the time-independent Hamiltonian [1]. In the preceding paragraphs, we have described ground-state quantum computation using a single qubit case for clarity. A real quantum computation, of course, requires more than one qubit. Fortunately, the basic idea behind the single qubit case can be extended to an arbitrary number of qubits [1].

1.1 Energy Barrier to Decoherence

Ground-state quantum computation is formulated to handle the same algorithms designed for time-dependent quantum computation. How are the two schemes distinct? When it comes to actually constructing a quantum computer, there could be an important advantage to computing with a time-independent ground-state $|\phi\rangle$ rather than a time-dependent state $|\psi(t)\rangle$. Unpredictable, time-dependent perturbations of a quantum computer will cause decoherence to a time-dependent state $|\psi(t)\rangle$. However, such perturbations will have to overcome an energy barrier to take a time-independent system out of the ground-state $|\phi\rangle$ into the first excited state. We therefore say that ground-state quantum computation "energetically" defends against decoherence. It is important to note that the energy barrier does decrease with the number of computational steps in the algorithm. Thus, the defense becomes less effective as the computation gets longer. However, the decrease is only algebraic [2]. It is also worth mentioning that, while the ground-state $|\phi\rangle$ resists time-dependent decoherence, it will be degraded by time-independent perturbations to the system's Hamiltonian. Just as in the time-dependent case, defects in the implementation will have to be removed by thorough testing and refinement. Aside from the energy barrier to decoherence, there are several other features that distinguish ground-state quantum computer implementation from time-dependent quantum computer implementation. First of all, a ground-state quantum computer must have a precise Hamiltonian on a large Hilbert space. This is more difficult than dealing with the smaller Hilbert space associated with time-dependent quantum computation. One the other hand, the Hamiltonian is time-independent in the ground-state case. It only needs to be re-tuned during the programming of a new computational algorithm. After the time-independent Hamiltonian has been set up, a ground-state quantum computer will function properly only if the system is in its ground-state $|\phi\rangle$ and not in some excited state. To ensure this, it may be necessary to anneal the system. This step is analogous to preparing the initial state in a time-dependent quantum computer, but it is likely to be more difficult because of the large Hilbert space involved. To extract the results of a ground-state quantum computation, it is necessary to measure certain amplitudes of the ground-state. For example, in the single qubit case discussed in Sec. **??**, the amplitudes $\langle 0, i = \mathrm{N}|\phi\rangle$ and $\langle 1, i = \mathrm{N}|\phi\rangle$ would contain the results of the computation. The measurement procedure may or may not be subtle depending on the form of the computational algorithm [1,2].

2 Ground-State Qubit Implementation

2.1 Finding a Suitable Implementation

To design a realistic ground-state quantum computer, it is natural to try to adapt one of the creative ideas for time-dependent quantum computer realization that appear in the literature. In considering which ideas might be adapted, one must take into account the fact that the ground-state case requires (i) a large Hilbert space and (ii) a precise Hamiltonian on the large Hilbert space. With these considerations in mind, the highly developed Josephson junction technology seems particularly well suited to our goal. Large arrays of Josephson junctions are commonly fabricated, so it well known how to construct a device with a large, quantum mechanically coherent Hilbert space. The Hamiltonian of the Hilbert space can be carefully tuned if SQUIDs are used for Josephson junction links. With a Josephson junction array, there are many routes for establishing the 2N dimensional Hilbert space we need for a qubit. For example, one could extend a design from the time-dependent case [3,4] and allow Cooper pairs to tunnel between 2N superconducting islands rather than 2 superconducting islands [5]. Alternatively, it may be more practical for us to work with current eigenstates. We will pursue this latter idea in the remainder of this paper, outlining the design of a ground-state persistent current qubit.

2.2 Josephson Junction Ground State Qubit

The building blocks of the qubit are superconducting loops with three Josephson junctions each (Fig. 1). These loops are adapted from the time-dependent case [6]. One of the three junctions in the loop has a large coupling energy E_J. We denote the gauge invariant phase difference across this junction by γ_a. The other two junctions have smaller coupling energies e_J. We denote the phase differences across these by γ_b and γ_c. A magnetic flux of $\Phi = \Phi_0/2 = (h/2e)/2$ pierces the loop. Because the loop's wavefunction must be single-valued, the phase differences satisfy the constraint $\gamma_a + \gamma_b + \gamma_c = 2\pi\Phi/\Phi_0 + 2\pi m = \pi + 2\pi m$, where m is an integer. If we assume that the capacitances of the junctions are negligible, the Hamiltonian of the loop is $H = E_J(1 - \cos\gamma_a) + e_J(1 - \cos\gamma_b) + e_J(1 - \cos\gamma_c)$. Taking into account the constraint, we find that this Hamiltonian has two degenerate minima. The two minima give two degenerate ground states, one with clockwise current and the other with counterclockwise current. For each state, the magnitude of the current is fixed by a parameter Γ, where $\sin(\Gamma/2) = e_J/2E_J$. The phase differences associated with each state appear in Fig. 1.

To construct a ground-state qubit, we need to establish a 2N dimensional Hilbert space. This is achieved by connecting 2N constituent loops together with a linking loop, as in Fig. 2. The linking loop does not change the form of the Hamiltonian of the constituent loops. It just imposes a constraint on phase differences, as is indicated on the bottom of Fig. 2. This constraint has been chosen [7] so that it is compatible with states of the form

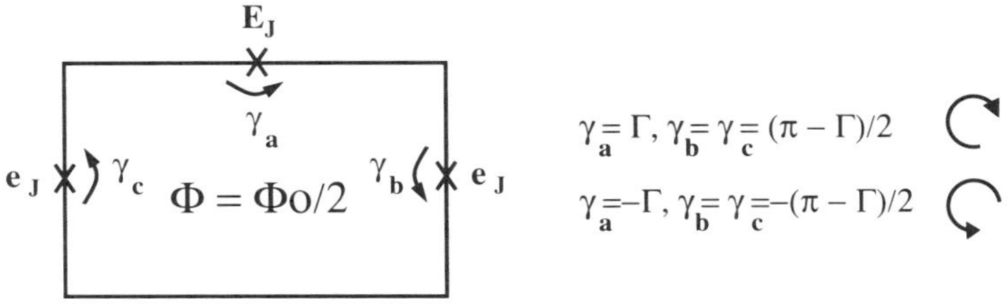

Figure 1. A component of a qubit. The superconducting loop possesses two degenerate states, which are indicated on the right side of the figure. The states depend upon the parameter Γ, where $\sin(\Gamma/2) = e_J/2E_J$.

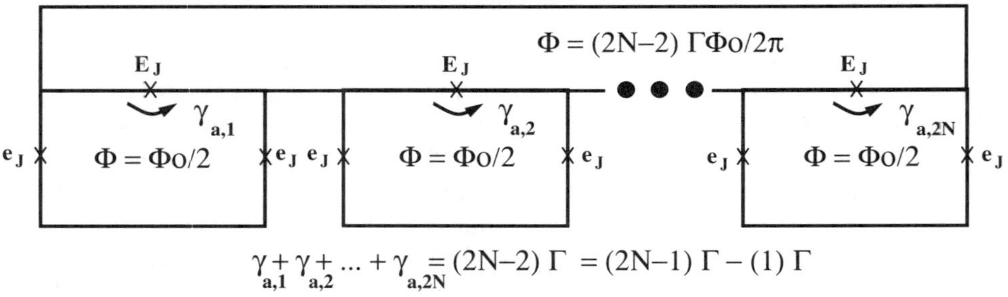

Figure 2. A single ground-state qubit. The qubit consists of 2N constituent loops joined by a linking loop. It possesses 2N degenerate ground-states that establish the 2N dimensional Hilbert space of the qubit. In each such state, (2N-1) constituent loops have counterclockwise current and one has clockwise current.

$\{\gamma_{a,j_0} = -\Gamma\ ;\ \gamma_{a,j} = \Gamma$ for $j = 1, ..., j_0 - 1, j_0 + 1, ..., 2\mathrm{N}\}$. Here, γ_a, j is the phase difference across junction E_J of loop i. The integer j_0 can take any value between 1 and 2N, so we find that the structure has 2N degenerate ground states. Physically, these ground states each have 2N -1 loops with current flowing clockwise and 1 with current flowing counterclockwise. There are 2N possibilities for the loop with the counterclockwise current, which yield the 2N degenerate ground states. With the proper number of ground states in place, the structure has the potential to act as a ground-state qubit.

In summary, we have described ground-state quantum computation and some of its strengths and weaknesses. Potential implementations were discussed and the outline of a Josephson junction ground-state qubit was presented. It is hoped that this outline conveys the idea behind a possible ground-state quantum computer design despite the absence of a description of qubit gates.

References

1. A. Mizel, M. W. Mitchell, and M. L. Cohen, *Phys. Rev.* A Rapid Comm., in press. Available at http://xxx.lanl.gov/abs/quant-ph/9908035.
2. A. Mizel, M. W. Mitchell, and M. L. Cohen, submitted for publication. Available at http://xxx.lanl.gov/abs/quant-ph/0007001.
3. A. Shnirman, G. Schon, Z. Hermon, *Phys. Rev. Lett.* **79**, 2371 (1997).
4. Y. Makhlin, G. Schon, A. Shnirman, *Nature* **398**, 305 (1999).
5. I am grateful to D. Dixon for his insights on this possible implementation.
6. J. E. Mooij *et al.*, *Science* **285**, 1036 (1999).
7. This argument requires that we fix e_J, E_J so $n\Gamma \neq 2\pi m$ for any integral $n < 2N$.

SEMICONDUCTOR QUANTUM INFORMATION SYSTEMS – A PROGRESS REPORT

KANG L. WANG*, HONG-WEN JIANG, AND ELI YABLONOVITCH

University of California, Los Angeles, Los Angeles, CA 90095-1594 Email: wang@ee.ucla.edu

Quantum information suggests a new approach to revolutionize computing capability beyond today's CMOS circuits. Thus it has a potential to go beyond the Moore's scaling limit in increasing the functionality of CMOS. In this paper, we will discuss our proposed scheme for implementation using SiGe nanostructures for computing and information processing. Our proposed system uses an electron spin as a qubit, in which the control of g factor and thus addressing of a specific qubit is done by a nanometer scale field effect transistor. We will describe g factor engineering in SiGe quantum well nanostructures and quantum dot systems. Several other alternative approaches using SiGe will be discussed. For quantum memory, we propose the use of a single P dopant atom in a single device or a quantum dot. Likewise, for quantum information systems, possible schemes of transfer of an electron spin to the polarization of a single photon from SiGe to InAs as well as challenges will be discussed. We will also describe our recent progress in realizing the basic elements for constructing a quantum information system. In characterization of electron spin resonance, we have developed an electronic detection method for spins using magneto-resistance measurement. The detection method was demonstrated in GaAs HEMT devices and a large Overhauser effect was also evident. In view of the large Overhauser effect, which will likely impede our ESR control in quantum computing, we will discuss and contrast the issue of the use of III-V and SiGe for quantum computing. This magneto-resistance measurement technique may be scaled down to the detection of a single spin. In fabrication, a new scheme of controlled placement of self-organized structures will be discussed. In this scheme, cooperative, well-organized Ge quantum dots were fabricated for potential realization of dot arrays for qubits.

1 Introduction

Solid-state implementation of quantum computing has a potential of convenient scale up in comparison to those of atomic traps, liquid NMR and others. The basic prerequisites for a quantum computer are: a quantum unit (qubit) with a decoherence time significantly less than the computation cycle [1], unitary rotations of the qubits and the ability to control interaction between the qubits [2]. Ge and SiGe layered nanostructures have been proposed for implementation in quantum information computation [3]. The layered structures are illustrated in Fig. 1, in which an electron spin with a donor atom is used to represent a qubit [4], although the donor is not really required for qubit operation in our proposed approach. There is one electron under one gate or one field effect device. The structure similar to a HEMT structure with proper band offsets for the device operation to transfer one electron from one region with a g factor to another region of a different g value and the g factor of SiGe layers can be tuned by adjusting the Ge content near 0.8 depending on the strain and orientation, where the transition from the L and X valley of the conduction band minima occurs. We refer to this type of device as "Field Effect Transferred Spinor" or "**Transpinor**" or electron spin resonance (**ESR**) transistor. In this scheme, there is one electron under one gate or under one field effect device, which is slightly different from a previous one proposed by this group in that the donor, although that may be desirable for different reasons, is not required in the structure. In the case of a qubit consisting of a donor, the qubit is represented by both an electron and a donor, very much like that of Kane who proposed the use of a donor nucleus in a pure Si nanostructure as a qubit and the use of the electron-mediated interactions for qubit preparation and elementary operations of quantum computing.

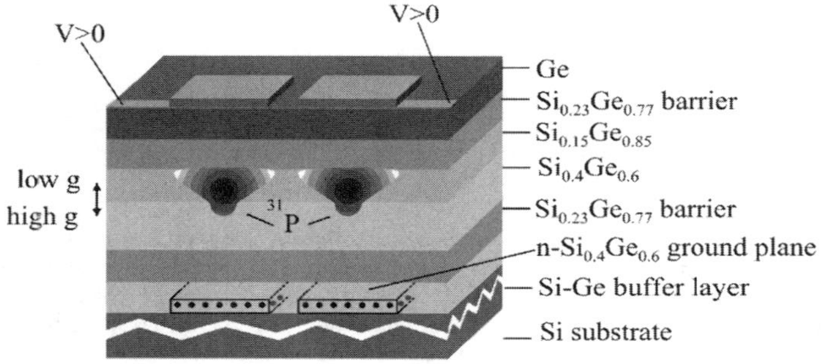

Figure 1. Transpinor or ESR transistor for a qubit. The structure consists of different layers having different g-factor numbers. The electron under each gate can be controlled by applying gate voltage to be in ESR for qubit preparation. The interaction for the adjacent two qubits can be done by applying the gate voltage simultaneously onto the two gates.

The large change of the g-factor can be seen from Fig. 2 in which the band alignment of the SiGe layered structure is illustrated. For (111) oriented substrate, for example, the D layer is shown to have a Ge content x=0.6, the T (top) layer has x=0.85 and the two B (barrier) layers have x=0.77. (It is also feasible to use other orientations with smaller changes of the g-factor.) The Ge contents are chosen such that the band alignment is shown in Fig. 3 and the electron is confined by the two barriers while it can move freely from the D to T layer under a gate bias without a barrier.

The g factor has not been measured for SiGe. We have estimated the g factor using the following equations:

$$g_{\parallel} = 2 - \frac{2}{3} \cdot \frac{\Delta_{SO}^{Ge,Si}}{E_c^{L,X} - E_v^{L,X}} \cdot \left(\frac{m}{m_{\perp}} - 1 \right) \qquad (1)$$

$$g_{\perp} = 2 - \frac{2}{3} \cdot \frac{\Delta_{SO}^{Ge,Si}}{E_c^{L,X} - E_v^{L,X}} \cdot \left(\frac{m}{m_{\parallel}} - 1 \right) \qquad (2)$$

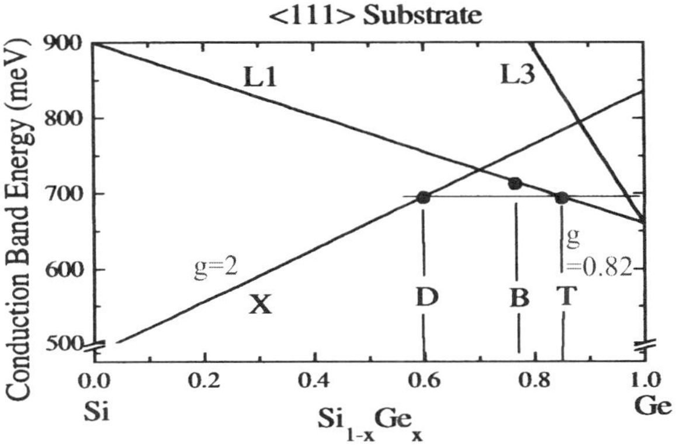

Figure 2. Band alignment of the conduction bands for SiGe alloys grown on <111> Ge substrate. Due to the strain, the 4-fold degenerate L valleys are split. The x valleys remain degenerate due to the symmetry.

Accurate estimate of these values is not easy as there are no reliable effective mass numbers available for the band far away from the Brillouin zone. We have, however, used the recent theoretical calculation to provide the guideline number for the preparation of device structures. Experimental work needs to be done to determine the g factor.

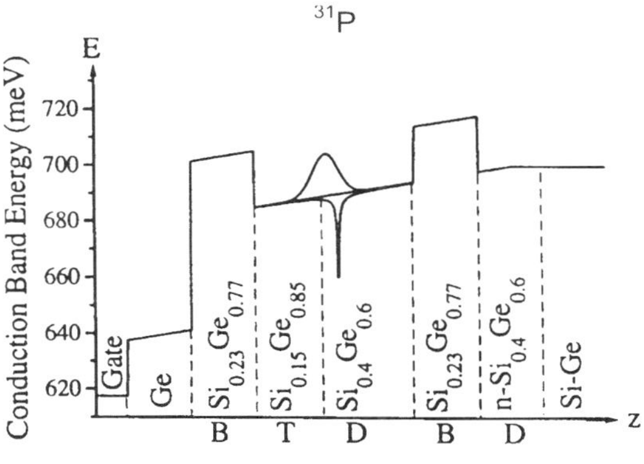

Figure 3. Band alignment of the conduction bands for the layered structures grown on <111> Ge substrate. The electron can move from the D to T layer freely by the gate voltage. Two barrier (B) layers are used to confine the electron.

The addressing for a given qubit is done similarly to those of CCD or DRAM by applying the gate voltage to move the electron wave function to the layer, where the g factor is such that the spin resonance frequency is in tune with the applied microwave frequency for performance of the elemental operations of computation ($E= q\mu B$). The entanglement of a set of two qubits can be done similarly by putting the wave functions of two adjacent qubits to have sufficient overlap for exchange interaction by applying the gate bias to the two gates to pull each electron from the D layer to the T layer as shown in Fig. 3. The ^{31}P donor will be primarily used for transferring the electron spin to the nuclear spin for qubit storage.

2 Experimental Progress

In our initial experimentation for characterization of ESR and nuclear spin resonance as well as the test of the gate control, we used AlGaAs/GaAs HEMT structures. The experiments are among the first steps towards the realization of spin resonance transistor, its use as a qubit, and the elemental operations for quantum computing. This paper also reports our progress towards these objectives.

2.1. Magneto-resistance detected ESR

We have detected the electron spin resonance by resistance with a variety of semiconductor heterostructures. It is well known that the conduction change of the resistance due to conducting (diffusion) or localized (hopping) electrons or holes in semiconductors can be used to effectively detect the ESR.

As shown in Fig. 4, the magneto-resistance was measured for the AlGaAs/GaAs device. The top figure illustrates the typical resistance under a high magnetic field. The bottom shows the resistance under an additional applied microwave signal. A small ESR signal was observed.

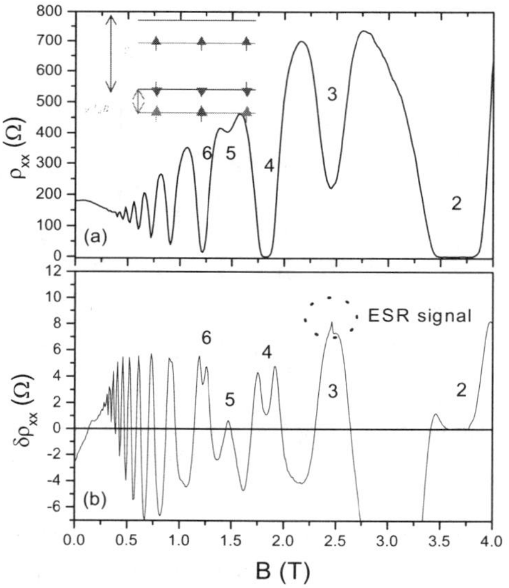

Figure 4. Magneto-resistance of ESR signal. The top diagram is a typical quantum Hall spectrum. The set shows the cyclotron and spin splits. The bottom curve was obtained by modulating the applied microwave signal in ESR resonance and the small ESR signal was detected with a lock-in amplifier. The signal shown was in resonance with ν=3. The signal

can be improved with the use of high Q microwave cavity. Indeed, ESR signal was detected even without the use of the lock-in amplifier.

2.2. *Gate controlled g-shift*

In the proposed Si-Ge structure, the g-factor is controlled by moving the electron wave function of the electron from one epitaxial layer to another by electrical gating. This allows local manipulation of electron spins, being essential in concepts for the quantum logic operations.

We indeed have demonstrated recently with a gated GaAs/AlGaAs heterostructure that the ESR resonance frequency can be tuned by means of a gate that electrostatically adjusts the weighting of the electron wave function between the two hetero-structure layers of different g-factors (i.e., GaAs has g=-0.44, while AlGaAs has g=+0.4). Figure 5 shows that the electrically detected electron spin resonance spectra at a sequence of gate voltages with the DC magnetic field sweep. Front and back gates produce opposite g-factor shifts verifying that g-factor is being controlled. We found that for a front gate, the resonant peak moves progressively to higher magnetic field when the amplitude of the bias gate voltage is increased. In contrast, the peak shifts towards lower field for the back gate case. The g-factor calculated from the resonance DC magnetic field is shown in Fig. 6, for which the front and back gate biases were used. It is clearly seen that the electric field applied to the gate pulls the electron wave function toward the AlGaAs layer, changing the g-factor. The result indicates the feasibility of the g-factor control by the gate.

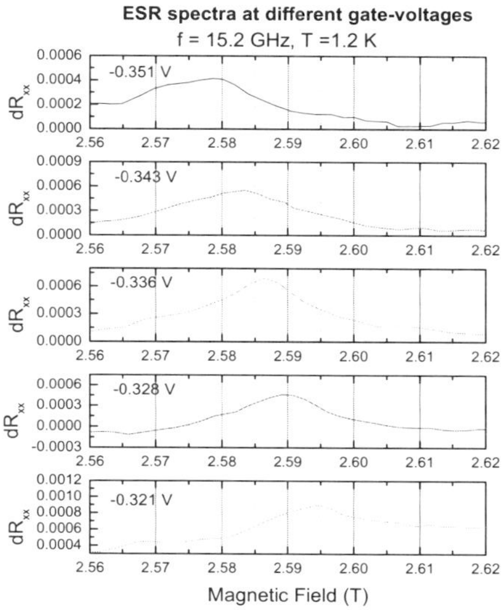

Figure 5. Gate control of ESR. With applied gate voltage, the ESR resonance signal shifts, indicating the change of the g-factor. The structure used was the same AlGaAs/GaAs. The applied microwave was at 15.2 GHz.

It is anticipated that gate controlled single qubit operations are possible in our proposed Si-Ge device. It is important to note here that since over 95% of the nuclear isotopes for the host Si or Ge materials have zero spin, the inhomogeneous broadening caused by hyperfine interactions from isotopic impurities should be vanishingly small. Of course, the narrow line width, which reflects the long decoherence time, is highly desirable for quantum computing. Furthermore, the narrower ESR line width will allow the gate to shift the qubit in and out the resonance with an even much smaller variation of the voltage.

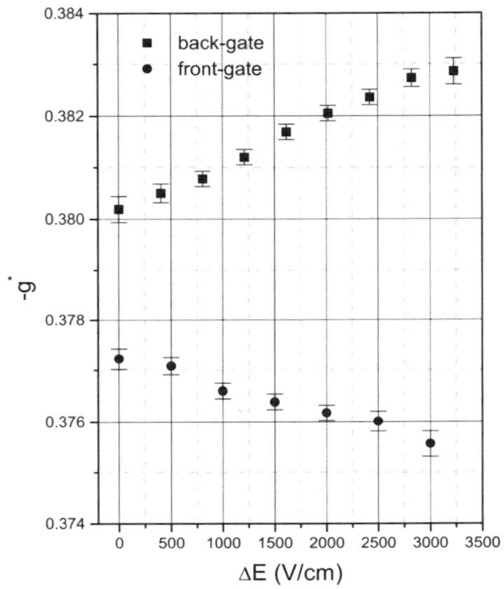

Figure 6. G-factor obtained from the ESR signal in Fig. 5. The g-factor was plotted as a function of the gate applied field. The change of the g-factor comes from the wave function penetrating into or pulling away from the AlGaAs layer.

2.3. Nuclear Spin resonance – Overhauser Effect

The residual nuclear spin may impose a serious problem for tuning ESR to qubit operations for III-V compounds due to the hyperfine interaction. The effect referred to as Overhauser effect tends to cause the ESR resonance energy to continuously change and make the steady state tuning of ESR for elementary operations of qubits extremely difficult, if not impossible. We have demonstrated in our ESR measurement that the polarization of nuclear spins, which in turn result in the net magnetic field, shifts the ESR signal. Fig. 7 shows the ESR signal vs. time after driving electron spins to saturation to flip and polarize the nuclear spins of As and Ga. The decay of the nuclear polarization, which exhibits as the shift of the ESR signal, is shown in Fig. 8. This continuous change indeed makes the tuning for qubit operations difficult. From this change, we can deduce the nuclear spin relaxation time to be on the order of 360 s. Similarly, the Overhauser effect for Ga and other isotopes was observed. The large nuclear spin polarization, or the Overhauser effect will be a critical problem in the use of III-V compounds. Thus Si/Ge should be more desirable for the implementation of quantum computers.

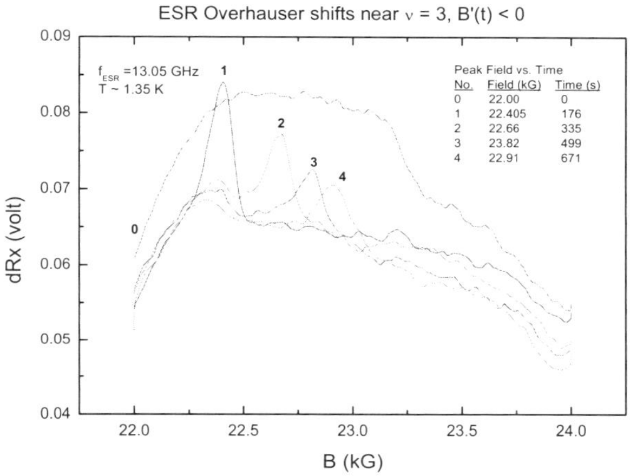

Figure 7. Magnetic field scans after the electron spins are saturated by the ESR microwave signal. The ESR shift shown in each trace is the result of the polarization of the nuclear spins caused by the electron spin relaxation.

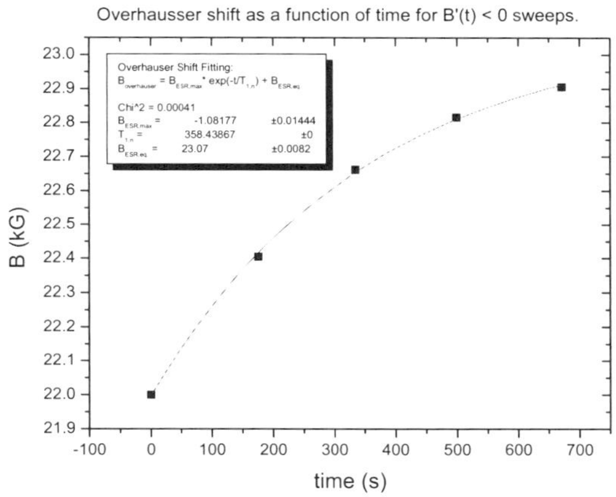

Figure 8. Shift of ESR as a function of time after the saturation of electron spins. The plot is obtained from Fig. 7 by plotting the shift vs. t. The change gives a T_1 of about 360 s.

3 Quantum dot Implementation of Transpinor Structures

Similar device concepts can be applied using the quantum dots. There are several physical reasons why the Ge quantum dots are very promising for application in quantum computing. Firstly, there are some theoretical evidences [5-7] that certain germanium rich SiGe alloys give the type I band offset of the conduction band with a coherently strained Ge film grown on top. Thus it is highly probable that electrons can be effectively localized in the quantum dots to serve as qubits. An electron trapped into the dot will be restricted in its orbital motion to the low-angular momentum states that will drastically suppress the spin-orbit interaction [6], and thereby effective decoupling of the electron spin from the environment can be achieved. Secondly, because of the spatial confinement, the electron g-factor is a sensitive function of the electric field [6] that allows efficient tuning of the electron spin in and out of resonance with a global microwave field by a local electric field. Thirdly, chemical impurity concentration is a few orders lower than in III-V compounds.

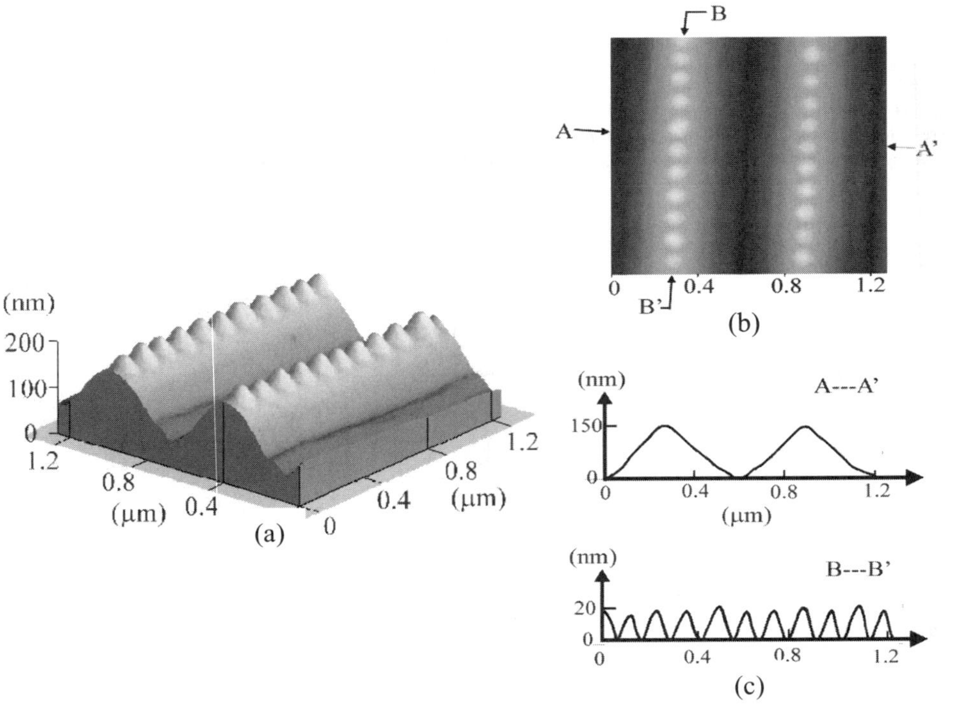

Figure 9 (a) 3-D AFM image of the self-assembled Ge dots on the <110>-oriented Si stripe mesas with a window width of 0.6 μm. The top plateaus have been fully reduced to the ridges after about 120 nm Si deposition. Self-aligned and well-spaced 1-D arrays of the Ge dots are formed on the ridges of the Si mesas after a subsequent deposition of 10 ML Ge. (b)

The 2-D image of the dot arrays in (a). (c) shows the cross section of the mesas (line AA') and one array (bottom) of the dots (line BB'). The sidewalls are {113} facets.

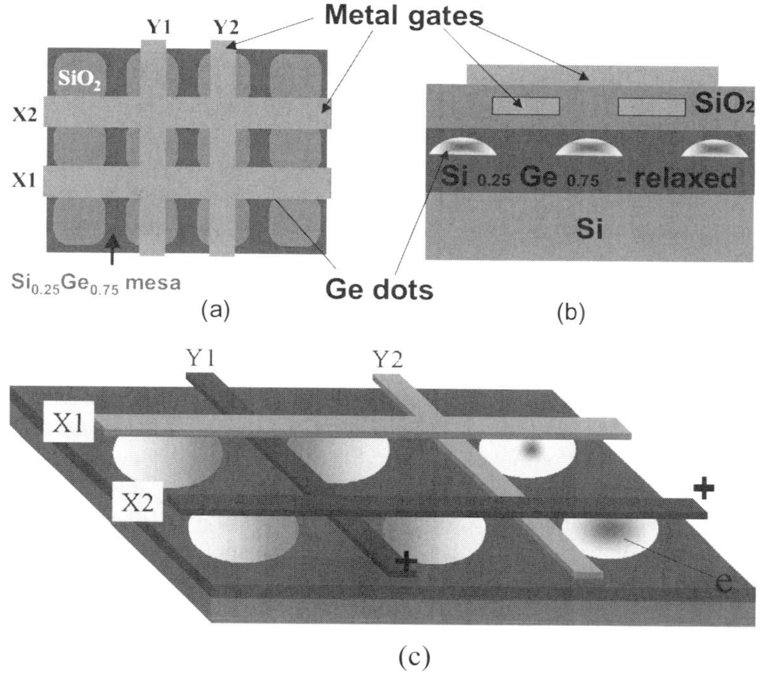

Fig. 10. Illustration of a quantum computation device using Ge quantum dots in linear arrays. The signal of interaction passes through in the x direction. As shown is a set of linear arrays. a) Cooperatively assembled Ge quantum dot growth on $Si_{0.25}Ge_{0.75}$ mesa by selective epitaxial growth technique. b) X cross sectional view of the device. (c). Showing the exchange interaction between two electrons being pulled by applying a high gate voltage to the X1 and the Y1 gate as indicated by +. Each qubit is tuned by the same means except applying a weaker potential to the X1 and Y1/Y2 lines to change the g factor.

If one P or other ½ spin atom can be incorporated into a single quantum dot, it is foreseeable that the nuclear spin of the dopant atom can be used as quantum memory. In such case, the electron spin can be transferred to the nuclear through hyperfine interaction. It is anticipated that nuclear spin has a much longer coherent time (hours) and thus may be used as quantum memory very much like DRAM of today's computers.

As shown in Fig. 9, we have demonstrated Ge quantum dots grown on Si mesas formed by selective epitaxial growth using patterned structures. The dots are quite uniform, the separation of the dots is about 0.1μm, and the distance between the rows is about 0.6 μm, which is limited by the mask we used. The arrangement mechanism is the cooperative interaction of neighboring dots via diffusion and substrate deformation induced by dots – referred to as Cooperative self-assembly [7]. We have achieved controlling the size of the dots to within 3% and their positioning to within

1 nm. For the dot array, we have conceived a simple structure as illustrated in Fig. 10, in which the signal passes by the x direction. We can arrange them in a rectangular grid with a 100 nm period. With the current technology of metal lithography, we should be able to produce 30 nm metal gates in the middle between two neighbor quantum dots (designated as Y1 and Y2 in the figure). One can control the spatial position of the electron wave function by applying the electric field to the gate. The basic operations can be performed by addressing the x and y lines. It is also possible to construct an x-y random access array similar to that of DRAMs.

For quantum information, the spin information must be transferred to the polarization of a photon. We have also proposed a system in which InAs can be deposited on Si/Ge and the electron spin information can be transferred to the electron spin of InAs before injection in to the p-type region to produce a polarized photon. Photon entanglement can be done.

Acknowledgement

The authors acknowledged partial support of DARPA, ARO, and NSA.

References

1. J. Preskill, Proc. R. Soc. London A **454**, 385 (1998).
2. D.P. DiVincenzo, G. Burkard, D. Loss, E.V. Sukhorukov cond-mat/9911245, 16Nov, 1999.
3. Rutger Vrijen, Eli Yablonovitch, Kang Wang, Hong Wen Jiang, Alex Blandin, Vwani Roychowdhury, Tal Mor, and David Di Vincenzo, "Electron-spin-resonance transistors for quantum computing in silicon-germanium heterostructures," *Physical Review A*, **62**, pp. 012306-1 – 012306-10, June 2000; R. Vrijen, E. Yablonovitch, K. L. Wang, H. Jiang, A. Balandin, V. Roychowdhury, T. Mor and D. DiVincenzo, quant-ph/9905096v2, 11 Jun, 1999.
4. B. Kane, Nature **393**, 133 (1998).
5. E.L. Ivchenko, A.A. Kiselev, and M. Willander, Solid State Comm. **102**, 375 (1997).
6. J.P. Gordon and K.D. Bowers, Phys. Rev. Lett. **1**, 368 (1958).
7. G. Jin, J. L. Liu, and K. L. Wang, Appl. Phys. Lett. **76**, 3591(2000).

CONSTRUCTION OF A SILICON-BASED SOLID STATE QUANTUM COMPUTER

A. S. DZURAK,[1,2] R. G. CLARK,[1,3] A. R. HAMILTON,[1,3] D. N. JAMIESON,[1,4]
G. J. MILBURN,[1,5] M. NEILSEN,[1,5] S. PRAWER,[1,4] H. RUBINSZTEIN-DUNLOP[1,5]
AND M. Y. SIMMONS[1,3]

[1] *Centre for Quantum Computer Technology, University of New South Wales, Sydney 2052, Australia*

[2] *School of Electrical Engineering & Telecommunications, University of New South Wales, Sydney, Australia*

[3] *School of Physics, University of New South Wales, Sydney 2052, Australia*

[4] *School of Physics, University of Melbourne, Parkville 3052, Australia*

[5] *Department of Physics, University of Queensland, Brisbane 4072, Australia*

We discuss progress towards the fabrication and demonstration of a prototype silicon-based quantum computer, being undertaken at the Centre for Quantum Computer Technology. The devices are based on a precise array of ^{31}P dopants embedded in ^{28}Si. Fabrication is being pursued via two complementary pathways – a 'top-down' approach for near-term production of few-qubit demonstration devices and a 'bottom-up' approach for large-scale qubit arrays. The 'top-down' approach employs ion implantation through a multi-layer resist structure which serves to accurately register the donors to metal control gates and single-electron transistor (SET) read-out devices. In contrast the 'bottom-up' approach uses STM lithography and epitaxial silicon overgrowth to construct devices at an atomic scale. Techniques for qubit read-out, which utilise coincidence measurements on novel twin-SET devices, are also presented.

1 Introduction

Solid-state architectures offer the promise of large-scale integrated quantum computer (QC) processors on a single chip. Of the solid-state schemes proposed thus far, silicon-based QC has the added advantage of being compatible with metal-oxide-semiconductor (MOS) technology that has dominated conventional computer technology for the past four decades. The Centre for Quantum Computer Technology has as its central aim the development of Si-based QC architectures, with particular emphasis on the proposal by Kane[1], which uses the nuclear spin of phosphorus atoms (spin $I=\frac{1}{2}$) as qubits embedded in isotopically-pure silicon-28 ($I=0$) – see figure 1.

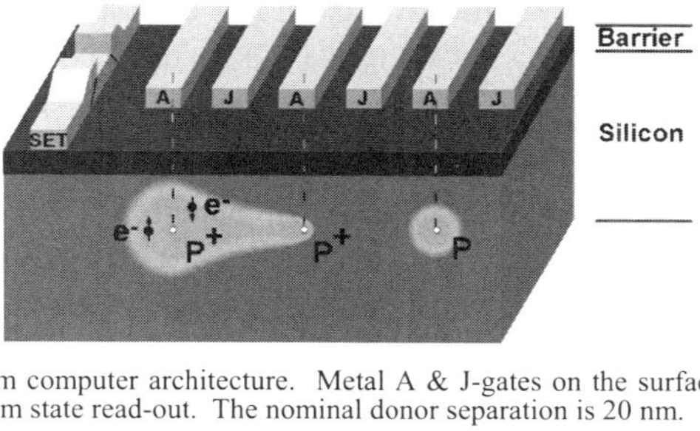

Figure 1. Silicon-based quantum computer architecture. Metal A & J-gates on the surface control the ^{31}P qubits below, while the SET is used for quantum state read-out. The nominal donor separation is 20 nm.

The development of a device such as that shown in Fig. 1, with critical dimensions in the range 5-10nm, requires significant advances in nanotechnology, along with new experimental and theoretical developments in quantum electrical measurement and quantum information. In order to achieve these challenging goals the Centre for Quantum Computer Technology is organised into a number of parallel research programs, spread over the partner institutions, as depicted in Fig. 2.

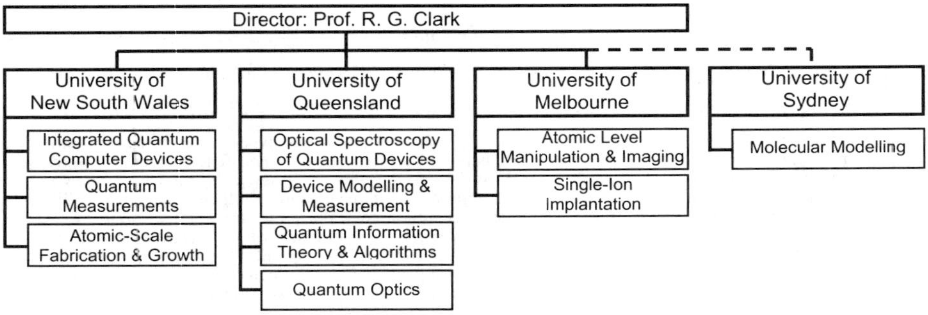

Figure 2. Program organisational chart for the Australian Centre for Quantum Computer Technology.

The nanofabrication and materials issues relevant to the architecture of Fig. 1 are addressed jointly by nodes at the University of New South Wales (UNSW) and the University of Melbourne. Electrical measurement of prototype devices is undertaken at UNSW while a program in optical studies of relevant nanostructures is being pursued at the University of Queensland (UQ). Modelling of the prototype devices and quantum measurement process is coordinated from UQ, as is research into quantum information theory and algorithms. A new program in molecular dynamics modelling of nanofabrication processes has recently been established with the University of Sydney. In addition to silicon-based architectures, the Centre is also investigating a number of other promising QC paradigms, with particular emphasis on linear optics QC. Beyond Australia, the Centre maintains strong collaborative links with teams in the USA, including those at Los Alamos National Laboratory (LANL), the University of Maryland and UCSB.

Here we focus on progress at UNSW in the nanofabrication of structures for prototype Si QC devices, together with the development of novel devices for spin read-out. Details of the other Centre programs are provided through a series of papers elsewhere in these Proceedings.

2 Parallel Fabrication Pathways

Two parallel pathways are being explored at the Centre for the fabrication of Si QC devices. In order to achieve few-qubit prototypes in the shortest possible time a *top-down* approach[2] is being used, which employs established semiconductor process techniques such as electron-beam lithography and ion-implantation. Although at present this approach is designed for small-qubit devices, we are exploring techniques which could extend this approach to the fabrication of large qubit arrays. The top-down approach is discussed in detail in section 3.

The key limitations of the top-down approach are related to the ion implantation process itself, namely damage to the Si lattice and the straggle of ^{31}P donors from their desired locations. An alternate pathway which avoids both of these is a *bottom-up* approach, in which STM lithography and epitaxial silicon overgrowth are used to construct devices at the atomic scale, thus producing low-disorder devices with ultra-high-precision placement of ^{31}P donors. Although necessitating the development of new technologies, this approach has already led to significant results, including the demonstrated placement of ordered PH_3 molecules on a Si surface[3]. Despite the technological challenges and longer development time, the bottom-up approach promises the fabrication of large-scale quantum processors with the low crystalline disorder required to ensure reliable operation. The approach and results to date are presented in section 4.

3 Top-Down Approach: Few-Qubit Prototypes

In the Si QC architecture shown in Fig. 1, a single-electron transistor (SET) is used to read out the electron spin state of the two nearest donors. Realisation of the structure in Fig. 1 requires not only the ability to form ordered arrays of ^{31}P atoms, but also the capability to accurately align these atoms to their control gates and readout SETs. Variations on this configuration have been suggested for proposals that utilise both nuclear-spin and electron-spin qubits[4,5]. However the need for donor alignment to surface features remains the same for all architectures. The top-down approach described here achieves all of these requirements.

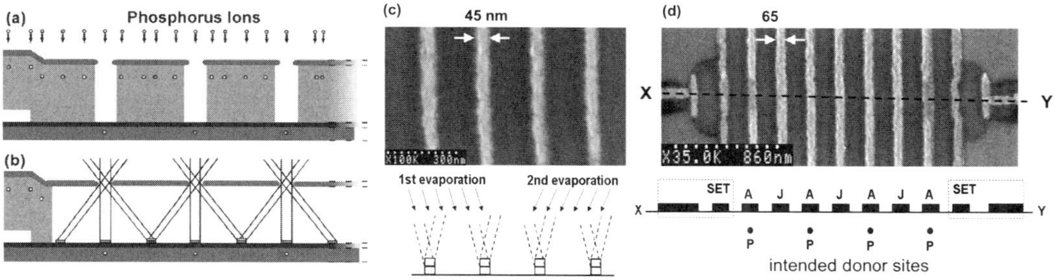

Figure 3. *Top-down* multilayer resist fabrication process and results. (a) Resist profile for ion implantation step. (b) Cavity profile formed after further development of lower resist layer, showing triple-angle metal evaporation. (c) SEM image of Al gates fabricated using double-angle evaporation. The two metallisations sit exactly above one another as shown in the schematic. (d) A- & J-gate and twin-SET pattern for a four donor device produced using multi-angle metallisation.

Central to the top-down approach is the use of multi-layer electron-beam-sensitive resists. Fig. 3 is a schematic cross-section of a tri-layer resist structure used in the process. The trenches in Fig. 3(a) correspond to lines exposed in the resist structure using electron beam lithography (EBL) and then developed using wet development or dry etching techniques. The resulting resist structure acts as a mask for the implantation of low energy (1-20 keV) $^{31}P^{+}$ ions as shown in the figure. The middle layer of resist is chosen to have a higher sensitivity to EBL exposure than that of the resist above it, so that upon further wet development the middle resist is removed leaving a cavity below the self-supporting top layer (see Fig. 3b). The surface control gates (A- and J-gates in the Kane architecture) are then deposited using metal evaporation from above, through the patterned top-layer, from a number of different angles. In Fig. 3(b) three different angles are used for the evaporations. A similar type of shadow evaporation process[6] is commonly used for the fabrication of Al/Al_2O_3 SETs and is therefore ideally suited to our application, since both SETs and Al control gates can be produced using the same process.

The evaporation at normal incidence forms the A-gates, which sit directly above the ^{31}P donors. Self-alignment of the A-gates to the donors is therefore achieved by using the same EBL exposure to define the pattern for the ion implantation and the A-gate. The gates may be quite long and are connected to external circuitry but the ^{31}P donors are constrained to a straight line perpendicular to the gates, which is defined using EBL in a third resist. This perpendicular cross-pattern technique is not evident in Fig. 3 but full details of this and all other aspects of the multilayer resist process can be found in Buehler *et al.*[7] elsewhere in these Proceedings.

Figures 3(c) & (d) shows gate metallisation trials achieved using this processes. Fig. 3(c) shows metal gates fabricated using shadow evaporation from two accurately controlled angles, such that the two lines sit precisely on top of one another, producing a linewidth of only 45nm. Fig. 3(d) shows the complete gate and readout SET metallisation for a four-donor device. This device has four A-gates and three J-gates in between, together with a SET at either end. It is just possible to distinguish the island regions of the SETs, which are about 200nm long and connected to the source and drain regions by thin (1-3 nm) Al_2O_3 tunnel barriers[7]. The darker region in the image corresponds to the area where the cavity existed, showing that it is possible to support such a structure over more than 2 μm. The gate width in this device is 65nm, leading to an inter-donor spacing of 400nm. While such a spacing is sufficient for devices based on electron spin[5], for nuclear spin devices[1,4] the gate widths will need to be reduced by an order of magnitude, which is close to the limits of EBL[8]. Note that the device shown in Fig. 3(d) does not incorporate donors. The inclusion of the ion implantation step in the process is currently underway.

4 Bottom-Up Approach: Scale-Up to Many-Qubit Arrays

The precision of donor arrays produced by the top-down approach will be limited by the straggle inherent to ion implantation. Whilst this should be surmountable for few-qubit devices, recent calculations[9] indicate that deviations from regularity of the ion array as small as a few nm could impair the operation of a large-scale Si QC. A solution to this problem is a *bottom-up* fabrication approach, which utilises recent advances in scanning probe technology. This approach employs a strategy[3] outlined in Fig. 4, which utilises a resist composed of a layer of hydrogen atoms that terminate a silicon surface. An STM tip is used to selectively desorb individual H atoms, exposing the underlying silicon surface in the required array. Such STM induced hydrogen desorption[10] previously has been proposed[11] for the assembly of atomically-ordered device structures.

The first step in fabricating a P qubit array is to prepare a clean, defect free Si surface. Fig. 4(b) shows an optimally-prepared[12] Si(001)2×1 surface consisting of rows of σ-bonded Si-Si dimers. Fig. 4(c) shows an image of a low defect density Si(001)2×1 surface after exposure to atomic H at 600 K. A near uniform coverage of the mono-hydride phase has occurred, where one H atom bonds to each Si atom. Sites of di-hydride, where two H atoms bond to each Si, and tri-hydride, a mixture of the other two phases, are also indicated. We have found that all three phases act as an effective resist during subsequent exposure to phosphine, thus the resist barrier depicted in Fig. 4(a) has been realised.

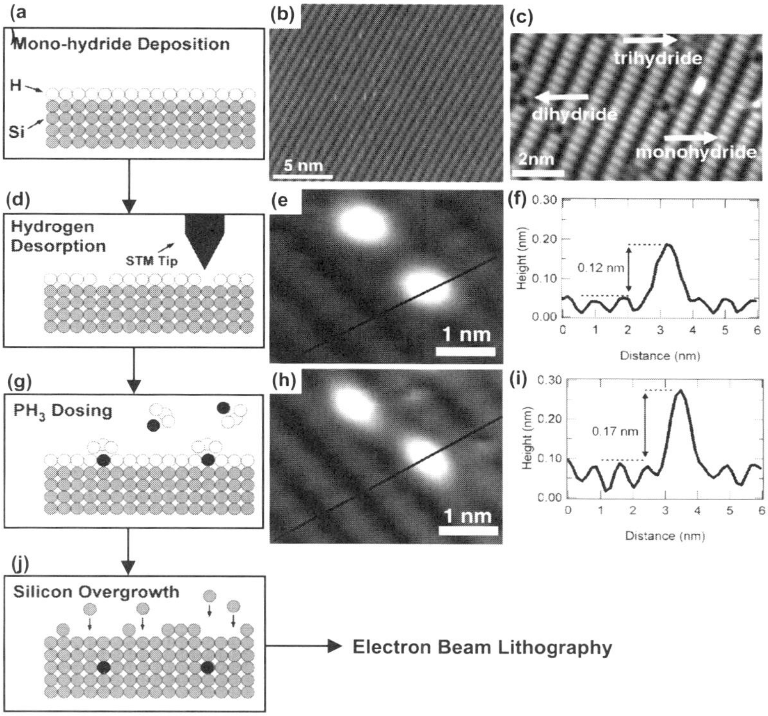

Figure 4. *Bottom-up* fabrication process [taken from Ref. 3]. (a) A Si(100)2×1surface is passivated with a monolayer of hydrogen. (b) Filled state STM image of a clean, very low defect density Si(100)2×1surface. (c) Fully hydrogen terminated Si(100)2×:H surface. (d) A voltage applied to an STM tip is used to desorb hydrogen at a single Si site. (e) STM image of two desorption sites, where each bright region represents the protusion through the monohydride resist of a single Si dangling bond. (f) STM height profile taken along the line marked in (e). (g) Exposure of the STM-patterned surface to PH_3 gas leads to adsorption at each exposed site. (h) STM image of the same two desorption sites in (e) following exposure to PH_3. The bright regions now correspond to PH_x molecules bonded to the exposed Si. (i) STM height profile taken along the line marked in (h), showing a characteristic 0.05 nm increase compared with the profile in (f), due to the presence of the PH_x. (j) Low temperature silicon overgrowth encapsulates the phosphorus.

Following STM lithography it is necessary to adsorb the individual phosphorus atoms or qubits using phosphine gas, as depicted schematically in Fig. 4(g). This process is aided by the strong sticking coefficient (very close to unity) of PH_3 on a clean silicon surface[13]. Figure 4(h) is an experimental demonstration of this step, showing the two desorption sites of Fig. 4(e) immediately after exposure to PH_3. The effectiveness of the H resist as a barrier to PH_3 adsorption is demonstrated by the uniform H coverage after PH_3 dosing except at the previously desorbed sites. Comparison of the line profiles in Figs. 4(f) and 4(i) shows a characteristic increase of ~0.05 nm in the height of the protrusion after phosphine dosing. This increase, which was seen reproducibly at a number of sites, confirms the adsorption of a PH_3 molecule and corresponds to the difference between the exposed silicon dangling bond and the adsorbed phosphine.

The final step, currently underway, is the encapsulation of the phosphorus qubits in a crystalline lattice of silicon-28, as depicted in Fig. 4(j). The main difficulty in this step is to ensure the P atoms incorporate into the Si crystal and remain in an ordered array. The most direct route to achieving this is to desorb the H resist from the surface by heating to ~700 K followed by epitaxial Si growth over the P array. A possible concern with this level of heating is the potential to induce lateral surface diffusion of the P atoms in the array. However it is known that the phosphorus atom in the PH_2 molecule, with a single bond to the silicon, incorporates into the silicon surface over the lower temperature range of 550-650 K[13]. Following this 1-3 nm of epitaxial silicon can be grown at low temperatures[14] down to 300 K to encapsulate the phosphorus, followed by elevated temperature growth at ~500 K to maintain crystallinity[14].

5 Quantum State Read-Out

To read out the nuclear spin state (|0> or |1>) of a target qubit (^{31}P donor) in the Kane architecture it is necessary to employ an additional ^{31}P donor which must be initialised to a known nuclear spin state (say |0>). At the low temperatures (T<1 K) and moderate magnetic fields (B~2 T) required in the Kane design, all donor electron spins will initially be in the |↓> state[15] so that the two-donor electron system will be in a |↓↓> spin triplet. By appropriate settings on the A- and J-gates it is possible to adiabatically transform the |↓↓> state into the |↑↓ – ↓↑> singlet state if, and only if, the target qubit was initially in the |0> state[1]. Qubit read-out then reduces to the process of differentiating between the two-electron |↓↓> and |↑↓ – ↓↑> states.

The difficult problem of single electron-*spin* detection can be reduced to the much more tractable problem of single electron-*charge* detection by using the architecture shown in Fig. 1 for many qubits, or Fig. 5(a) for a single qubit (two donor) system. By asymmetric biasing of the gates and/or SETs above the donors an electric field can be created which will encourage electron transfer from one donor to another (Fig. 5a). The change in local electric potential which results from such a single electron transfer can be detected with a single electron transistor (SET) device, which acts as an ultra-sensitive electrometer[16]. Since only the |↑↓ – ↓↑> singlet state forms a stable two-electron bound state on a P donor in Si (referred to as a D^- state), the detection (or converse) of this state constitutes qubit read-out[1]. Although this discussion has focused on the Kane nuclear spin architecture[1], the method would apply equally well to electron spin qubits[5].

To *simulate* two-donor read-out we have designed and fabricated an all metal twin-SET/double-dot system[17] (Figs. 5b & 5c). In this design the two P atoms are replaced by two metal islands (double-dot) connected by a tunnel junction. A similar type of construction has been employed in the read-out mechanisms of quantum cellular automata[18]. Applying a differential voltage $+V_A$ and $-V_A$ to the two control gates A1 and A2 causes single electrons to tunnel from one metal island to the other, just as electrons are transferred from one P donor to the other in Fig. 5(a). When an electron tunnels from the left dot to the right dot, SET-1 senses the departure of an electron, while SET-2 simultaneously senses the arrival of an electron. The charges $Q_1 = -Q_2$ on the two dots thus change in discrete steps as V_A increases (Fig. 5e). This is most clearly seen in the transconductance $\partial G/\partial V_A$ which shows peaks corresponding to single-electron transfers. Spurious signal rejection is accomplished by using both SETs as shown in Fig. 5(f). Charge transfer in the double-dot is detected by simultaneous and opposite peaks on the twin-SET signals. Noise due to charge movements elsewhere in the device can be either a peak in just one SET output (I) or simultaneous but equidirectional peaks in both SETs (II), dependent on the position of the charge trap. Correlating the outputs of both SETs thus allows for a significant reduction in unwanted charge noise, which could otherwise lead to read-out errors in a QC.

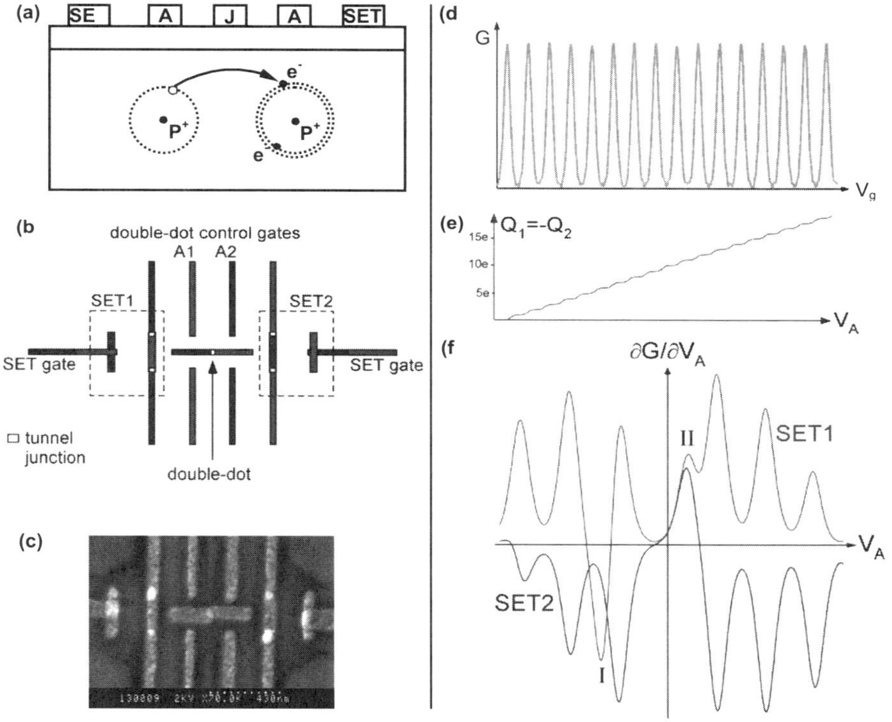

Figure 5. SSQC read-out process and twin-SET simulation device. (a) A bound two electron state on a P donor following single electron transfer will only occur if the electrons form a spin-singlet state. The success of this transfer is monitored by two SETs on the surface. (b) Schematic and (c) SEM image of a twin-SET device, fabricated by double-angle evaporation of Al. (d) Coulomb blockade oscillations obtained on a typical SET device. (e) Charge $Q_1 = -Q_2$ on the two dots at the centre of the device as a function of control gate voltage V_A. (f) Calculated signal outputs of both SETs. Peaks represent controlled single electron transfer events. I and II indicate random charge noise.

6 Conclusions

We have presented two complementary fabrication strategies, and initial experimental results, for construction of a Si-based solid state quantum computer, together with devices to simulate qubit read-out in such architectures. The *top-down* process utilises industrial production techniques such as ion-implantation and electron-beam lithography and initial results demonstrate the fabrication of surface gate and SET arrays for multi-qubit devices. The next step, currently underway, is controlled single-ion implantation. The *bottom-up* process, while less suited to high-throughput production, has the potential of creating highly regular large qubit arrays. Initial results demonstrate the accurate placement of phosphorus-bearing PH_3 molecules on a patterned Si(001)2×1:H surface, making such arrays a possibility. The next stage entails encapsulation by a Si overlayer. The demonstration of prototype devices will require sensitive SET electrical read-out, via a spin-dependent single-electron tunneling process. We have designed an all-metal twin-SET device to simulate this read-out process and have fabricated test devices ready for measurement. There remain significant challenges in both the nanofabrication and quantum electrical measurement of Si-based QC architectures, however, the initial results presented here offer considerable hope that large-scale solid state quantum computing may one day be realised.

Acknowledgements

This work was funded by the Australian Research Council, and the US National Security Agency through the Army Research Office.

References

1. B.E. Kane, *Nature* **393**, 133 (1998).
2. T.M. Buehler, R.P. McKinnon, N.E. Lumpkin, R. Brenner, D.J. Reilly, L.D. Macks, A.S. Dzurak and R.G. Clark, submitted to *Appl. Phys. Lett.* (2001).
3. J.L. O'Brien, S.R. Schofield, M.Y. Simmons, R.G. Clark, A.S. Dzurak, N.J. Curson, B.E. Kane, N.S. McAlpine, M.E. Hawley and G.W. Brown, to appear in *Phys. Rev. B* (2001).
4. B.E. Kane, *Fortschr. Phys.* **48**, 1023 (2000); LANL preprint *quant-ph*/0003031.
5. R. Vrijen, E. Yablonovitch, K. Wang, H.W. Jiang, A. Balandin, V. Roychowdhury, T. Mor and D. DiVincenzo, *Phys. Rev. A* **62**, 012306 (2000).
6. T. Fulton and G. Dolan, Phys. Rev. Lett. 59, 109 (1987).
7. T.M. Buehler, R.P. McKinnon, N.E. Lumpkin, R. Brenner, A.S. Dzurak, R.G. Clark, D. Jamieson and J. McCallum, *in these Proceedings*.
8. W. Chen and H. Ahmed, *Appl. Phys. Lett.* **62**, 1499 (1993).
9. B. Koiller, Xuedong Hu and S. Das Sarma, LANL preprint *cond-mat*/0106259.
10. C. Thirstrup, M. Sakurai, T. Nakayama and M. Aono, *Surf. Sci.* **411**, 203 (1998).
11. J.R. Tucker and T.-C. Shen, *Solid State Electronics* **42**, 1061 (1998).
12. S.R. Schofield, J.L. O'Brien, M.Y. Simmons, R.G. Clark, A.S. Dzurak, B.E. Kane, M.E. Hawley, G.W. Brown and N.J. Curson, *in these Proceedings*.
13. D.S. Lin, T.S. Ku and R.P. Chen, *Phys. Rev. B* **61**, 2799 (2000).
14. D.J. Eaglesham, H.-J. Gossman and M. Cerullo, *Phys. Rev. Lett.* **65**, 1227 (1990).
15. We adopt the convention of Ref. [1] in using $|0>$ and $|1>$ to represent nuclear spin states and $|\downarrow>$ and $|\uparrow>$ to represent electron spin states.
16. *Single Charge Tunneling*, edited by H. Grabert and M. Devoret (Plenum, New York, 1992).
17. R. Brenner, T.M. Buehler, R.P. McKinnon, D.J. Reilly, A.R. Hamilton, A.S. Dzurak, R.G. Clark, N.E. Lumpkin and G.J. Milburn, *in these Proceedings*.
18. I. Amlani, A.O. Orlov, R.K. Kummamuru, G.H. Bernstein, C.S. Lent and G.L. Snider, *Appl. Phys. Lett.* **77**, 738 (2000).

FABRICATING AN ALL-EPITAXIAL SILICON QUANTUM COMPUTER

J. R. TUCKER

Department of Electrical and Computer ENGINEERING, University of Illinois, 1406 West Green Street, Urbana, IL 61801, USA

T.-C. SHEN

Department of Physics, Utah State University, Logan, UT 84322, USA

Scalable silicon quantum computers will require a material perfection that has never been attempted. Ground state wavefunctions for conduction electrons orbiting individual phosphorous donors must be polarized electronically and coupled to nearest neighbors with great precision. Elimination of all randomizing influences can be achieved only with a fully epitaxial structure; and we believe that output circuitry must also be integrated into the qubit arrays in order to achieve the uniformity needed for large-scale integration. A process that could potentially accomplish this will be outlined, based on scanning tunneling microscope (STM) removal of individual hydrogen atoms from the H-terminated silicon surface followed by phosphine dosing and ultra-low-temperature overgrowth. Self-ordering of PH_3 molecules onto extended areas of bare silicon should permit patterning of planar single-electron transistors along with P-donor qubits in the same lithographic step. Initial plans for an experiment to characterize exchange coupling under gate control will be described.

1 Introduction

Silicon-based quantum computers[1,2] are made possible by the large Bohr diameter for bound electron wavefunctions on individual P-atom donors, and the astonishing perfection of single-crystal silicon. Whether quantum logic of this type can be implemented in practice will be the focus of intense research over the next several years. Those efforts will also extend conventional silicon technology toward single electrons and individual spins, with many unforeseen possibilities in addition to a scalable quantum computer.

This larger framework is familiar to us, because much of our research over the past ten years has been devoted to developing scanning tunneling microscope (STM) patterning techniques for atom-scale electronic devices. Prior work had laid a good foundation by characterizing the properties of hydrogen-terminated silicon surfaces at atomic resolution[3], and by demonstrating desorption of hydrogen at nanometer dimensions using the STM's low-energy electron beam.[4] We and our colleagues saw an opportunity to employ this single-monolayer 'Hydrogen resist' for atom-resolved patterning on silicon, and set about developing the required techniques. Atomic resolution was first demonstrated by desorbing individual hydrogen atoms to expose bare silicon dimers on a Si(100)-2x1:H surface, as shown in Fig.1, using a tip voltage well below the ~4.5eV tungsten workfunction.[5] Systematic studies were then performed to characterize STM electron-stimulated desorption over the entire experimental range from ~1 to 12V, leading to a detailed theoretical interpretation of the desorption mechanisms in both tunneling and field emission regimes and the discovery of a large isotope effect for deuterium vs. hydrogen.[6] Atomic resolution on 'Hydrogen resist' can only be obtained in the tunneling regime below ~4V, where wavefunction coupling between particular tip and sample states limits the electron beam diameter. Voltage pulses can also be applied in tunneling mode to eliminate hydrogen atoms from an individual silicon dimer at any preselected location, creating two bare silicon dangling bonds and a reaction site small enough to accommodate only one adsorbate atom or molecule.

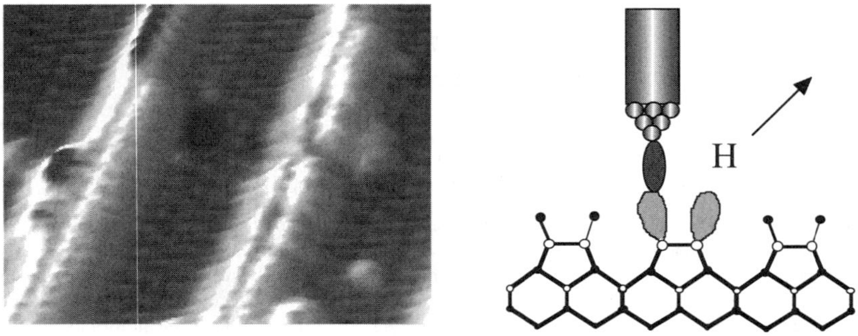

Figure 1. Lines of bare Si dangling bonds written 30Å apart along the dimer rows of a Si(100)-2x1 surface by STM-induced desorption of hydrogen. Sharp nodal lines of the bare Si-Si π-antibonding state are clearly seen, and hydrogen atoms are removed only from the left side of each dimer near the top of this image.

Our initial attempts at fabrication demonstrated selective oxidation of ~4nm-wide lines[7], and utilized these 'native oxide' patterns as masks in experiments on aluminum chemical vapor deposition (CVD).[8] Uniform CVD nucleation proved very difficult to realize at these dimensions, but we had some success in defining atom-scale metal patterns by direct evaporation of aluminum at ~0.1ML onto H-terminated surfaces containing ~2nm-wide STM-exposed lines of bare silicon at room temperature.[9] These experiments, and subsequent efforts to fabricate a 'single-atom wire' using this technique[10], indicated a need for some type of self-ordering to eliminate gaps and clusters on the atomic scale; and it occurred to us that common dopant precursor molecules such as PH_3 might be ideal for this purpose. By coincidence, our proposal for fabricating electronic structures in silicon based on overlap of individual P-donor wavefunctions[11] was published simultaneously with Kane's proposal for a silicon-based quantum computer.

2 Patterning Atom-Scale Circuits

Atom-resolved studies of phosphine adsorption onto the bare Si(100)-2x1 surface[12] show very strong self-ordering of individual PH_3 molecules on alternating dimers within a single row at room temperature. The adsorption sites are staggered between adjacent rows to form a ~1/4ML c(4x2) structure (with ~5nm-size domains) at saturation coverage. On the other hand, phosphine does not adsorb onto H-terminated surfaces at room temperature; and substrates must be heated above ~500°C to thermally evolve hydrogen and expose bare Si dangling bonds in order to sustain common CVD reactions involving PH_3, SiH_4, AsH_3, B_2H_6, and related molecular species. In a future industrial process, conventional e-beam sources might be employed *in situ* to selectively pattern the 'Hydrogen resist' and expose a 2D circuit pattern for selective deposition of self-ordered ~1/4ML P-donor arrays, with STM used only in those areas requiring atomic resolution or attachment of individual precursor molecules. Figure 2 illustrates one such pattern that might be employed in fabricating a planar single-electron transistor (SET).

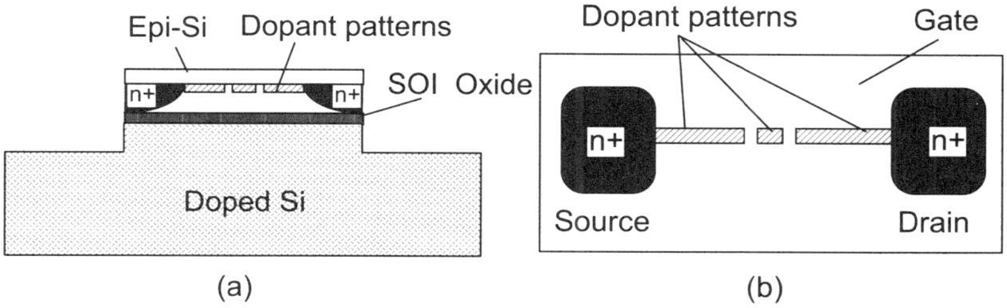

Figure 2. (a) Side and (b) top view schematic for a SET on SOI substrate utilizing the 2D dopant patterns fabricated *in situ* by STM. The mesa and source/drain regions are prefabricated *ex situ* by conventional processes. The heavily-doped Si substrate will be used together with a top-gate (not shown) to control wavefunction overlap across the lateral tunnel junctions.

The critical step in transforming these patterns into useful electronic devices will be to overgrow and activate all of the selectively deposited donor atoms inside the silicon crystal while limiting redistribution to the atomic scale. The planar density of a ~1/4ML c(4x2) donor sheet is $\sim 1.7\times10^{14}cm^{-2}$, a factor of ~100 greater than in conventional δ-doped layers. Prospects for successful overgrowth appear excellent, however, in view of recent progress in low-temperature silicon molecular beam epitaxy (LT-SiMBE). Self-ordered 2x1 boron adlayers have been overgrown at temperatures as low as ~350K, and full activation is seen in Hall effect up to ~1/2ML coverage with only ~4Å spread in the growth direction.[13] More recent experiments with Sb donors[14] have yielded similar results. Surface hydrogen up to 1/2ML is found to have no discernible effect on LT-SiMBE[15], so it appears that these techniques can be successfully employed in overgrowing the patterned PH_3 precursor layers described here.

The overgrown ~1/4ML P-donor patterns are intended to function as 'metallic' electrodes in the planar SET structure sketched in Fig.2. Nearest-neighbor distances in a c(4x2) precursor adlayer are only 7.7Å, much smaller than the ~50Å average Bohr diameter for isolated donor bound states. Strong wavefunction overlap should thus yield high conductivity within the dopant plane. At low temperatures, the conduction electrons will be tightly confined within ~10Å in the vertical (growth) direction by the large electric fields of a ~1/4ML donor sheet. The resulting 2-dimensional electron gas (2DEG) is expected to populate only the lowest subbands for normal effective masses $m_\perp^*=0.916m_o$ and $m_t^*=0.190m_o$, respectively. However, the characteristic distance for lateral wavefunction spreading at the edges of these highly-conducting 2D patterns will still be set by the Bohr radius, so lateral tunnel junctions should correspond to ~10nm-wide gaps whose impedance levels can be accurately controlled. Through use of silicon-on insulator (SOI) device templates with pre-implanted contact regions, we hope to produce seamless atom-scale integrated circuits of overlapping donor wavefunctions with no barriers to the outside world.

Full activation of patterned 2D donor sheets offers a unique opportunity to eliminate the random offset charges that plague all single-electron devices today. The large 1/f noise of metal-oxide tunnel barriers should also be greatly reduced, yielding much greater sensitivity in detecting sub-electron charge displacements with SET electrometers. If realized, these extraordinary electrical properties could provide ideal output circuitry for a silicon-based quantum computer, imposing the smallest possible perturbation on individual donor qubits.

3 Controlling The Exchange Interaction

Perhaps the most critical step toward building a silicon quantum computer will be to demonstrate electronic control over the exchange interaction between isolated donor pairs, and characterize the quantum states of the resulting two-electron system. Figure 3 illustrates our conceptual approach to this problem. In addition to the SET circuit sketched in Fig. 2, STM pulses will also be used to create attachment sites for two individual PH_3 precursor molecules near the Coulomb island. A second set of planar electrodes will be patterned (contacts not shown) to apply a small ac electric field along the inter-nuclear axis, polarizing the spin singlet state of the coupled donor-pair. The magnitude of this in-plane electric field, E_x^{ac}~1kV/cm, will be limited by the relatively small binding energy of the 2D donor sheets, and we anticipate using an ac bias of ~20mV across a ~200nm-wide gap in order to polarize the coupled donor-pair.

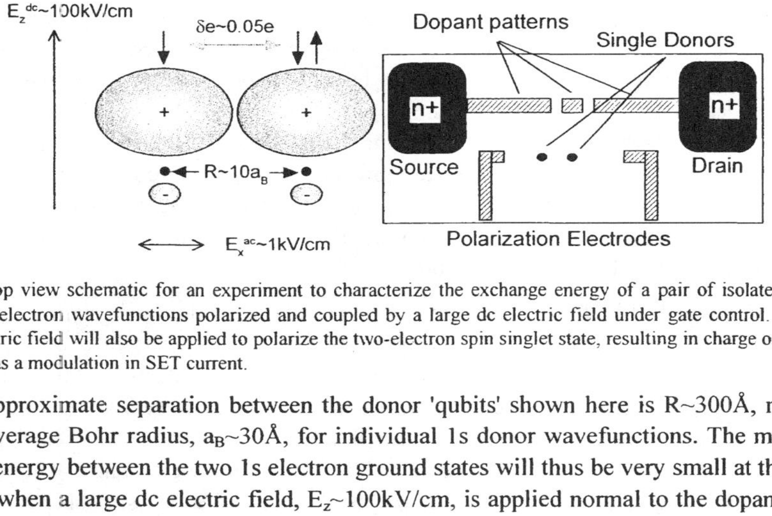

Figure 3. Top view schematic for an experiment to characterize the exchange energy of a pair of isolated donors with ground state electron wavefunctions polarized and coupled by a large dc electric field under gate control. A smaller in-plane ac electric field will also be applied to polarize the two-electron spin singlet state, resulting in charge oscillations that are detected as a modulation in SET current.

The approximate separation between the donor 'qubits' shown here is R~300Å, much larger than the average Bohr radius, a_B~30Å, for individual 1s donor wavefunctions. The magnitude of exchange energy between the two 1s electron ground states will thus be very small at this distance. However, when a large dc electric field, E_z~100kV/cm, is applied normal to the dopant plane, the individual ground states will become highly polarized with a significant $2p_z$ component. For the sample configuration shown in Fig.2, this dc electric field would be established between the heavily doped SOI substrate and a common metal top-gate (not shown). Since the characteristic decay length for $2p_z$ states is $2a_B$, twice as large as for 1s states, it should be possible to establish gate control over wavefunction overlap and exchange energy in a convenient experimental range, with J~0.2meV for the parameters described here. This value of exchange energy corresponds to J/kT~40 at a measurement temperature T~50mK, and the singlet to triplet transition would be expected at an applied magnetic field B_z~2T.

The basic output measurement, as outlined in Kane's proposal, is a determination of the two-electron quantum state for coupled donor-pairs via electronic polarization. For anti-parallel spins in the singlet state, a transverse electric field will push both electrons toward a doubly-occupied D^- configuration on one of the donor sites. Charge transfer will not occur in the triplet state, however, since double occupancy is statistically forbidden for electrons having the same spin. A nearby SET electrometer can potentially detect the presence or absence of this sub-electron charge transfer, through the polarization induced onto its Coulomb island.

According to a simplified Hubbard model, an ac electric field E_x^{ac}~1kV/cm can be expected to set up a charge oscillation δe~0.05e between the two donor sites for the experimental parameters described here. Because these two donors are placed asymmetrically with respect to the SET in Fig.3, a somewhat smaller polarization charge would be induced onto the Coulomb island, resulting in modulation of SET current at the applied frequency. Conventional SET electrometers have sensitivities in the range 10^{-3} to 10^{-5}e/√Hz, limited by large 1/f noise of Al-oxide tunnel junctions. Epitaxial SETs with silicon tunnel barriers are expected to have greatly reduced 1/f noise, so detecting charges of order 10^{-2}e should not be a problem. By employing lock-in and derivative techniques as the magnetic field is increased slowly, we expect to observe the narrow singlet-triplet transition of a single donor-pair in the presence of much larger background components of ac polarization. The value of magnetic field at the transition provides a direct measure of exchange energy; and the entire phase space for coupled donor-pairs could be mapped out as a function of gate voltage if this approach is successful.

4 Conclusion

Controlling the wavefunction coupling between individual P-atom donors in silicon may hold the key to large-scale quantum computation, but it is a very difficult undertaking. Our approach emphasizes the benefits of a fully integrated, all-epitaxial structure for several reasons. First, the SET output circuitry needed to detect the results of quantum logic can be incorporated along with donor qubits in the same STM patterning process. Major difficulties in registration can thereby be avoided. More importantly, growing the entire structure into single-crystal silicon can greatly reduce the density of defects and random polarization charges that might otherwise deform electron wavefunctions on the individual donor 'qubits', precluding high levels of integration. SiGe heterolayers can eventually be incorporated into the overgrowth for ESR tuning structures and epitaxial gate dielectrics. Given the already advanced state of silicon materials technology, there is room for serious optimism.

Acknowledgements

This work was supported in part by the National Security Agency (NSA) and Advanced Research and Development Activity (ARDA) under Army Research Office (ARO) contract number DAAD19-00-1-0407 and by the U.S. National Science Foundation under grant number DMR-9875129.

References

1. B.E. Kane, "A silicon-based nuclear spin quantum computer", Nature **393**, 133 (1998).
2. R. Vrijen, E. Yablonovitch, K. Wang, H.W. Jiang, A. Balandin, V. Roychowdhury, T. Mor, and D. DiVincenzo, "Electron spin resonance transistors for quantum computing in silicon-germanium hetero-structures" Phys. Rev. **A62**, 12306 (2000).
3. For a review, see J. J. Boland, "Scanning tuneling microscopy of the interaction of hydrogen with silicon surfaces", Adv. Phys. **42**, 129 (1993).
4. R. S. Becker, G. S. Higashi, Y. J. Chabal, and A. J. Becker, "Atomic scale conversion of clean Si(111):H-1x1 to Si(111)-2x1 by electron-stimulated desorption", Phys. Rev. Lett. **65**, 1917 (1990).
5. T.-C. Shen, C. Wang, G. C. Abeln, J. R. Tucker, J. W. Lyding, Ph. Avouris, R. E. Walkup, "Atomic scale desorption through electronic and vibrational excitation mechanisms", *Science* **268**, 1590 (1995).
6. T.-C. Shen and P. Avouris, "Electron stimulated desorption induced by the scanning tunneling microscope", Surf. Sci. **390**, 35 (1997).
7. T.-C. Shen, C. Wang, J. W. Lyding and J. R. Tucker, "Nanoscale oxide patterns on Si (100) surfaces", Appl. Phys. Lett. **66**, 976 (1995).
8. T.-C. Shen, C. Wang, and J. R. Tucker, "The initial stage of nucleation and growth of Al on H/Si(100)-1x1 by dimethylaluminum hydride vapor deposition" Appl Surf. Sci. **141**, 228 (1999).
9. T.-C. Shen, C. Wang, and J. R. Tucker, "Al nucleation on monohydride and bare Si(001) surfaces: atomic scale patterning", Phys. Rev. Lett. **78**, 1271 (1997).
10. Hashizume, T., et al. Interaction of Ga adsorbates with dagling bonds on the hydrogen terminated Si(100) surface. Japan J. Applied Physics 1996; 35: L-1085-1088.
11. J. R. Tucker and T.-C. Shen, "Prospects for atomically ordered device structures based on STM lithography", Solid State Electronics **42**, 1061 (1998).
12. Y. Wang, M. J. Bronikowski, and R. J. Hamers, "An atomically resolved STM study of the interaction of phosphine with the silicon (001) surface", J. Phys. Chem. **98**, 5966 (1994).
13. B. E. Weir, L. C. Feldman, D. Monroe, H.-J. Gossmann, R. L. Headrick, and T. R. Hart, "Electrical characterization of an ultrahigh concentration boron delta-doping layer", Appl. Phys. Lett. **65,** 737 (1994).
14. P.H. Citrin, D.A. Muller, H.-J. Gossmann, R. Vanfleet, and P.A. Northrup, "Geometric frustration of 2D dopants in silicon: surpassing electrical saturation", Phys. Rev. Lett. **83**, 3234 (1999).
15. M. Copel and R. M. Tromp, "H coverage dependence of Si(100) homoepitaxy", Phys. Rev. Lett. **72**, 1236 (1994).

PHOSPHORUS QUBIT ARRAYS IN SILICON

J. L. O'BRIEN,[1,2*] S. R. SCHOFIELD,[1,2] M. Y. SIMMONS,[1,2] R. G. CLARK,[1,2] A. S. DZURAK,[1,3] N. J. CURSON,[1,2] B. E. KANE,[4] N. S. MCALPINE,[2†] M. E. HAWLEY[5] AND G. W. BROWN[5]

[1] *Centre for Quantum Computer Technology, University of New South Wales, Sydney 2052, Australia*
[2] *School of Physics, University of New South Wales, Sydney 2052, Australia*
[3] *School of Electrical Engineering & Telecommunication, University of New South Wales, Sydney 2052, Australia*
[4] *Laboratory for Physical Sciences, University of Maryland, College Park, MD 20740, USA*
[5] *Los Alamos National Laboratory, Los Alamos, NM 87545, USA*

We outline a detailed strategy for the construction of a solid state phosphorus in silicon quantum computer and demonstrate the first step towards this goal - the fabrication of atomically-precise arrays of P atoms on a silicon surface. Using a monolayer hydrogen resist to passivate a silicon surface we apply pulsed voltages to a scanning tunneling microscope tip to selectively desorb individual hydrogen atoms atoms with atomic resolution. Exposure of this surface to the phosphorus precursor phosphine results in precise placement of single phosphorus atoms on the surface. The fabrication of an atomically precise array of phosphorus atoms in silicon represents a significant advance towards the construction of a silicon quantum computer

1 Introduction

Solid state architectures for the realization of a quantum computer are attractive for their potential scalability. In particular, a proposal by Kane[1] to use the nuclear spin of phosphorus atoms (nuclear spin $I=\frac{1}{2}$) as qubits embedded in isotopically-pure silicon-28 ($I=0$) holds the promise of well isolated qubits and easy integration with existing silicon technology. Fabrication of such devices, however, requires atomic scale manipulation which is an immense technological challenge, particularly with semiconductor surfaces.

The architecture proposed by Kane is shown in Fig. 1, where the phosphorus qubits are 20 nm apart. At low temperatures the donor electron remains bound to the phosphorus nucleus and surface A gates control the hyperfine interaction between nuclear and electron spins, enabling polarization of the two. Coupling between adjacent donor electrons is achieved using separate surface J gates, enabling an electron mediated interaction between qubits. The fabrication of sensitive aluminum-aluminum/oxide single electron transistor detection systems, specifically for the spin read-out of this quantum computer, is detailed elsewhere.[2,3]

The ability to form a precise array of P atoms in a silicon matrix lies at the heart of the Kane strategy for a silicon quantum computer. In section 2 we detail a hydrogen resist/scanning tunneling microscope (STM) strategy for fabricating such a structure and in section 3 we demonstrate the atomically precise placement of phosphine molecules on the silicon surface using this technique. Finally, in section 4, we discuss the implications of these results for silicon based quantum computation.

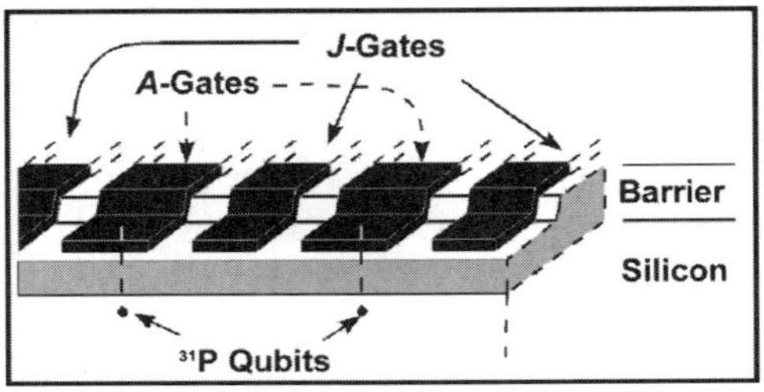

Figure 1. A schematic of the Kane quantum computer architecture taken from Ref. 1, showing two P atoms incorporated into isotopically pure ^{28}Si with surface metal gates separated by an electrical barrier. The P qubits are $\sim$ 20 nm apart to allow adjacent electron wavefunctions to overlap.

2 Fabrication Strategy

Atomic scale manipulations of metal atoms on metal surfaces using an STM tip to move individual atoms has been achieved previously,[4] however, atomic scale rearrangement of individual atoms in a semiconductor system is problematic due to the strong covalent bonds involved. In Fig. 2 we outline a strategy for fabricating the phosphorus in silicon device shown in Fig. 1 using an STM and a H-resist technique developed specifically for the construction of the Kane quantum computer.

A clean low defect density Si(001)2x1 surface is prepared by annealing to just below the melting point of silicon ($\sim$1500 K). This surface is then exposed to high purity atomic H formed by passing molecular H_2 gas over a tungsten filament heated to $\sim$1900 K. Careful control of the sample temperature during this procedure ensures that a monohydride phase is formed with a single H atom bonded to each silicon atom. Full details of this process can be found elsewhere.[5]

An STM tip is used to selectively pattern the H terminated surface by application of controlled voltage pulses between the tip and sample to remove H atoms, thus exposing the underlying silicon surface.[6] The remaining H then acts as a resist layer during subsequent exposure to phosphine (PH_3) gas so that PH_3 molecules only adsorb at the sites where silicon is exposed, much like a resist in conventional lithography. By forming sufficiently small desorption sites, it is possible to achieve the adsorption of single PH_3 molecules required for the Kane architecture and this is demonstrated in section 3.

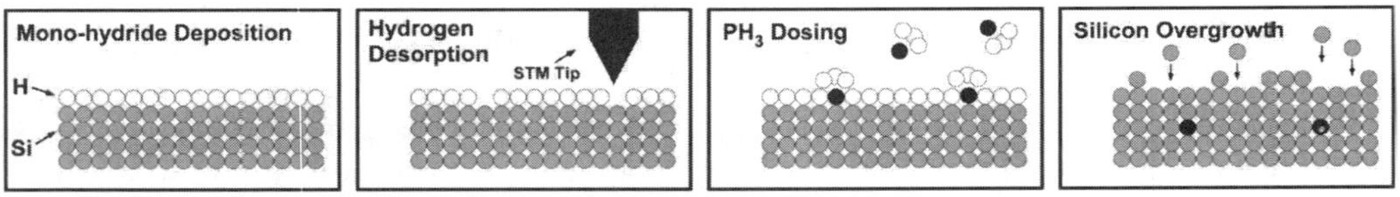

Figure 2. Schematic of the process to fabricate the structure shown in Fig. 1. An STM tip is used to selectively desorb H with atomic resoultion from a H terminated surface, allowing phosphine (PH_3) to adsorb only at these sites. Finally the array of P atoms is encapsulated by epitaxial silicon growth.

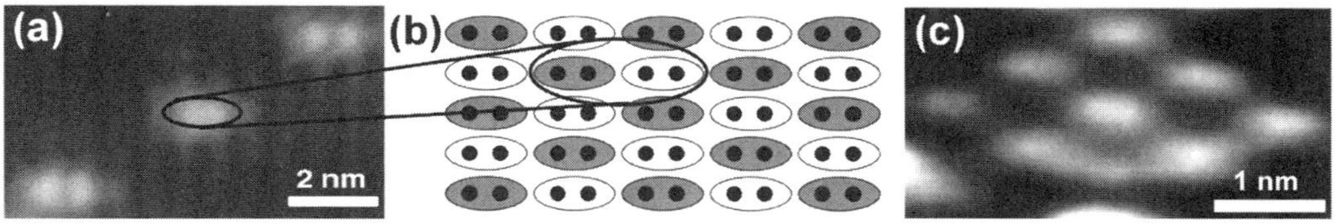

Figure 3. (a) STM image of three hydrogen desorption sites on a Si(001)2x1 hydrogen terminated surface. (b) Schematic of Si(001)2x1 surface where the shaded dimers indicate the c(4x2) bonding structure of PH_3. (c) STM image of PH_3 bonded to a clean silicon surface in a c(4x2)structure.

3 Formation of phosphorus arrays

Fig. 3a shows an STM image of three H desorption sites <1 nm in diameter with a pitch of ~4 nm on a Si(001)2x1:H surface. In this image rows of dimers formed by the pairing of adjacent silicon atoms can be seen running from top to bottom. The sites where H has been desorbed from this surface appear as bright protrusions since the electron density at these sites extends out of the surface due to the exposed silicon dangling bonds.

To calculate the required desorption site size in order that only one PH_3 molecule bonds at that site, we consider the bonding structure of PH_3 on a clean silicon surface. Fig. 3c shows an image of the clean Si(001)2x1 surface after exposure to a saturation dose of PH_3 gas, where the sticking coefficient of PH_3 on the silicon surface is $S = 1$ (Ref. 7). The PH_3 bonds to the surface with a c(4x2) surface periodicity with a PH_3 molecule bonding to every other dimer,[8] shown schematically in Fig. 3b. Here each black dot represents a silicon atom, each of which forms a dimer with it neighbor (indicated by surrounding ellipses), and the shaded dimers are those where a PH_3 molecule bonds. Relating this structure back to

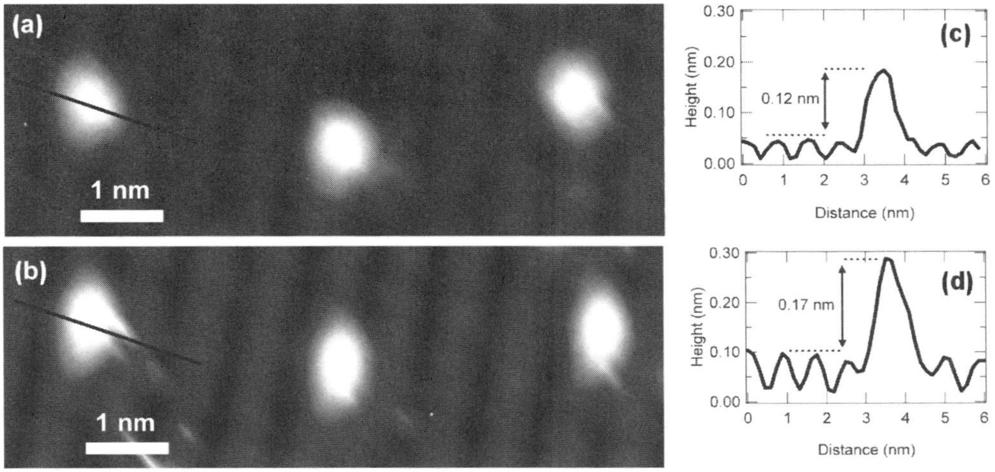

Figure 4. STM image of three desorption sites before (a) and after (b) PH_3 dosing. Corresponding line profiles, as indicated by the black lines in the images, are shown in (c) and (d).

Fig. 3a we see that the STM induced desorption sites are sufficiently small to allow only one PH_3 molecule to bond at each of these sites.

Fig. 4a shows a higher resolution image of three desorption sites resulting from the removal of single H atoms (as determined from the underlying atomic structure). Fig. 4b shows the same three sites after exposure to high purity PH_3 gas. At these atomic sites, and all others studied, a characteristic elongation of the site is observed after exposure to PH_3. Analysis of line profiles through the left hand site as indicated in the images and shown in Figs. 4c and 4d, reveal a height increase of ~ 0.05 nm in the height of the protrusion after PH_3 dosing. This increase, observed at all adsorption sites over several images, has been calibrated against an atomic step edge on the same surface before and after PH_3 dosing. We conclude that these changes are the result of a single PH_3 molecule bonding at each site. A more detailed discussion of the surface bonding at these adsorption sites is described in Ref. 9.

4 Conclusions

We have outlined a detailed strategy for the fabrication of an atomic array of phosphorus atoms in silicon for the quantum computer proposal by Kane.[1] A hydrogen resist technique has been used to directly form an atomic scale pattern on the Si(001)2x1:H surface using an STM tip. This patterning of a H resist can be exploited to promote the adsorption of *single* PH_3 molecules at predefined locations on the silicon surface. A detailed program to encapsulate the P arrays by epitaxial silicon overgrowth is underway to complete the qubit architecture. These results represent a significant step towards the realization of a silicon quantum computer and are directly relevant to other silicon quantum computer architectures[10].

Acknowledgments

We would like to thank I. Andrienko, T. Kamins, G. J. Milburn, S. Prawer, C. Thirstrup and S. Williams for useful discussions. This work was funded by the Australian Research Council, and the US National Security Agency through the Army Research Office.

References

[*] e-mail: job@phys.unsw.edu.au
[†] Present address: School of Mathematical and Physical Sciences, University of Newcastle, Callaghan, New South Wales, 2308, Australia

1. B. E. Kane. Nature **393**, 133 (1998).
2. T. M. Buehler *et al. presented at IQC01*, (2001).
3. R. Brenner *et al. presented at IQC01*, (2001).
4. M. F. Crommie, C. P. Lutz, and D. M. Eigler. *Science*, **262**, 218 (1993).
5. S. R. Schofield *et al. presented at IQC01* (2001).
6. C. Thirstrup, *et al. Surf. Sci.*, **411**, 203 (1998) and references therein.
7. M. L. Yu, D. J. Vitkavage and B. S.Meyerson. *J. Appl. Phys.*, 59:4032–4037, 1986.
8. Y. Wang and M. J. Bronikowski and R. J. Hamers *J. Phys. Chem.* **98**, 5966 (1994).
9. J. L. O'Brien *et al. to appear in Phys. Rev. B Rapid Comm.* (see cond-mat/0104569).

10. R. Vrijen *et al.* *Phys. Rev. A*, **62**, 012306 (2000).

CONTROL OF NON-RESONANT EFFECTS IN A NUCLEAR SPIN QUANTUM COMPUTER WITH A LARGE NUMBER OF QUBITS

G.P. BERMAN[†], G.D. DOOLEN[†], D.I. KAMENEV[†], AND V.I. TSIFRINOVICH[‡]

[†] *Theoretical Division and CNLS, Los Alamos National Laboratory, Los Alamos, NM 87545*

[‡] *IDS Department, Polytechnic University, Six Metrotech Center, Brooklyn, New York 11201*

We discuss how to simulate simple quantum logic operations with a large number of qubits. These simulations are needed for experimental testing of scalable solid-state quantum computers. Quantum logic for remote qubits is simulated in a spin c hain. Analytical estimates are presented for possible correlated errors caused by non-resonant transitions. A range of parameters is given in which non-resonant effects can be minimized.

1 Introduction

Several proposals for scalable solid-state quantum computers have been recently published [1,2,3]. The simulation of quantum information processing with a large number of qubits will be necessary to test these computers. We show in this paper how to implement the Control Not (CN) gate between remote qubits in a spin chain using a sequence of electromagnetic pulses. The errors generated by non-resonant effects are estimated analytically and computed numerically. We give the probability of error as a function of the Rabi frequency and the number of qubits.

The t wo-level approximation for an individual transition used in this paper for analysis of the system (see also Ref. [4]) allows one to simulate the quantum dynamics in a system with an enormous number (2^L, where L is the number of qubits) of quantum states. This technique is based on selection of states generated as a result of the resonant or near-resonant transitions, while other transitions are neglected.

In Sec. I we give the Hamiltonian of the nuclear spin quantum computer. In Sec. II we use the resonance approximation to show that each quantum state interacts resonantly only with a single state. This allows us to decrease from exponential in L to polynomial in L the number of differential equations in the system. In Sec. III we use the solution of the problem in the resonance approximation to estimate error probabilities in the implementation of a CN gate caused by finite detuning from exact resonance.

2 Dynamics of the spin chain

We consider a chain of identical nuclear spins placed in a nonuniform high external magnetic field B, with a uniform gradient. The nuclear magnetic resonance (NMR) frequency for the kth spin is $\omega_k = \gamma_n B_k$, where γ_n is the nuclear gyromagnetic ratio and B_k is the z-component of the magnetic field at the location of the kth spin. The gradient of the magnetic field (in the direction of the chain) provides a shift of ω_k by the value $\delta\omega$ between the neighboring spins. (For the physical parameters see Ref. [4].)

The Hamiltonian of the spin chain in an external radio-frequency (rf) field is,

$$H_n = -\sum_{k=0}^{L-1} \omega_k I_k^z - 2J \sum_{k=0}^{L-1} I_k^z I_{k+1}^z + V_n = H_0 + V_n, \tag{1}$$

where J is the Ising interaction constant and I_k^z is the operator of the z-component of spin 1/2. The operator, V_n, describing the interaction of spins with the *rf* pulses, can be written in the form, [5] $V_n = -(\Omega_n/2)\sum_{k=0}^{L-1}[I_k^- \exp(-i(\nu_n t + \varphi_n) + I_k^+ \exp(i\nu_n t + \varphi_n)]$, where Ω_n, ν_n and φ_n are the Rabi frequency, the frequency and the phase of the nth pulse; $I_k^\pm = I_k^x \pm I_k^y$. Each quantum state of the spin chain can be described as a superposition of eigenstates of H_0, for example, $|00\dots00\rangle$, $|00\dots01\rangle$, and so on, where the state $|\dots 0_k \dots\rangle$ corresponds to the direction of the kth nuclear spin along the direction of the magnetic field (spin up), and the state $|\dots 1_k \dots\rangle$ corresponds to the spin being in the opposite direction (spin down). The wave function, Ψ, of the spin chain can be written in the interaction representation as a linear combination of the individual states, $\Psi(t) = \sum_p C_p(t)|p\rangle \exp(-iE_p t)$, where E_p is the energy of the state $|p\rangle$. The contribution of each quantum state to the wave function, $\Psi(t)$, is given by the coefficient $C_p(t)$. The quantity $|C_p(t)|^2$ is the probability of finding the spin chain in the state $|p\rangle$ at time t.

The Schrödinger equation for the coefficients $C_p(t)$ has the form (we put $\varphi_n = 0$), $i\dot{C}_p(t) = \sum_{m=0}^{2^L-1} V_{pm}^n \exp[i(E_p - E_m)t + ir_{pm}\nu_n t]C_m(t)$, where $r_{pm} = \mp 1$ for $E_p > E_m$ and $E_p < E_m$, respectively. $V_{pm}^n = -\Omega_n/2$ for the states $|p\rangle$ and $|m\rangle$ connected by a single-spin transition. $V_{pm}^n = 0$ for all other states.

3 The approximate solution

The Schrödinger equation can be numerically integrated only for a spin chain with small enough number of spins since the number of states increases exponentially with L. The problem can be simplified when the Rabi frequency, Ω_n, is much less than the difference, $\delta\omega$, between the NMR frequencies of the neighboring spins, $\Omega_n \ll \delta\omega$. Suppose that a pulse is resonant with the kth spin in the chain, $\nu_n \approx \omega_k$. Then one can admit that this pulse affects in some approximation only the kth spin in the chain and does not interact with other spins. In this case, only the term with the energy $E_{m'} \approx E_p + \nu_n$ effectively contributes to the right-hand side of the Schrödinger equation, so that the states $|m'\rangle$ and $|p\rangle$ are connected by a single-spin flip of the kth spin. In this case, we reduce the Schrödinger equation to the set of t wo differential equations,

$$i\dot{C}_m(t) = -\frac{\Omega_n}{2}\exp[i(E_p - E_m - \nu_n t]C_m(t),$$

$$i\dot{C}_p(t) = -\frac{\Omega_n}{2}\exp[-i(E_p - E_m - \nu_n t]C_p(t), \tag{2}$$

where $E_p > E_m$, and $|p\rangle$ and $|m\rangle$ are t wo stationary states which are connected by a single-spin transition of the kth spin, $\nu_n \approx \omega_k$.

The solution of Eq. (2), for initial conditions, $C_m(t_{n-1}) = 1, \quad C_p(t_{n-1}) = 0$, after the action of the nth pulse has the form,

$$C_m(t_n) = \left[\cos(\lambda_n \tau_n/2) + i\frac{\Delta_n}{\lambda_n}\sin(\lambda_n \tau_n/2)\right] e^{\left(\frac{-i\tau_n \Delta_n}{2}\right)},$$

$$C_p(t_n) = i\frac{\Delta_n}{\lambda_n}\sin(\lambda_n\tau_n/2)\,e^{\left(it_{n-1}\Delta_n+i\frac{\tau_n\Delta_n}{2}\right)}, \quad (3)$$

where t_{n-1} and t_n are the beginning and the and of the nth pulse, τ_n is the duration of the nth pulse, $\Delta_n = E_p - E_m - \nu_n$ (we do not indicate the dependence of Δ_n on the indices p and m), $\lambda_n = \sqrt{\Omega_n^2 + \Delta_n^2}$ is the precession frequency in the frame rotating with the frequency ν_n. Suppose that at t_{n-1} the system was in the state $|m\rangle$. In the case of the exact resonance ($\Delta_n = 0$) and for $\Omega_n\tau_n = \pi$ (a π pulse) Eqs. (3) describe the resonant transition from the state $|m\rangle$ to the state $|p\rangle$. If the frequency of the next pulse also satisfies the resonance condition, $\Delta_{n+1} = E_l - E_p - \nu_{n+1} = 0$, then the system with probability equal to unity will transform to the state $|l\rangle$, and so on.

4 Errors in creation of entangled state for remote qubits

The advantages given by the t wo-level approximation described above are obvious: (a) we solve the problem by using the discrete map (3) instead of integrating differential equations; (b) we can consider only the states with large enough probabilities and neglect all other states in a controlled way; (c) we can estimate errors caused by non-resonant effects and minimize them by choosing optimal parameters.

In this paper we will estimate the error in implementation of a unitary operation for remote qubits (which is a particular case of the well-known Control-Not (CN) gate) using a sequence of π-pulses. This CN gate is defined as the unitary operator, U, with the following properties:

$$U|0_{L-1}0_{L-2}\dots0_10_0\rangle = e^{i\varphi_1}|0_{L-1}0_{L-2}\dots0_10_0\rangle, \quad (4)$$

$$U|1_{L-1}0_{L-2}\dots0_10_0\rangle = e^{i\varphi_2}|1_{L-1}0_{L-2}\dots0_11_0\rangle, \quad (5)$$

where φ_1 and φ_2 are known phases. The target qubit (0) should change its state only when the control qubit ($L-1$) is in the state $|1\rangle$. The operator, U, can be used to create an entangled state for remote qubits: $U(\alpha|0_{L-1}...0_0\rangle + \beta|1_{L-1}...0_0\rangle) = \alpha e^{\varphi_1}|0_{L-1}...0_0\rangle + \beta e^{\varphi_2}|1_{L-1}...1_0\rangle$. To accomplish the (4b) operation in the system described by the Hamiltonian (1), we choose a sequence of π-pulses with resonant frequencies (for which all $\Delta_n = 0$). In this case, the state $|10\dots00\rangle$ transforms to the state $|10\dots01\rangle$, with probability equal to unity, by the following scheme: $|1000\dots0\rangle \to |1100\dots0\rangle \to |1110\dots0\rangle \to |1010\dots0\rangle \to |1011\dots0\rangle \to |1001\dots0\rangle \to \dots \to |100\dots11\rangle \to |100\dots01\rangle$. The sequence of pulses which realizes this protocol has the following form: $\nu_1 = \omega_{2L-2}$, $\nu_2 = \omega_{2L-3}$, $\nu_3 = \omega_{2L-2} - 2J$, $\nu_{2L-4} = \omega_6$, ..., $\nu_{2L-4} = \omega_0 - J$, $\nu_{2L-3} = \omega_1$. If we apply the same protocol to the state $|00\dots00\rangle$, then with some probability the system will remain in this state because all transitions are non-resonant with the detuning $\Delta_n = \pm J, \pm 2J$. Since $\Delta_n \neq 0$ these non-resonant transitions have the probabilities (see Eq. (3)),

$$\varepsilon_n = (\Omega_n/\lambda_n)^2\sin^2(\lambda_n\tau_n/2). \quad (6)$$

In this paper we study the errors generated during the operation (4) under the condition that the transitions in Eq. (**??**) are resonant. The operator U in Eqs. (4) and (**??**) can be written as a product, $U = U_{2L-3}U_{2L-4}\dots U_1U_1$, of operators of individual pulses U_n, where $n = 1, ,2, \dots, 2L-3$. We take the Rabi frequency to be the same, $\Omega_n = \Omega$, for all pulses. Then the error, ε_n, generated by the nth pulse is defined only by the detuning, Δ_n.

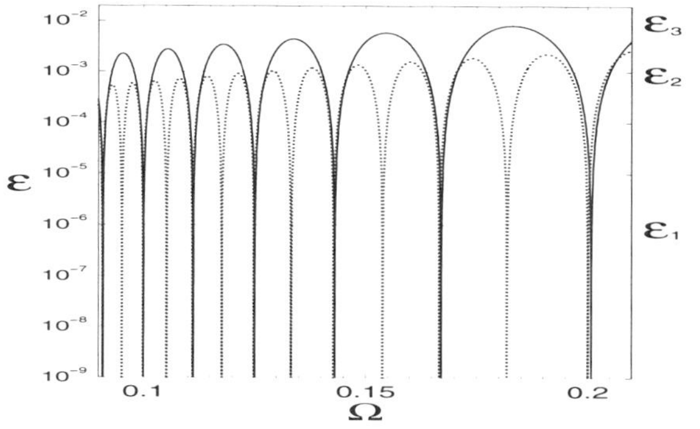

Figure 1. Dependence of ε (solid line) and ε' (dotted line) on Ω. $J = 1$.

The values of detuning, $|\Delta_n|$, are the same for all pulses, $|\Delta_n| = J$, except for the third pulse, where $|\Delta_3| = 2J$. We denote $\varepsilon_n \equiv \varepsilon$ for $n \neq 3$ and $\varepsilon_{n=3} \equiv \varepsilon'$.

Each unwanted state generated by the nth pulse can produce other unwanted states under the action of the other $(n + m)$ pulses. This causes the generation of a hierarchy of different states. Instead of one ground state, which should be the result of (4a), we generate a series of different unwanted states.

Since the dynamics is generated by the discrete map (3), probability of error can be estimated analytically. Suppose that we can measure only the states with probability $P \geq P_0$, where P_0 can be defined by the experimental conditions. Then, if ε, $\varepsilon' < \varepsilon_1 = P_0$, we shall not find in the system any unwanted states and the CN gate is realized with probability equal to unity. In Fig. 1, we plot ε (solid line) and ε' (dotted line) as functions of Ω. We suppose that $\varepsilon_1 = P_0 = 10^{-6}$. One can see from Fig. 1 that $\varepsilon(\Omega)$ and $\varepsilon'(\Omega)$ are less than ε_1 only in very narrow regions of the values of Ω (in the vicinity of $\Omega_k = |\Delta_n|/\sqrt{4k^2 - 1}$, $k = 1, 2, \ldots$ [4,5]). Hence, high precision is required to implement the CN gate, (4), (??), without error.

Let us consider now in detail the case of arbitrary ε: $0 \leq \varepsilon \leq 1$, and explain how the errors can be estimated analytically in the operation (4). The first pulse U_1 generates one unwanted state $U_1|0000\ldots0\rangle \rightarrow |0100\ldots0\rangle$ with the probability ε. The probability to remain in the ground state is $1 - \varepsilon$. The second pulse, U_2, generates an additional unwanted state from the ground state, $U_2|0000\ldots0\rangle \rightarrow |0010\ldots0\rangle$, with probability $\varepsilon(1-\varepsilon)$, and transforms the state $|0100\ldots0\rangle$ to the state $|0110\ldots0\rangle$ without generating additional states, since the second pulse is resonant to this transition. The probability of remaining in the ground state is now $(1 - \varepsilon)^2$. The total number of states after the action of two pulses is three. The probabilities of all generated states after the action of the U_3 pulse are:

$\|0000\ldots0\rangle$:	$(1-\varepsilon)^2(1-\varepsilon')$,
$\|0100\ldots0\rangle$:	$\varepsilon'(1-\varepsilon)^2$,
$\|0010\ldots0\rangle$:	$\varepsilon(1-\varepsilon+\varepsilon^2)$,
$\|0110\ldots0\rangle$:	$\varepsilon(1-\varepsilon^2)$.

(Note, that above expressions can be considered only as estimates because we estimated probabilities instead of complex amplitudes.) One can continue to calculate the states generated by subsequent pulses and estimate their probabilities in a similar way. However, for ε, $\varepsilon' \ll 1$, one can omit the contribution from the higher order terms in ε, and neglect the states with the probabilities less than P_0.

Suppose, that $\varepsilon' \approx \varepsilon$ and $\varepsilon_1 < \varepsilon < \varepsilon_2 \equiv \sqrt{P_0}$. In this case, each pulse generates only one unwanted state from the ground state. Then, only

$$N_1(L) = 2L - 3 \tag{7}$$

unwanted states will be generated with probabilities (with the accuracy up to ε^2) ε, $\varepsilon(1-\varepsilon)$, $\varepsilon(1-2\varepsilon)$, ..., $\varepsilon(1-(2L-4)\varepsilon)$. After the action of $2L-3$ pulses there are $L-1$ unwanted states of the form,

$$|011\dots11\rangle,\ |0011\dots11\rangle, \dots, |00\dots011\rangle,\ |00\dots01\rangle, \tag{8}$$

with the target spin in the state $|1\rangle$, and $L-2$ unwanted states of the form,

$$|011\dots110\rangle,\ |0011\dots110\rangle,\ \dots,\ |00\dots010\rangle, \tag{9}$$

with the target spin in the state $|0\rangle$. As one can see from Eqs. (8) and (9), already in the first order of ε many states have correlated error with many qubits in excited states. The estimate of the total probability of unwanted states is

$$P_1(\mathcal{E}) = \varepsilon \sum_{n=0}^{2L-4} (1 - n\varepsilon) \approx \mathcal{E}\left(1 - \frac{1}{2}\mathcal{E}\right), \tag{10}$$

where $\mathcal{E} = (2L-3)\varepsilon$.

The probability of unwanted effect of finding a target qubit in the state $|1\rangle$ while the control qubit is in the state $|0\rangle$ (i.e. the probability of the process $U|0\dots0\rangle \to |0\dots1\rangle$) is,

$$\mathcal{P}_1(\varepsilon) = \varepsilon + \varepsilon \sum_{n=0}^{L-3} (1 - (2n+1)\varepsilon) \approx \Gamma(1-\Gamma), \tag{11}$$

where $\Gamma = (L-2)\varepsilon$.

In Figs. 2 (a), (b) we plot the probabilities, P_1 and $\mathcal{P}_1$, as functions of the number of spins, L, in the chain, for $\Omega = 0.0906$, so that $\varepsilon_1 < \varepsilon(\Omega) = 4.78 \times 10^{-5} < \varepsilon_2 = 10^{-3}$ ($\varepsilon'(\Omega) = 3.23 \times 10^{-5}$). The number of unwanted states in our calculations was exactly equal to N_1 in Eq. (7). From Figs. 2 (a) and (b) one can see that the analytical estimates agree with the results of numerical calculations.

In a similar way one can calculate the probabilities of unwanted states for the case $\varepsilon_2 < \varepsilon < \varepsilon_3 \equiv (P_0)^{1/3}$. In this situation one should take into account states with the probabilities up to ε^2. Each of these states is generated as a result of two successive non-resonant processes. However, since the ground state generates only unwanted states with the probabilities of order ε, the total probability of unwanted states will be given again by Eq. (10). But the number of generated states increases and at the condition, $\varepsilon_2 < \varepsilon < \varepsilon_3$, it is proportional to L^2.

In Fig. 3 we compare the numerical results with the analytical estimates for the total probability of unwanted states given by Eq. (10) when the value ε is relatively large. From Fig. 3 one can see that formula (10) correctly describes the behavior of the system for large value of ε. When L becomes large enough, one should also include into consideration in Eq. (10) the terms of order $(L\varepsilon)^3$.

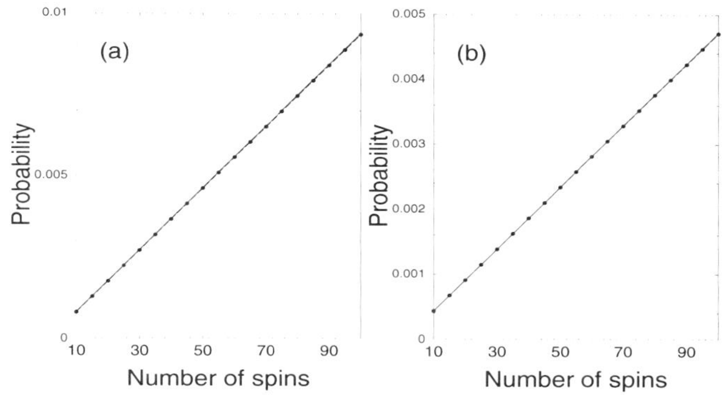

Figure 2. (a): The total probability, P_1, of unwanted states in the execution of the CN logic gate as a function of the number of qubits, L, in the spin chain. Solid line: analytical estimate (10), filled circles connected by dashed line: numerical results. (b) Probability, $\mathcal{P}_1$, of observing the target spin in the state $|1\rangle$, while the control qubit is in the state $|0\rangle$, after the CN operation; solid line: analytical estimate (11), filled circles connected by dashed line: numerical results. $J = 1$, $\Omega = 0.0906$.

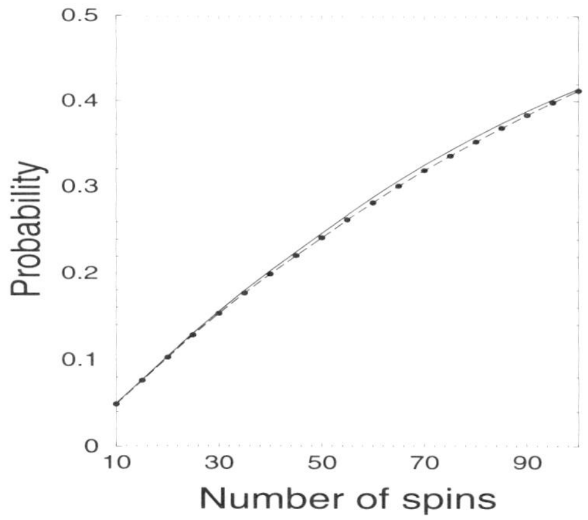

Figure 3. The total probability of unwanted states for the case of relatively large ε, when $\varepsilon_1 < \varepsilon < \varepsilon_2$. Solid line: analytical estimate (10), filled circles connected by dashed line: numerical results. $J = 1$, $\Omega = 0.20844$, $\varepsilon(\Omega) = 2.98 \times 10^{-3}$, $\varepsilon'(\Omega) = 2.41 \times 10^{-3}$.

The probabilities of unwanted states are shown in Figs. 4 (a) and (b) in two different scales. Two types of states are clearly seen. The states in Fig. 4 (a) are generated as a result of one non-resonant transition, so the probabilities of these states are proportional to ε. On the other hand, the states in Fig. 4 (b) are generated by two successive non-resonant transitions, and they have the probabilities of the order ε^2.

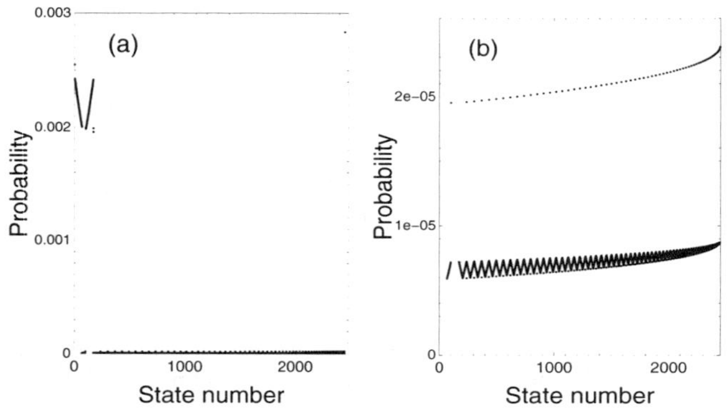

Figure 4. The probabilities of generated states in two different scales. $L = 70$. All other parameters are the same as in Fig. 3.

5 Conclusion

We have demonstrated how one can simulate a simple logic operation for a large number of qubits. We analyzed the probability of correlated errors in the implementation of a CN gate between remote qubits in a spin chain with a large number of qubits. In the limit when the Rabi frequency Ω is much less than the difference between the NMR frequencies of neighboring spins, it is shown that the probability of error, P, is mainly defined by the small parameter ε defined by (5) and by the number of qubits in the spin chain. It is demonstrated that at definite values of Ω the value of ε is small and there is no correlated error in the implementation of the CN gate. The total probability of unwanted states at other values of ε is estimated analytically and computed numerically.

Since the Hamiltonian of the system only allows transitions between the states connected by a single spin transformation, and because the electromagnetic field interacts resonantly only with one spin in the spin chain, the probability of error cannot accumulate in a definite single unwanted state. The probability of each unwanted state is always less than the value of the small parameter of the problem $\varepsilon(\Omega)$. The considered approach can be useful for experimental testing of scalable solid-state quantum computers.

The paper was supported by the Department of Energy (DOE) under contract W-7405-ENG-36 and by the National Security Agency (NSA), and by the Advanced Research and Development Activity (ARDA).

References

1. B.E. Kane, Nature, **393**, 133 (1998).
2. R. Vrijen, E. Yablonovitch, K. Wang, H.W. Jiang, A. Balandin, V. Roychowdhury, T. Mor, and D. Divincenzo, Phys. Rev. A **62**, 2306 (2000).
3. G.P. Berman, G.D. Doolen, P.C. Hammel, and V.I. Tsifrinovich, Phys. Rev. B **61**, 14694 (2000).
4. G.P. Berman, G. D. Doolen, G. V. Lòpez, and V. I. Tsifrinovich, Phys. Rev. A **61**,

2305 (2000).
5. G.P. Berman, G. D. Doolen, R. Mainieri, and V. I. Tsifrinovich, *Introduction to Quantum Computers* (World Scientific, Singapore, 1998).

A ROLE FOR ION IMPLANTATION IN QUANTUM COMPUTING

DAVID .N. JAMIESON, STEVEN PRAWER, IGOR ANDRIENKO, DAVID A. BRETT, VICTORIA MILLAR CHANGYI YANG and ALBERTO CIMMINO

Centre for Quantum Computer Technology, Microanalytical Research Centre
School of Physics, University of Melbourne, Victoria 3010, Australia

The Kane solid state quantum computer aims to exploit phosphorous-31 nuclear spins as qubits embedded in a silicon matrix. We propose to create this array using ion implantation. Since the implantation of the ions is essentially random, the yield of usefully spaced atoms is low. This problem can be reduced if the passage of a single ion can be registered so those usefully positioned ions can be identified. Or, the arrival of randomly timed ions implanted though a movable mask can be registered by electrical transients induced in the substrate. The former can be accomplished by implantation of the ions through a thin surface layer consisting of resist. Changes to the chemical and/or electrical properties of the resist will be used to mark the site of the buried ion. For chemical changes, the latent damage can be developed and the atomic force microscope (AFM) used to image the changes in topography. Alternatively, the electrical transients in the substrate can be detected if the substrate contains an electric field which separates the electron-hole pairs induced by ion impact. This technique is well established with MeV ions. These methods are discussed here as potential construction tools for the Kane quantum computer.

Keywords: Quantum Computer, Ion implantation, single ion nanomachining, ion tracks

1. Introduction

Various approaches to the construction of a quantum computer have recently been reviewed by DiVincenzo[1]. The solid state quantum computer proposed by Kane[2], which is based on an array of $I = \frac{1}{2}$ ^{31}P ions embedded within a silicon matrix, is the focus of this paper. A key problem with the construction of this device is how to construct the array of ^{31}P ions with the required precision, with a period of around tens of nanometres, in the silicon matrix. Ion implantation may be used to deliver the ions to the correct positions. Ion implantation traditionally delivers single ions at random times and positions at a depth determined by the ion energy. In the Kane quantum computer, the depth of the ^{31}P atoms must be of the order of 20 nm corresponding to around 13 keV implantation energy. The mean spacing, d (in nm) of a given ion dose D (in ions/cm^2) is $d = 10^7 / \sqrt{D}$. To achieve the required spacing the dose needs to be in the range from 10^{10} to 10^{13} ions/cm^2. Additional factors such as ion straggling, sputtering of the substrate and diffusion following damage annealing are likely to reduce the yield of useful qubits.

Realisation of suitable architectures under these constraints is a daunting technical challenge. Vrijen et al[3] have introduced a scheme where ^{31}P ions are randomly implanted through a mask containing an array of apertures above cells in which ^{31}P qubits are to be located. It was suggested that pre-existing circuitry could be used to determine the presence or absence of 0, 1 or more ^{31}P atoms arriving in the cells. Poisson statistics shows that the yield of usefully placed qubits is likely to be low. The testing of devices can be made considerably more efficient if it can unambiguously demonstrated that an ion has entered a particular cell. This can be done "off-line" if, during implantation, the ion impact sites are recorded by a suitable resist. The impact sites are then imaged, after implantation, by a suitable microscope and cells successfully loaded with a qubit are then wired for incorporation in a device. It can also be done "on-line" by implantation though a mask that confines the implanted ions to the desired locations in the substrate. It is the purpose of this paper to discuss ways these proposals could be accomplished. We focus on several promising strategies based on existing technology of ion irradiation sensitive resists and solid state particle detectors.

2. Registration by thin resists

If the ions are implanted through a thin resist layer, ion passage creates latent damage in the resist material. The latent damage can be developed to produce topological contrast on the surface. Alternative resists record the passage of a single ion from changes in the electrical properties to identify the ion impact site. In either case a search of the implanted region could reveal, in the random spatial distribution of ion strikes, a suitable array of implanted ions where further control gates could be added or where pre-existing cells have received an ion. Without a means of identifying suitable ion impact sites, the task of screening unsuccessful devices

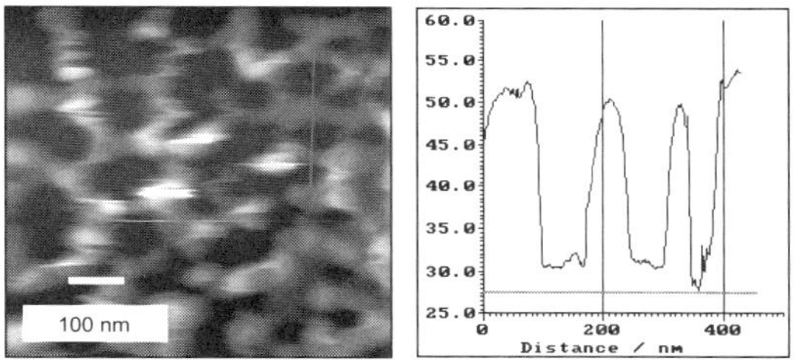

Fig. 1. STM image and height profile of 3 MeV He ion-irradiated and developed PMMA. Ion dose was $10^9/cm^2$; dark is the Si substrate; light is the residual PMMA.

would be prohibitive. Considerable previous work has demonstrated the detection of single ion impacts by topological changes in a resist (see a recent review[4]). In the case of polymethylmethacrylate (PMMA) the diameter of the latent damage track of a MeV ion is between 2 and 25 nm. These scales are favourable for the Kane proposal. In fact the characteristics of 50 nm thick PMMA films for registering single ions have been investigated and some recent results are available[5]. Although the experiments were done with MeV He ions, the results are promising show evidence for single ion registration (see the example in fig. 1.) and will be repeated in the near future with 10 keV ^{31}P.

The properties of C_{60} may also be exploited to register single ion impacts from changes in electrical conductivity. Upon single ion impact resistive C_{60} molecules disintegrate into its constituent 60 carbon atoms which are conducting. Conductivity contrast scans can then identify the ion impact sites. The results of low dose ion irradiation of C_{60} films have been presented[6] showing single ion impact sites can be registered. However it was found to be very difficult to obtain a sufficiently uniform film so that the ion-induced changes in contrast were visible. Improved C_{60} films, or other organic films with similar properties[7], show promise. Although an inevitable drawback with implantation through a resist is that knock-on atoms from the resist may co-implant into the substrate and introduce contamination

3. Registration with electrical detection

While the use of thin resists looks promising, the requirement of a thin surface layer will contribute to undesirable straggling which will reduce the yield of working devices. With electrical detection of the ion arrival in the substrate, the yield could be considerably improved at the expense of greater complexity of the fabrication method.

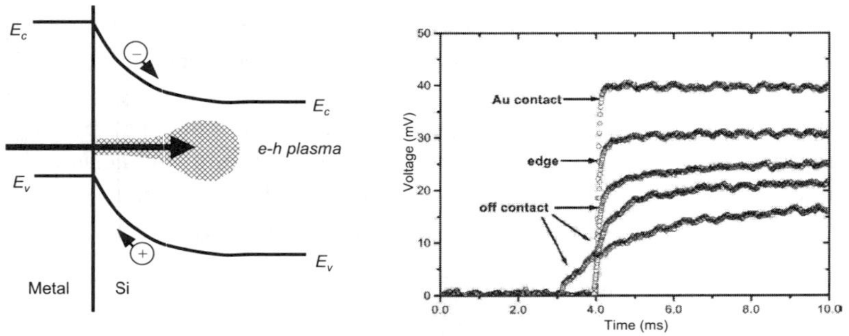

Fig. 2. Charge injection from MeV ion implantation into a gold Schottky barrier on Si and the resulting voltage transients from on, at the edge and just off the contact.

As an energetic ion dissipates its kinetic energy when it slows to a stop inside a solid, much of the energy appears as ionization in the form of electron hole (e-h) pairs[8]. These e-h pairs may be separated by an electric field within the specimen and thus collected by suitable electrodes, usually applied to the surface, where a current transient (see Fig. 2) can be recorded. This is the basis of a new technique for sensitive, spatially resolved, characterisation of defects in semiconductors[9]. The electric field necessary to collect the charge may arise from an externally applied field across the surface electrodes, or from a built-in field arising from the depletion region of a p-n junction, or from the band bending arising from a

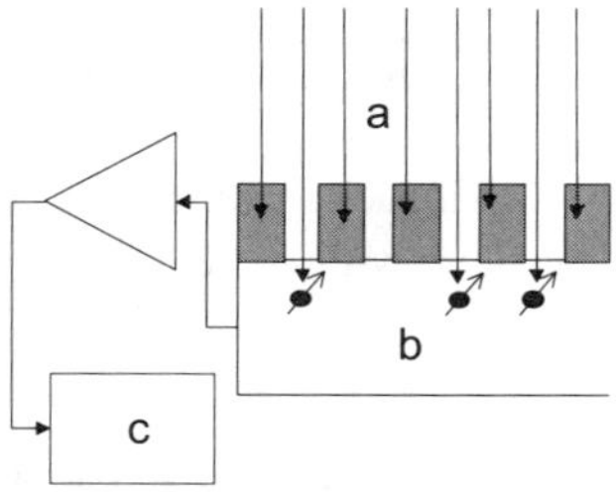

Fig. 3. Ion implantation (a) through a pre-patterned mask into a substrate (b) with ion counting circuitry (c)

metal/semiconductor Schottky barrier. Of these alternatives, the use of an external electric field applied to a very pure, highly insulating substrate, a suitable environment for the Kane computer, is the most desirable. The technology for the detection of the charge transients is well established for MeV ion irradiation and could be adapted to the lower level transients produced by keV ions[10]. Standard nuclear instrumentation can then be employed to count the ions successfully implanted into the substrate through the patterned mask.

4. Conclusion

Both PMMA and C_{60} show considerable promise for the registration of the passage of single ions with the ion track showing strong contrast with the unirradiated film. Those qubits likely to be active, can then be readily identified from an appropriate surface scan. Electrical registration can be used to precisely count the ions arriving in the substrate through a pre-patterned mask. Combining the two methods will allow verification that the array of ^{31}P ions have actually been delivered to the substrate in the correct locations.

Acknowledgements

We acknowledge stimulating discussions with our collaborators in the Special Research Centre for Quantum Computer Technology, in particular Prof Robert Clark, Prof Gerad Milburn, Dr Andrew Dzurak and Dr Nancy Lumpkin.

References

1. D.P. DiVincenzo, arXiv:quant-ph/0002077v2 6Mar 2000
2. B.E. Kane, Nature 393 (1998) 133
3. R. Vrijen, E. Yablonovitch, K. Wang, H.W. Jiang, A. Balandin, V. Roychowdhury, T. Mor and C. DiVincenzo, Phys. Rev. A62 (2000) 12306.
4. D.N. Jamieson, S. Prawer, I. Andrienko, D.A. Brett and V.E. Millar, Nucl. Instr. Meth. B, in press (2001).
5. D.A. Brett, D.N. Jamieson, A. Cimmino, I. Andrienko and S. Prawer, Studies of PMMA for the ion beam route into quantum computing, IQC01, these proceedings.
6. V.E. Millar, I. Andrienko, A. Cimmino and S. Prawer, Study of C60 as a thin film resist in the Kane quantum computer, IQC01, these proceedings.
7. I. Andrienko, A. Cimmino, D.N. Jamieson and S. Prawer, Demonstration of single ion impact on the Si(001) surface using a self-assembled organic monolayer of 1,3,5,7-Cyclooctatetraene, IQC01, these proceedings.
8. M.B.H. Breeese, D.N. Jamieson and P.J.C. King, Materials analysis with a nuclear microprobe, John Wiley and Sons, New York (1996).
9. J.S. Laird, R.A. Bardos, D.N. Jamieson and G.J.F. Legge, submitted to Nucl. Instr. Meth. B (2001).
10. C. Yang, D.N. Jamieson, S.M. Hearne, P. Spizzirri and S. Prawer, Ion implantation into electrically active substrates for quantum computer fabrication, IQC01, these proceedings.

SIMULATING THE FABRICATION OF A SILICON-BASED QUANTUM COMPUTER: MOLECULAR DYNAMICS CALCULATIONS OF ION IMPLANTATION

N.A.MARKS, K.H. LEE, AND D.R. MCKENZIE

School of Physics A28, The University of Sydney NSW 2006, Australia

Molecular dynamics calculations are used to explore the feasibility of ion implantation as a means to construct phosphorus quantum bits in a silicon-based quantum computing device. The simulations employ the recently developed Environment Dependent Interaction Potential which describes bonded interactions more accurately and efficiently than comparable empirical methods. Implantation events with energies of 1 keV are considered, and the evolution of the damage cascade is followed. Transient enhanced diffusion during the post-implantation annealing cycle is small, but significant. Other diffusive processes associated with the silicon self-interstitial are also in close agreement with highly accurate quantum mechanical calculations.

1 Introduction

Ion implantation has been suggested as a fast-track, low-yield means of constructing the silicon-based quantum computation device proposed by Kane.[1] Electron-beam lithography has already been shown[2] to be capable of laying the patterned resist barriers necessary to align the overlying A-gates with the phosphorus implantation channel. In the proposed scheme a low-flux beam would implant phosphorus into silicon with on average one P atom passing into each resist channel. The A-gates which control the electron-nuclear coupling are of order 50 Å in width, and fix the precision of the phosphorus positioning technology to be ±25 Å, just ten times the interatomic distance in silicon.

Subsequent to implantation an annealing step restores the crystalline lattice. During this time of high-mobility the phosphorus atom will undergo diffusion, and if the lateral displacement exceeds 25 Å the qubit will be misaligned with the A-gate. While implantation and mobility are well understood for bulk doping, very little is known for the case of a single dopant due to the difficulty of characterization at the atomic scale. Atomistic simulation is therefore a most useful analytical technique, providing fine spatial and temporal resolution, and the goal of this work is to use simulations to determine the degree of lateral displacement during the implantation and annealing stages. In this work we use the most accurate empirical potential available for silicon in simulations of phosphorus implantation into (100) oriented silicon.

2 Molecular Dynamics Method

The molecular dynamics simulations in this study employ two empirical potentials to describe the interatomic interactions. The repulsive potential of Ziegler, Biersack and Littmark (ZBL)[3] describes atoms undergoing high-energy collisions, applying for pairs of atoms with a separation R_{ij} less than 1.35 Å. The ZBL energy is given by

$$U_{\mathrm{ZBL}} = \sum_{i<j} \frac{Q_i Q_j}{R_{ij}} \phi(Q_i, Q_j, R_{ij}) \tag{1}$$

where the Q_i denote the nuclear charges and ϕ is a damping function capturing the screening of the nuclear charges by the electrons.

Bonded interactions are described using the recently developed Environment Dependent Interaction Potential (EDIP)[4] which expresses the total energy as a sum of two-body (bond-streching) and three-body (bond-bending) contributions, in the manner

$$U_{\mathrm{EDIP}} = \sum_{ij} U_2(R_{ij}, Z_i) + \sum_{ijk} U_3(R_{ij}, R_{ik}, \theta_i, Z_i) \tag{2}$$

with the strength of both terms varying with the coordination Z_i, thus describing environmental dependence. This environmental dependence represents a significant improvement over the fixed-coordination Stillinger-Weber (SW)[5] potential which has been widely used in previous simulations of ion implantation. A linear transition potential connects the intermediate regime between EDIP and ZBL. A cubic spline cannot be used as in ZBL/SW simulations due to the environmental dependence which gives rise to a varying intersection point with coordination. Phosphorus atoms are modeled as "heavy" silicon since the 3p electron is ionized at room temperature.

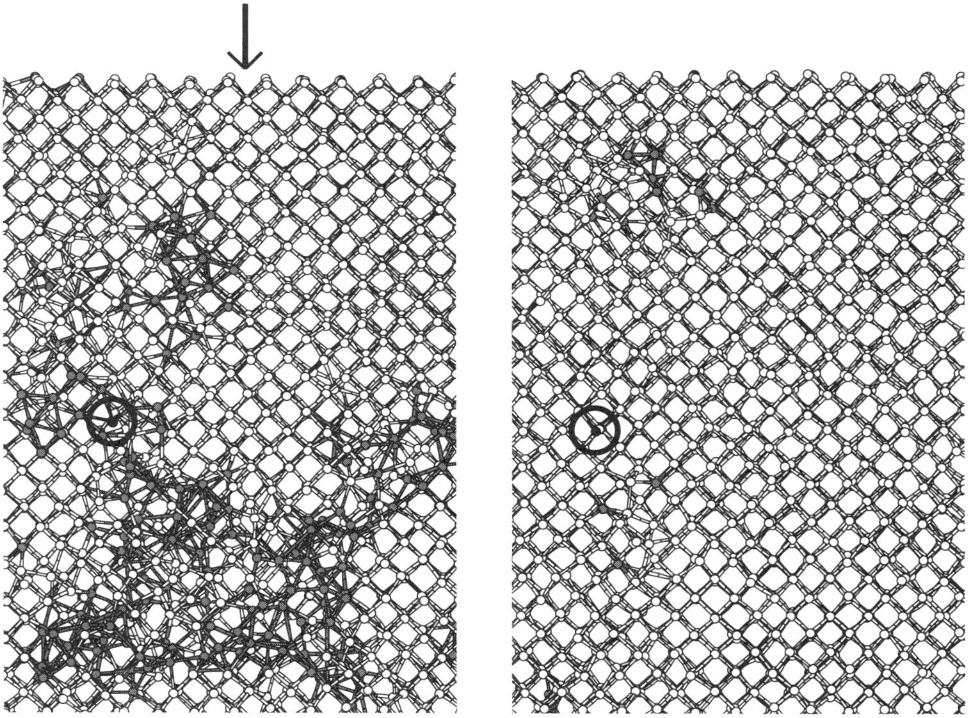

Figure 1. Left: Damage cascade produced by the implantation of a 1 keV atom at the position indicated by the arrow. At this point ≈10 ps has elapsed. Atoms in the damage region are shown in grey, while the final resting place of the implanted atom is indicated by the large circle. Right: Same system after annealing at 1250 K for 100 ps.

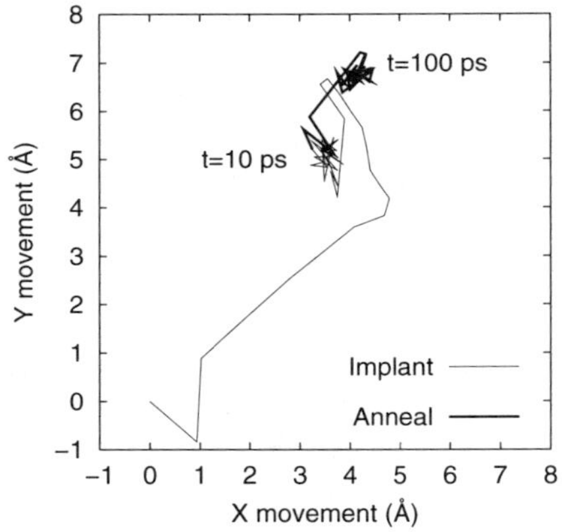

Figure 2. Projected trajectory of implanted atom.

Verlet's algorithm[6] is used to numerically integrate the equations of motion, periodic boundaries are used to eliminate edge effects, and velocity rescaling is used for temperature control. The integration timestep is less than one femtosecond and varies according to the maximum kinetic energy.

3 Results & Discussion

Five separate molecular dynamics studies have been completed using a (100) oriented silicon substrate and an impact energy of 1 keV. A typical simulation is shown in Fig. 1. The left hand image shows the situation after the implantation phase has ceased, while the right hand image shows the result of 100 ps of annealing at 1250 K. The point of impact of the implanted phosphorus atom is indicated by an arrow and the final resting point is indicated by a circle.

During the implantation phase a large number of silicon atoms are displaced from their lattice sites. Some of this damage self-anneals due to the transient temperature spike associated with the atomic displacements, leaving the residual damage indicated in the left-hand image in Fig. 1. This damage is substantially reduced by the 1250 K anneal.

The trajectory of the implanted atom in Fig. 1 is shown in projection in Fig. 2, with time as a parameter along the trajectory. The thin line indicates the lateral displacement during the impact phase, while the thick line indicates the effect of annealing. The yield of devices with a given size of A-gate is directly determined by the distribution in projected distances from the origin. In this instance the extent of the displacement is sufficiently small for the qubit to be functional.

Simulations at slightly different angles of incidence show that the mobility of the atom during the annealing step is strongly dependent on the position of the atom at the end of the implantation phase. In the case of an atom coming to rest outside the principal damage region the annealing mobility is very small and the atom moves only slightly during the anneal. If on the other hand the atom is surrounded by damage, it moves substantially further (e.g. 10 Å) due to the process of transient enhanced diffusion. This diffusion process occurs because of an interaction between the phosphorus and a silicon defect, and involves the "kicking out" of an atom in a lattice site by an interstitial atom.

To confirm that the diffusion process is well described, we calculated the diffusion constant for a silicon self-interstitial at a range of temperatures. The process was found to be strictly activated, with an activation energy of 0.33 eV, in good agreement with the value of 0.30 eV obtained by an accurate *ab initio* calculation[7], and much improved over the 0.90 eV value obtained with the Stillinger-Weber potential.[8]

4 Conclusion

Preliminary simulations of 1 keV phosphorus implantation suggest that the lateral displacement is sufficiently small for the A-gate to align with the phosphorus qubit. Further simulations will build an accurate probabilistic picture of the phosphorus motion during implantation and annealing, and provide insight into the relevant physical processes.

References

1. B. E. Kane, *Nature* (1998) **393**, 133.
2. T.M. Buehler, R.P. McKinnon, N.E. Lumpkin, R. Brenner, A.S. Dzurak, R.G. Clark and D.N. Jamieson and J. McCallum, *International Conference on Experimental Implentation of Quantum Computation,* Poster WP6.
3. J. F. Ziegler, J. P. Biersack and U. Littmark, *The stopping and range of ions in matter*, 1985, Pergamon (New York).
4. J. F. Justo, M. Z. Bazant, E. Kaxiras, V. V. Bulatov and S. Yip, *Phys. Rev. B* (1998) **58**, 2539.
5. F. Stillinger and T. A. Weber, *Phys. Rev. B* (1985) **31**, 5262.
6. L. Verlet, *Phys. Rev. B* (1967) **159**, 98.
7. P.E. Blochl, E. Smargassi, R. Car, D. B. Laks, W. Andreoni and S. T. Pantelides, *Phys. Rev. Lett.* (1993) **70**, 2435.
8. G. H. Gilmer, T. Diaz de la Rubia, D. M. Stock and J. Jaraiz *Nucl. Instr. Meth. B* (1995) **102**, 247.

LONG-TERM CHARGE OFFSET NOISE IN COULOMB-BLOCKADE DEVICES

WILLIAM H. HUBER, STUART B. MARTIN AND NEIL M. ZIMMERMAN

National Institute of Standards & Technology, 100 Bureau Drive, Gaithersburg, MD 20899-8112
http://www.eeel.nist.gov/811/femg/set.html

While Coulomb-Blockade devices have shown promise as qubit systems or qubit state detectors, integration of these devices is severely hampered due to the low-frequency charge offset (Q_0) drift in time. We note that an extrapolation of the 1/f noise to lower frequencies cannot account for the large ($\delta Q_0 \sim 1$ e) changes. We have investigated possible approaches towards reducing the Q_0 noise in $Al/Al_2O_{3-x}/Al$ based Coulomb-Blockade (CB) devices. In $Al/Al_2O_{3-x}/Al$-based CB devices, the maximum stability was typically 1 day to 3 days, nearly independent of processing techniques or annealing treatments. In our best result, $\delta Q_0 < 0.1$ e (except for a single excursion) for 12 days. Encasing devices in a continuous metal vault (a nano-Faraday cage) did not however improve the Q_0 noise. We also discuss one specific, identified cause of the low frequency ($< 10^{-3}$ Hz) Q_0 noise. By extending conventional, two-level system theory (previously used to explain the behavior of macroscopic disordered materials), the noise is shown to be caused by atomic structural relaxations of nearby insulating regions.

1 Introduction

The variety of Coulomb-Blockade (CB) based phenomena observed over the past 15 years rely on the generally accepted postulate that similar quasiparticles (e.g. electrons in a normal metal or Cooper pairs in a superconductor) have nearly the same electronic charge e^*. Therefore the electrostatic energy U of a quasiparticle island is similarly quantized, $U \sim e^{*2}/2C$, where C is the sum capacitance of the system. While usually negligible, the energy can dominate the dynamics of a device at low enough temperatures if the quasiparticle island weakly couples to the surrounding environment[1].

By exploiting the CB phenomenon, a number of potentially useful devices have been or are being developed such as a Single Electron Tunneling Transistor (SETT)[1] an electron Pump[2] and a CB based single photon source[3]. In addition, CB devices may also factor strongly in a solid state-implemented quantum computer -- assuming one could be built -- either directly as a qubit[4] or indirectly as a qubit state detector[5].

At present the number of useful CB devices has remained relatively low, partly because these devices are strongly affected by nanometer length scaled movements of small packets of charge in insulators located either near or in the devices, a phenomenon called the Charge-Offset (Q_0) noise. First investigated soon after the initial CB confirmation, the power spectral density of the noise is typically found to be approximately inversely proportional to frequency[6] between $\sim 10^{-1}$ Hz and 10^3 Hz, similar to the noise observed in other small-scale electronic devices. The Q_0 noise deleteriously affects a CB device in several ways. Above 10^{-1} Hz, the Q_0 noise limits the sensitivity of CB based charge electrometer, and thus the sensitivity of recently proposed detectors for single nuclear spin qubit systems[5]. In addition, recent evidence strongly suggests that the Q_0 noise at very high frequencies ($\sim 10^9$ Hz) is chiefly responsible for decoherence in recently demonstrated Single Cooper Pair Box qubits[7].

In $Al/Al_2O_{3-x}/Al$-based CB devices, we have found that Q_0 drifts by $\delta Q_0 \sim 1$ e, on a typical time scale of 1 to 3 days. Thus over ~ 1 week the device has switched several times between the ON state (e.g. maximum current in a SETT) and the OFF state. *This long-term drift cannot be attributed to a low-frequency extrapolation of the 1/f noise measured at higher frequencies.* Indeed, a simple estimate shows that one would need to wait $10^{3,000,000}$ years for $1/f$ charge noise with typical amplitude to yield $\delta Q \sim 1$ e. While the low frequency Q_0 noise of individual CB devices can be compensated, it becomes increasing difficult to implement this partial solution as the number of CB devices increases because the number of compensation circuits is proportional to the system size and the time between compensation circuit readjustment is inversely proportional to system size. Therefore the low frequency ($< 10^{-3}$ Hz) Q_0 drift prevents integrating large numbers of CB devices.

Interestingly, while the low frequency Q_0 drift presents such a serious problem, it has received far less attention than its higher frequency counterpart, chiefly because practitioners have generally assumed that the *1/f* spectra, a ubiquitous feature above ~ 0.1 Hz, extends down to $f = 0$. Thus conventional wisdom has until now dictated that the 10 Hz noise is sufficient to characterize the noise over *all* frequency regimes.

We have undertaken an investigation of the charge offset noise over a broad frequency range (~ 10^{-5} Hz to 10^{2} Hz) in $Al/Al_2O_{3-x}/Al$-based CB devices. Our effort has focused on reducing the Q_0 noise and concurrently understanding what is driving the phenomenon. In $Al/AlO_x/Al$-based CB devices we were unable to reliably reduce the low frequency Q_0 noise using a variety of different approaches. Therefore it seems that $Al/Al_2O_{3-x}/Al$-based CB devices will likely not be sufficiently stable to permit integrating large numbers of CB devices. We have however recently found that the Q_0 drift in a new class of silicon based single electron tunneling devices is several orders of magnitude lower, thus pointing the way towards integration of a large number of CB devices[20].

• Thermal vs. e-beam evaporation
• Annealing in Forming gas
• Annealing in inert gas
• Using Ozone to oxidize tunnel junction insulators
• Thermal cycling (4-300 K)
• Nano-Faraday cage with O_2
• Nano-Faraday cage with O_3
• Cooling device with or without electrical short protection

Table 1. Attempts to reduce the low-frequency Q_0 noise.

In addition, we have discovered that the low frequency ($< 10^{-3}$ Hz) Q_0 noise often behaves quite differently than the noise at higher frequencies ($> 10^{-1}$ Hz) because it is driven by a different mechanism. Thus a measurement of the 10 Hz noise in an $Al/Al_2O_{3-x}/Al$ based CB device may not reliably determine whether large scale integration of similarly fabricated CB devices is feasible.

2 Attempts to Reduce Charge Offset (Q_0) Noise & Drift

Until recently Al/Al_2O_{3-x}/Al-based CB devices were investigated principally because this material system has already demonstrated its utility in a variety of different applications. To fabricate the devices a standard double angle deposition process was utilized[8]. Early in the investigation, a thermal evaporator was used to deposit the Al films ($P_{base} \sim 2 \times 10^{-6}$ Torr; $P_{depo} \sim 2 \times 10^{-5}$ Torr; rate $\sim$ 2.0 nm/sec). We later switched to an e-beam evaporator ($P_{base} \sim 1 \times 10^{-7}$ Torr; $P_{depo} \sim 5 \times 10^{-7}$ Torr; rate $\sim$ 0.2 nm/sec) but were unable to detect any difference in the charge offset noise level between the two methods.

While imperfections in the insulator almost certainly cause the Q_0 noise, its exact location (and moreover its microscopic form) remains somewhat controversial[9-13]. One faction maintains that the noise must originate in the tunnel junctions because the electric fields are much more intense. Another faction maintains that the noise originates outside the tunnel junctions because the volume of insulating material and its associated imperfection density is likely much larger. Experimental evidence has implicated both sources, possibly owing to the different techniques used to fabricate the devices.

Our early work did not attempt to isolate the location of the noise, but instead attempted to improve all insulators near or in the CB devices. We first operated under the assumption that the Q_0 noise in our Al/Al_2O_{3-x}/Al-based CB devices was driven by charge traps which were similar in nature to charge traps SiO_2. Borrowing a technique for charge trap passivation[14] in SiO_2, we annealed the SETTs in Forming gas for 30 min at 350 °C. Unfortunately however, annealing in a reducing environment did not significantly improve the low frequency Q_0 noise. We also considered the possibility that the Q_0 noise was caused by oxygen deficiencies at the Al/Al_2O_{3-x} interface. In the magnetic tunnel junctions this problem is particularly deleterious, as unoxidized aluminum reduces the magnetoresistance[15]. To improve the interface, magnetic tunnel junctions can be annealed in an inert environment[15]. While the Al/Al_2O_{3-x}/Al tunnel junctions contain only a single metallic element (unlike magnetic tunnel junctions), we suspected that annealing in an inert environment might also sharpen the interface, possibly because the annealing temperatures are well below the aluminum/oxygen phase separation temperature but high enough to permit significant atomic motion. We therefore annealed the devices in a relatively inert environment (N_2), also at 350 °C for 30 min. Again however, we were unable to reliably reduce the low-frequency Q_0 noise. Recent evidence has suggested that by switching from O_2 to an O_2/O_3 mixture, the interface state trap density in Si can be reduced[16] and the quality of large area Nb Josephson junctions similarly improved[17]. We thus attempted to improve the small area tunnel junctions in the CB devices by switching from pure O_2 to (5% O_3)/O_2 but were again unsuccessful at reducing the charge offset noise. Finally we considered the possibility that the charge offset noise arose from mechanical stresses occurring, presumably, from thermal expansion mismatches between the various materials in or near the device. However thermally cycling the devices between 4 K and 300 K did not affect the long term charge offset noise.

More recently we screened noise sources more than $\sim$ 100 nm away by, for the first time, encasing SETTs inside 'nano-Faraday' cages. In these structures, the cage consists of two Al thin film plates which sandwich the SETT. While the sides of the nano-Faraday cage are open, the

SETT is still well-shielded from outside noise because the gap between the plates is an order of magnitude smaller than the distance between the SETT and the nearest plate edge. Because $Al/Al_2O_{3-x}/Al$ based CB devices generally cannot tolerate post-fabrication processing, the entire structure was fabricated with a single lithographic mask using an extension of the conventional angled deposition technique.

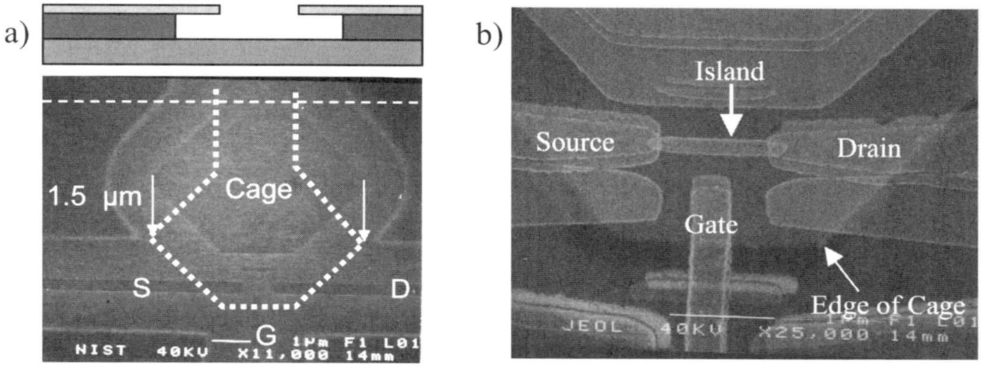

Figure 1. a) SEM Micrograph of lithographic mask used to fabricate nano-Faraday cage. A cross-section of the mask through the horizontal dashed line is shown above. The large hexagonal opening at the top of the mask is used to fabricate the two mask plates. b) Micrograph of SET transistor encased in nano-Faraday cage. Fabrication was stopped prior to depositing the top layers of the cage to permit viewing.

Figure 1a shows a micrograph of the lithographic mask. By depositing Al at a shallow angle the bottom plate of the cage is placed underneath the SETT device area. Al_2O_{3-x} is then e-beam evaporated from an Al_2O_3 charge in the presence of ~ 1 x 10^{-5} Torr O_2 to electrically isolate the cage from the SETT. The SETT is then fabricated using double angled Al evaporations at near normal incidence. To cap the device, Al_2O_{3-x} is again deposited before depositing the top layer of the cage.

Because this technique requires a large mask undercut to insure that the bottom of the cage lies comfortably underneath the SETT and because ~ 10 depositions are required to fabricate these structures, film stress and/or heating caused the masks to severely warp as shown in Fig. 2b. To stiffen the mask and provide the additional heat sinking, a 40 nm SiO layer was inserted between the bottom PMGI layer and the top PMMA layer. Additionally, a 20 nm Au layer was deposited prior to fabricating the cage bottom to decrease adhesion between deposited material and the mask. Combining these two techniques, mask warping was considerably reduced (Fig. 2c). Absent the cage top to permit examination, a SEM micrograph of a nano-Faraday caged SETT is shown in Fig. 1b.

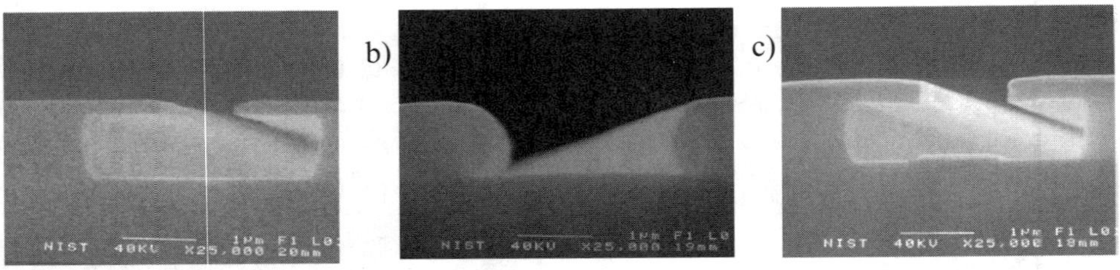

Figure 2. a) Cross-section of large undercut lithographic mask before deposition. b) Severe warping of conventional PMMA/PMGI bi-layer mask after depositing ~50 nm Al.. c) Decreased warping of PMMA/SiO/PMGI tri-layer mask after depositing ~ 20 nm Au/50 nm Al..

As anticipated, the nano-Faraday caged SETTs are far more robust than their uncaged counterparts. While conventional Al/Al_2O_{3-x}/Al-based CB devices are extremely sensitive to static electricity, the caged SETTs could be examined for several minutes using a SEM operating at 40 kV without destroying the tunnel junctions. In addition, caged devices could be left in ordinary tap water (which rapidly destroys uncaged devices) for over 24 hours without affecting the tunnel junction resistance. Thus encasement of Al/Al_2O_{3-x}/Al-based CB devices should likely permit limited post-fabrication processing. Unfortunately, however both the short and long term Q_0 noise in the caged devices was approximately a factor of two higher than uncaged SETTs. The additional noise was caused by defects in the deposited Al_2O_{3-x} film used to electrically isolate the SETT from the cage. Attempts to reduce the noise, including changing the background O_2 pressure, switching from O_2 to (5% O_3)/O_2 and varying the deposition rate did not unfortunately affect the charge offset noise level. Because the e-beam deposited Al_2O_{3-x} films are reliable noise sources however, they proved to be quite useful in determining the origin of a component of the charge offset noise.

3 Understanding the Charge Offset (Q_0) Noise Origin

While evaluating our attempts to reduce the charge offset noise in Al/Al_2O_{3-x}/Al based CB devices, we observed that the low frequency (< 10^{-3} Hz) Q_0 noise tended to decrease strongly over time. To parameterize the behavior, we measured the periodic transconductance of a SETT over extended periods of time to determine the phase shift and thus the charge-offset (Q_0). In Fig. 3a, Q_0 vs. time is shown for a conventional Al/Al_2O_{3-x}/Al based SETT at T ~ 30 mK. A drastic change of the Q_0 stability as a function of time is clearly evident. Unfortunately, we were unable to consistently and unambiguously observe the effect with SETTs fabricated using the conventional double angle deposition technique. To controllably introduce noise, Al_2O_{3-x} was deposited at well-defined locations relative to the Al/Al_2O_{3-x}/Al SETT using an extension of the nano-Faraday caged fabrication technique.

We can quantify the time dependence of the characteristic switching rate by determining the characteristic time Γ_Q^{-1}, before the charge offset deviates outside a predetermined window size (e.g. δQ_0 = 0.1 e). Introducing controlled noise sources permitted us to determine Γ_Q^{-1} due to a specific noise source.

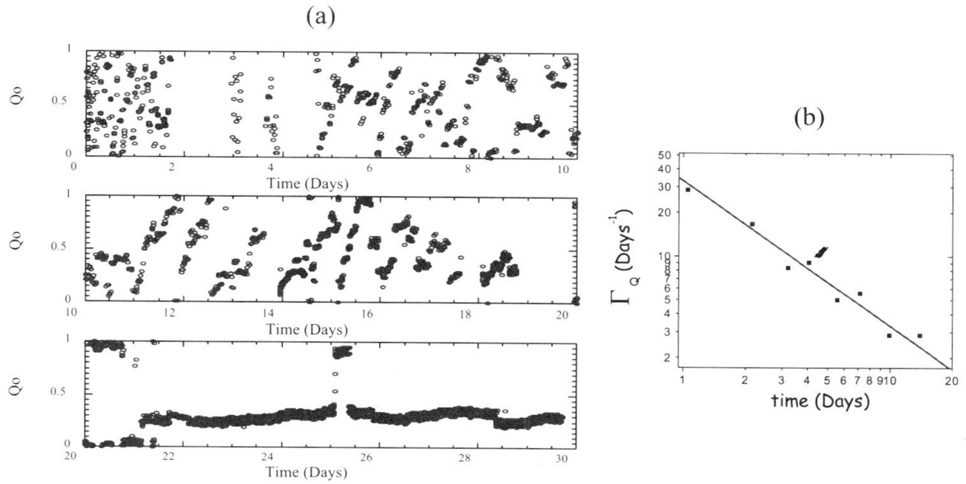

Figure 3. a) Dependence of the charge-offset (Q_0) over time in a SETT without a deliberate noise source. b) Characteristic switching rate Γ_Q as a function of time since fabrication for a SETT with a deliberate noise source. This response was also observed in a fraction of the devices without deliberate noise sources, such as in Fig. 3a.

In Fig. 3b the switching rate Γ_Q is shown as a function of time since fabrication for a SETT with a deliberate noise source. For the first several weeks, *the characteristic switching rate is inversely proportional to time since fabrication*. The inverse time dependence was observed in all $Al/Al_2O_{3-x}/Al$ based SETTs with deliberate noise sources and in most SETTs without deliberate noise sources, albeit usually less pronounced.

The Q_0 transient response was recently attributed to the same mechanism responsible for anomalous heat evolution in disordered materials[18]. For almost thirty years it has been recognized that disordered materials contain large numbers of two-level systems (TLS) which in general are broadly distributed in energy and relaxation time[19]. At temperatures high compared to the maximum TLS energy splitting Δ, the two states of the TLS are statistically equally populated and thus do not contribute appreciably to the specific heat. When a disordered material is quenched from high temperatures however, TLS with $\Delta > k_B T$ relax, releasing packets of heat which can be calorimetrically detected. Experiments performed over the past twenty years have demonstrated that in a broad range of materials with varying degrees of disorder, *the heat evolution is inversely proportional to time since quenching*[19].

By extending conventional TLS theory, we are able to explain the transient response of the charge offset noise[18]. It seems likely that the insulators surrounding the SET device contain large number of TLS (whose microscopic origin is not critical for the model). Each time a TLS relaxes, it emits a packet of energy. This process can change Q_0 in at least two ways. First, the atoms constituting the TLS, which likely move during the relaxation, may change their charge configuration. The charge reconfiguration will then induce a charge change on the central island which then changes the source drain current of the SETT. Second, the packet of energy from the relaxing TLS, likely in the from of a phonon, may travel some distance before being absorbed in another insulating region. The energy may then cause the insulator to rearrange which again would be detected using the SETT. Inserting rough estimates of the TLS density and characteristic volume, we have obtained relatively good agreement between theory and experiment[18].

At sufficiently long times, the $Al/Al_2O_{3-x}/Al$-based CB devices cease to significantly improve with age. For the majority of these devices, a Q_0 change larger than ~ 0.1 e is typically observed every 1 day to 3 days. In a single, non-repeatable instance, partially shown in Fig. 3a, *the Q_0 remained stable, except for a single excursion, for twelve days* - our best result. The $Al/Al_2O_{3-x}/Al$ Q_0 baseline noise is a surprisingly consistent value: it is insensitive even to high temperature anneals and processing modifications. Thus regardless of the baseline noise source, $Al/Al_2O_{3-x}/Al$ based tunnel junction systems will probably not be sufficiently stable for even moderate scale CB device integration.

Recent results have shown however that Silicon based SETTs fabricated using a pattern dependent oxidation process have an exceptionally small low frequency Q_0 noise[20]. One device drifted by $\delta Q_0 \sim 0.005$ e over several days. Further research on the mechanism responsible for the remarkably low charge offset noise is being pursued.

References

1. Single Charge Tunneling, Coulomb Blockade Phenomena in Nanostructures, ed. H. Grabert and M.H. Devoret (Plenum, New York, 1992).
2. M. W. Keller, J. M. Martinis, N. M. Zimmerman and A. H. Steinbach, Appl. Phys. Lett. **69**, 1804 (1996).
3. J. Kim, O. Benson, H. Kan and Y Yamamoto, Nature **397**, 500 (1999).
4. Y. Nakamura, Y. A. Pashkin and J. S. Tsai, Nature **398**, 686 (1999).
5. B. E. Kane et al., Phys. Rev. B **61**, 2961 (2000).
6. G. Zimmerli, T. M. Eiles, R. L. Kautz and J. M. Martinis, Appl. Phys. Lett. **61**, 237 (1992).
7. J. S. Tsai, this conference.
8. G. J. Dolan, Appl. Phys. Lett. **31**, 337 (1977).
9. N. M. Zimmerman, J. L. Cobb and A. F. Clark, Phys. Rev. B **56**, 7675 (1997).
10. A. J. Manninen and J. P. Pekola, Czech. J Phys **46 (Suppl S4)**, 2293 (1996).
11. A. B. Zorin et al., Phys. Rev. B **53**, 13682 (1996).
12. V. A. Krupenin et al., J. Appl. Phys. **84**, 3212 (1998).
13. D. Song, A. Amar, C. J. Lobb and F. C. Wellstood, IEEE Appl. Superconductivity **5**, 3085 (1995).
14. S. M. Sze in Physics of Semiconductor Devices, 2nd ed. (John Wiley, New York, 1981).
15. D. Song, private communication.
16. T. Nakanishi, S. Ohkubo and Y. Tamura, Fujitsu Sci. Tech. J. **32**, 128 (1996).
17. L. Fritch, H.-J. Köhler, F. Thrum, G. Wende and H.-G. Meyer, Physica C **296**, 319 (1998).
18. W. H. Huber and N. M. Zimmerman, submitted for publication.
19. A. Nittke, S.Sahling and P. Esquinazi in Tunneling Systems in Amorphous and Crystalline Solids, ed. P. Esquinazi (Springer Verlag, Heidelberg, Berlin, New York, Tokyo, 1998).
20. W. H. Huber, N. M. Zimmerman, A. Fujiwara and Y. Takahashi, private communication.

ADVANCED TECHNIQUES FOR SILICON SINGLE-ELECTRON DEVICES

YASUO TAKAHASHI, YUKINORI ONO, AKIRA FUJIWARA, KENJI SHIRAISHI, MASAO NAGASE, SEIJI HORIGUCHI, AND KATSUMI MURASE

NTT Basic Research Laboratories, 3-1, Morinosato Wakamiya, Atsugi-shi, 243-0198 Japan
E-mail: ytaka@aecl.ntt.co.jp

We have developed two novel methods for fabricating Si single-electron transistors (SETs). These methods exploit special phenomena that occur when a small laterally-or vertically-modulated Si layer on SiO_2 is oxidized. Hence, we call the methods pattern-dependent oxidation (PADOX) and vertical pattern-dependent oxidation (V-PADOX). The Si layer is converted into a small Si island with a tunnel barrier at each end. The conversion is a consequence of stress-induced bandgap narrowing and quantum size effect. Since the size of the resultant Si island is about 10 nm, the SETs operate at relatively high temperatures.

1 Introduction

The single-electron transistor (SET) is expected to be a key device for future extremely large-scale-integrated circuits because of its ultrtalow power consumption and small size. The most difficult aspect of fabricating SETs is the formation of a nanometer-scale island sandwiched between two small tunnel capacitors. Although many methods have been proposed, the operating temperature of SETs has been very low owing to the relatively large total capacitance of the island. This is because the minimum feature size of lithography is used for the formation of the tunnel barrier, not for the formation of the island. To overcome this problem, we have developed a method to attach two tunnel barriers to a very small Si island in a self-aligned manner. [1,2]

As a base material for SETs, any conductive material can be used. Among the many kinds of semiconductors and metals available, Si is the most promising because we can utilize fabrication technologies that are continuously advancing. The International Technology Roadmap for Semiconductors predicts the minimum feature size of 22 nm by 2014. One of the most reproducible and simple process technologies for Si LSIs is thermal oxidation of Si, which produces a very stable SiO_2 indispensable for achieving ultralarge-scale integrated circuits. The oxidation rate depends on mechanical stress developing as a result of volume expansion during thermal oxidation. [3,4] The influence of stress becomes striking especially for oxidation of nanometer-scale Si structures. We found that, by using stress-dependent oxidation, a small appropriately patterned Si layer can be converted into a small Si island furnished with tunnel barriers in a self-aligned manner. We call this conversion method pattern-dependent oxidation (PADOX), because the amount of oxidation at a particular point is modulated in accordance with the initial pattern. [1,5]

2 SETs fabricated by PADOX

The structure fabricated on a thin (001) SOI (Silicon On Insulator) layer contains a narrow and short Si wire oriented to the [110] direction as shown in Fig. 1(a). Typical wire width is about 30 nm. The wire height, or the thickness of the SOI layer, is also about 30 nm. The

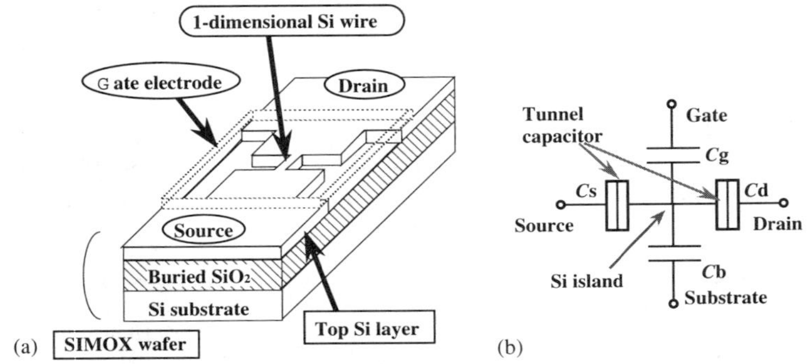

Figure 1. Initial device structure of the SET before PADOX (a), and equivalent circuit of the device (b). The small one-dimensional Si wire is converted to a small Si island. The Si island is almost completely surrounded by the gate electrode, substrate Si, and source and drain electrodes.

wire length is varied from 30 to 200 nm for obtaining SETs with different gate capacitances. [5] When such a Si structure is subjected to thermal oxidation, a huge amount of compressive stress is exerted on the middle part of the wire because this part is surrounded completely by newly grown SiO_2. It is reported that a compressive stress larger than 20000 atm accumulates in the newly grown SiO_2 when a 10-nm-diameter Si wire is formed by oxidizing a 30-nm-diameter Si wire. [6] This large stress causes about a 150 meV bandgap reduction, which cancels out an effective bandgap increase due to the quantum size effect of about 50 meV in a 5-10 nm Si wire. [7,8] As a result, a potential profile suitable for a SET is formed as schematically shown in Fig. 2. The two potential hills serve as a tunnel barriers: a tunnel capacitor is formed between the Si island and end of the wider Si layer. The great advantages of this method are that an island smaller than the initially defined size can be made due to the conversion from Si to SiO_2 and tunnel barriers are formed automatically at both ends of the wire. [7,8]

The typical conductance vs. gate-voltage (V_g) characteristics of a SET fabricated by PADOX are shown in Fig. 3(a). Here, the device temperature is 40 K. The initial width, length and height of the wire are 30, 70, and 30 nm, respectively. The thermal oxidation condition used for PADOX is 1000°C in a dry oxygen ambient. We have also successfully obtained SETs by using PADOX at 1100°C. [9] Figure 3(b) shows a contour plot of $\partial I/\partial V_d$ as a function of drain voltage (V_d) and V_g. Clear Coulomb-blockade diamonds are observed together with some exited states appearing as fine structures parallel to the edges of the diamonds. Total capacitance of less than 2 aF is obtained owing to the small island size.

3 SETs fabricated by V-PADOX

We have developed another method for fabricating SETs. In this method, the initial Si pattern is vertically modulated, hence V-PADOX. [2] The initial structure contains a fine trench formed in the middle of a Si wire on a (001) SOI substrate as shown in Fig. 4(a). The wire is oriented to the [110] direction as in the case for PADOX. The thickness of the Si layer under the trench is about 15 nm. Thermal oxidation in a dry oxygen ambient at 900°C converts the thin Si layer under the trench into two small islands embedded in

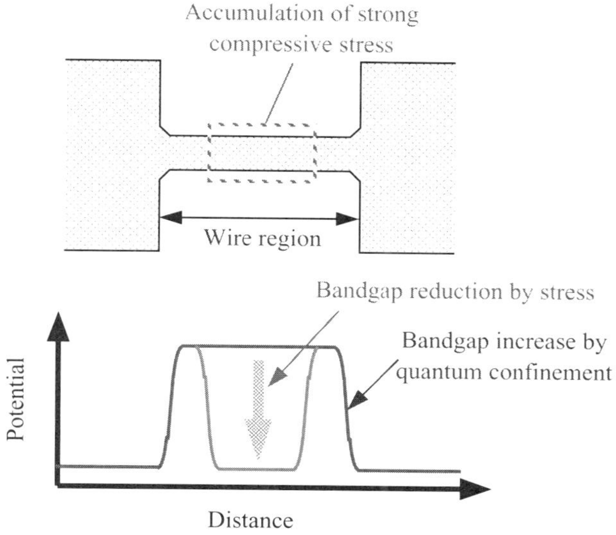

Figure 2. Schematic plane view of the device and potential diagram along the Si wire.

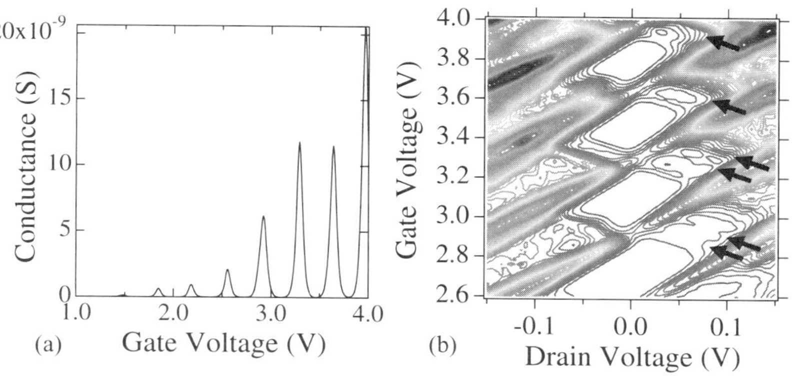

Figure 3. Conductance oscillations as a function of the gate voltage measured at 40 K and at a drain voltage of 1 mV (a) and contour plot of $\partial I/\partial V_d$ as a function of drain voltage V_d and V_g (b). Arrows indicate exited states.

both edges of the Si layer. This is because less oxidation occurs around the edges due to stress accumulation, which allows us to produce two tiny Si islands without the need for lithographic definition of the islands themselves. A tunnel barrier is formed at either side of the Si island due to almost the same mechanism as PADOX. The rest of the thin region under the trench is converted into SiO_2. A schematic structure of the device and an equivalent circuit are shown in Fig. 4(b) and (c). In V-PADOX, since the size of the

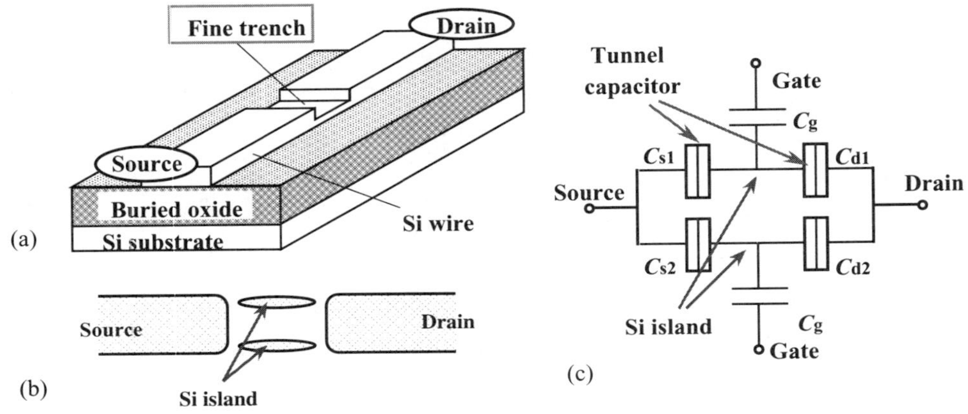

Figure 4. Initial structure of the SETs before V-PADOX (a), schematic structure of the SETs after V-PADOX (b), and the equivalent circuit of the device (c). W, L, d represent the designed wire width, trench length, and trench depth, respect.

SET island is mainly determined by the thickness and length of the Si layer under the trench, almost independently of the wire width, the size is less affected by the lithographic fluctuation.

V-PADOX forms twin SETs connected in parallel, as shown in Fig. 4(c). But, by modulating the two-dimensional pattern of the thicker Si layer, we can also connect the two islands in series as shown in Fig. 5. In this pattern, we attached a side gate electrode to each SET island formed at the edge of the small trench, which enables us to control the electrostatic potential of the two SETs individually.[10] As a result, the conductance of each SET can be measured independently. The typical gate-voltage dependence of the conductance, measured at 30 K, is shown in Fig. 6(a) for one of the twin-island SETs with a trench length of 50 nm and a thickness of 15 nm. A contour plot of $\partial I/\partial V_d$ as a function of drain voltage (V_d) is shown in Fig. 6(b), where Coulomb diamonds and exited states are also clearly observed.

4 Discussion

The starting patterns for PADOX and V-PADOX are different. As seen in Figs. 3(b) and 6(b), however, the contour plots of $\partial I/\partial V_d$ of the SETs fabricated by PADOX and V-PADOX are very similar. This strongly implies that the mechanisms of the island and tunnel barrier formation are the same for the two methods. In the initial stage of the oxidation process in V-PADOX, the two-dimensional Si layer under the trench is converted into two parallel wires because of slow oxidation around the edges due to stress accumulation. This forms two small wires automatically. And then, the compressive stress in the narrow wires converts them to small SETs. This is the same formation mechanism in PADOX.

Another difference between PADOX and V-PADOX is the oxidation temperature. The fact that the SETs fabricated by PADOX at 1100 and 1000°C operate in a similar manner implies that the oxidation temperature dose not play critical role in SET fabrication.

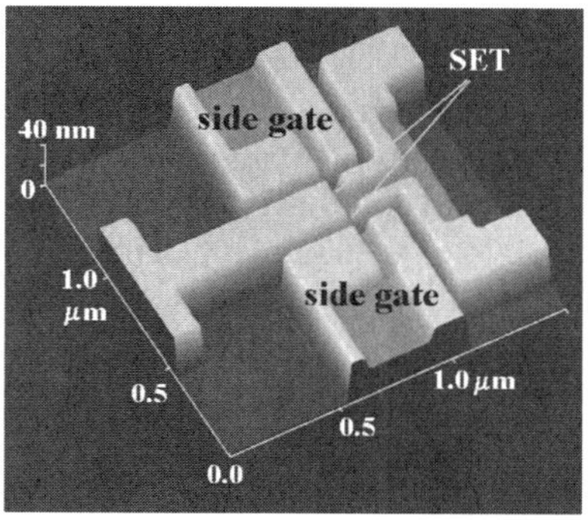

Figure 5. AFM (Atomic Force Microscopy) image of the device structure having two SETs connected in series.

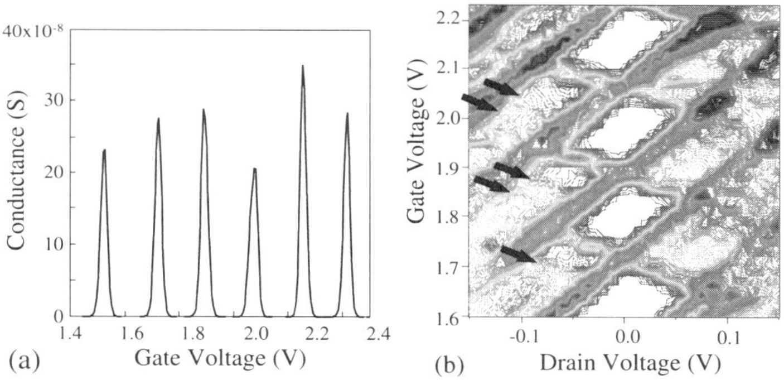

Figure 6. Conductance oscillations as a function of the gate voltage measured at 30 K and at a drain voltage of 1 mV (a) and contour plot of $\partial I/\partial V_d$ as a function of drain voltage V_d and V_g (b). Arrows indicate exited states.

However, the lower temperature causes slower oxidation rate and produces relatively strong stress in spite of the thinner oxide. Therefore, the V-PADOX device has a relatively thin SiO_2 layer, which means the SET has a relatively large gate capacitance, C_g, in comparison with source and drain capacitance, C_s and C_d. This results in larger voltage gain if the SET is used for a signal-processing device.

The most prominent feature of the SETs fabricated by PADOX and V-PADOX is that they work very stabely during thermal cycling between low and high temperature. After a one-year interval, the same characteristics were obtained. This feature is important for practical applications of SETs as memory devices, logic devices, and in quantum computing. In addition, the fabrication process is almost the same as that for conventional Si LSIs, which allows us to use SETs in combination with conventional Si devices. [11]

5 Conclusion

We have developed two special methods for fabricating Si SETs: PADOX and V-PADOX. Both methods utilize a special oxidation phenomenon that occurs when a very small Si structure is thermally oxidized and enable us to fabricate small SETs in a self-aligned manner. PADOX and V-PADOX are promising methods for SETs because of their flexibility and the very stable operation of the fabricated devices.

Acknowledgments

The authors thank Dr. Katsutoshi Izumi, Dr. Takahiro Makino and Dr. Takaaki Mukai for their continued encouragement and Dr. Michiharu Tabe, Dr. Hideo Namatsu, Dr. Kenji Kurihara, Kenji Yamazaki, Yasuyuki Nakajima, Kazumi Iwadate, Dr. Hiroyuki Kageshima, and Toru Yamaguchi for valuable discussions and their support in the experiments.

References

1. Y. Takahashi, M. Nagase, H. Namatsu, K. Kurihara, K. Iwadate, Y. Nakajima, S. Horiguchi, K. Murase, and M. Tabe, *Electron. Lett.* **31**, 136 (1995).
2. Y. Ono, Y. Takahashi, K. Yamazaki, M. Nagase, H. Namatsu, K. Kurihara, and K. Murase, *IEEE Trans. Electron Devices* **47**, 147 (2000).
3. D. Kao, J. P. McVittie, W. D. Nix, and, K. C. Saraswat, *IEEE Trans. Electron Device* **35**, 25 (1988).
4. M. Nagase, A. Fujiwara, K. Yamazaki, Y. Takahashi, K. Murase, and K. Kurihara, *Microelectronic Engineering* **41/42**, 527 (1998).
5. Y. Takahashi, A. Fujiwara, M. Nagase, H. Namatsu, K. Kurihara, K. Iwadate, and K. Murase, *Int. J. Electronics* **86**, 549 (1999).
6. H. I. Liu, D. K. Biegelsen, N. M. Johnson, F. A. Ponce, R. F. W. Pease, *J. Vac. Sci. Technol.* **B 11**, 2532 (1993).
7. K. Shiraishi, M. Nagase, S. Horiguchi, H. Kageshima, M. Uematsu, Y. Takahashi, and K. Murase, *Physica E* **7**, 337 (2000).
8. S. Horiguchi, M. Nagase, K. Shiraishi, H. Kageshima, Y. Takahashi, and K. Murase, *Jpn. J. Appl. Phys.* **40**, L29 (2001).
9. A. Fujiwara, Y. Takahashi, M. Nagase, and K. Murase, *Trans. Material Reserch Society Jpn.* **26**, to be published (2001).
10. Y. Ono, Y. Takahashi, K. Yamazaki, M. Nagase, H. Namatsu, K. Kurihara, and K. Murase, *Appl. Phys. Lett.* **76**, 3121 (2000).
11. Y. Takahashi, A. Fujiwara, K. Yamazaki, H. Namatsu, K. Kurihara, and K. Murase, *Jpn. J. Appl. Phys.* **38**, 2457 (1999).

QUANTUM MEASUREMENT WITH NANOMECHANICAL SYSTEMS

KEITH SCHWAB
Laboratory for Physical Sciences, 8050 Greenmead Drive, College Park, MD 20740, USA
E-mail: schwab@lps.umd.edu

The technology exists for preparing and measuring nanoscale mechanical systems at the quantum limit. A nanomechanical system has recently demonstrated the fundamental limit for heat flow through discrete channels. This universal energy transport limit is intimately related to the predicted maximum information rate per channel. I will discuss this measurement and experiments that are under preparation at LPS, using nanomechanical resonators with rf-SET read-out to achieve position detection limited by the uncertainty principle: the standard quantum limit. This system offers the possibility of implementing advanced measurement strategies, such as quantum non-demolition techniques, to beat this limit. In addition to revealing the fascinating collective quantum behavior of 10^{10} atoms, developing ultra-sensitive position sensors may be important for quantum computing applications such as single electron spin read-out with cantilevers, and provide an understanding of the interaction between delicate quantum systems and real read-out devices.

1 Introduction and Motivation

A central challenge for constructing quantum computing devices is read-out. A situation must be engineered where one can decouple a detector from the qubit to allow free evolution, and then couple the detector strongly in order to read-out the state of the qubit, all before the information is lost to the corrupting influence of decoherence. This is a very delicate experimental problem. The experiments proposed here with nanomechancial systems coupled to nanoelectronic readout intend to shed light on these problems.

A very exciting goal of this research is to observe the collective quantum properties of mechanical systems comprised of over 10^{10} atoms. Given the last advances in nanomachining,[1,2] advanced nanoelectronic detectors[3], and ultra-low temperature techniques it appears possible that the fully quantum behavior of mechanical systems may become apparent: resolution to the uncertainty principle, generation of superposition states, and entanglement with other quantum systems. To reach the quantum limit with a mechanical system has been a long discussed limit by the gravitational wave community[4,5] and realized to be an important chapter to the understanding of measurement.[6,7]

In order to be able to implement and develop advanced quantum measurement techniques, such as stroboscopic, squeezing,[8] and quantum non-demolition techniques, it is essential to have a resonator and position detector at the quantum limit. This short paper describes some of the considerations for achieving uncertainty principle limited detection of a nanomechanical systems, and gives a hint of what might lie ahead for this research. I will focus on the proposed system using an *rf*-SET and a nanomechanical device[9], and especially on the back-action of the measuring device on the resonator. Understanding the subtleties of both *rf*-SET read-out and the limits of nanomechanical resonators will be especially relevant to proposed schemes to read-out Cooper-pair qubits[10,11] and single electron spins[12].

2 Nanomechanics and the Standard Quantum Limit

If one performs two consecutive position measurements, the first measurement will effect the out come of the second measurement through the uncertainty principle 'kick' to the momentum from the first measurement. This will place a minimum uncertainty for two consecutive measurements of a freely ringing simple harmonic oscillator $\delta x_{SQL} = \sqrt{\frac{\hbar}{2m\omega_0}}$, which is referred to as the "Standard Quantum Limit" (SQL).[6] This fundamental consequence of $[x,p] = i\hbar$ has never been observed for a mechanical system.

In order to observe this physics, δx_{SQL} can be maximized by minimizing the product of $m\omega_0$. In previously demonstrated Si nanomechanical resonators with $w = 100nm$, $t = 200nm$, $l = 2.3\mu m$, $m \approx 10^{-16}kg$, $\omega_0 = 2\pi \cdot 10^7 rad/sec$, $Q = 10^4$, one expects $\delta x_{SQL} = 8 \cdot 10^{-14}m$, where as a resonant mass gravitational wave detector m=10^3kg and $\omega_0 = 2\pi \cdot 10^3 rad/sec$, $\delta x_{SQL} = 3 \cdot 10^{-21}m$. Nanoscale systems should reveal their quantum properties at a much larger length scale, facilitating observation.

Although the higher resonant frequency will decrease δx_{SQL}, it has an important advantage. A resonator will sit in the ground state when $k_B T < \hbar\omega_0$. For a resonator with $\omega_0 = 2\pi \cdot 10^7 Hz$, this 'freeze-out' occurs at $\approx 1mK$. Temperatures as low as $100\mu K$ are routinely achieved for macroscopic samples, but how easy it will be to cool nanomechanical systems to this temperature is an unexplored question (but probably not easy!).

There are penalties for considering such a small resonator. Because of the microscopic size of the device, efficient coupling to read-out amplifiers becomes very difficult. A solution to this problem is fabricating an amplifier (SET or SQUID) in close proximity to the nanomechanical system. In this way the parasitic impedances which divide away signals, are avoided.

With the development of the *rf*-SET[3] there has been proposal to use this device as a position detector[9]. This scheme is shown in Figure 1 and relies on the modulation of the gate capacitance by the motion of the mechanical system. In Ref. [9], the authors calculate the displacement sensitivity by evaluating both the coupling of the mechanical device to the *rf*-SET, dI_{DS}/dx, and the noise, given by the shot noise through the SET, $S_I \approx 2eI_{DS}$. The authors conclude that detection with sensitivity at the SQL is possible by providing sufficiently strong coupling,V_g, between the SET and the nanomechanical resonator.

Unfortunately, there is more to the story than just forward coupling; one must consider the back-action from the SET onto the mechanical resonator. Each time an electron tunnels onto the island of the SET, the potential of the island, U, changes by the Coulomb blockade voltage, $\delta U = E_c/e \approx 270\mu V$[a]. This noise is spread over a frequency range of $\approx I_{DS}/e \approx 30GHz$, typically. This picture is born-out in numerical simulation[14] where the power spectrum of the island potential fluctuations, S_U is found to have a white spectrum with amplitude: $S_U = \left(\frac{e}{2C_\Sigma}\right)^2 \frac{e}{I} \longrightarrow 1.5nV/\sqrt{Hz}$

This fluctuating voltage will produce a fluctuating force to the resonator, producing motion which in turn modulates the gate capacitance and is detected by the SET. Both noise sources can be included by using the engineering model of the SET as a perfect amplifier with input voltage noise S_V and current noise S_I[13]. The power spectrum of the position noise will be given by the sum of noise power contributions of the equivalent voltage

[a] I consider an optimized SET[13] $C_\Sigma = 0.3fF, C_g = 20aF, R_J = h/e^2, V_{DS} = e/2C_\Sigma$

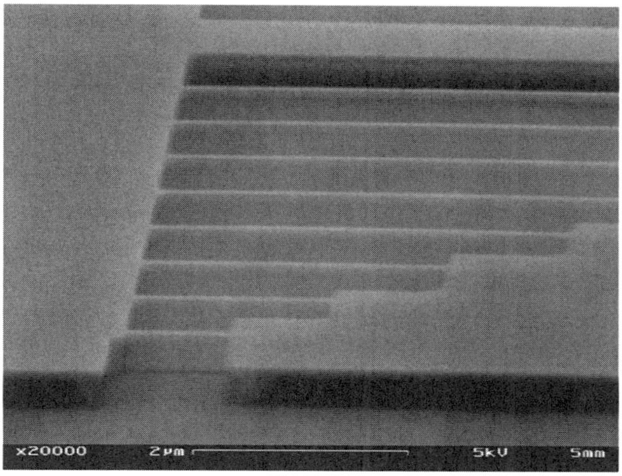

Figure 1. Scheme to couple a *rf*-SET to a nanomechanical resonator, where $d = 200nm$ is the gap between the resonator and the SET gate capacitance C_g, biased at voltage V_g. Metalization which defines the SET and the electrode on the resonator is shown as dark shading.

noise and the back-action current noise: $S_x^{total} = S_V \left(\frac{dx}{dV}\right)^2 + S_I \left(\frac{dx}{dI}\right)^2$ For the situation shown in Figure 1 it is found that $\frac{dx}{dV} = \frac{C_g}{V_g}\left(\frac{dC_g}{dx}\right)^{-1}$, and $\frac{dx}{dI}(\omega = \omega_0) = \left(\frac{QV_g}{m\omega_0^3 C_g}\frac{dC_g}{dx}\right)$, and using an optimized SET: $S_V = \frac{hC_\Sigma}{4C_g^2}$, $S_I = \frac{h\omega^2 C_g^2}{2C_\Sigma}$.

The total position noise power, S_x^{total}, will have a minimum value when $\left(\frac{dC_g}{dx}V_g\right)^2 = \frac{C_\Sigma m\omega_0^2}{\sqrt{2}Q}$ (when $V_g \approx 70mV$) giving $\sqrt{S_X} = \sqrt{\sqrt{2}\pi Q\frac{\hbar}{m\omega_0^2}} = 3.1 \cdot 10^{-15}m/\sqrt{Hz}$. The total rms amplitude of the mechanical system is given by $\delta x_{min} = \sqrt{S_x}\sqrt{\frac{\omega_0}{2\pi Q}} = \sqrt{\frac{\hbar}{\sqrt{2}m\omega_0}} = 1.2 \cdot \delta x_{SQL}$. Thus for only the most carefully engineered SET can one approach the SQL.

The situation may be improved by a more sophisticated position detector based on the Josephson electrometer[15,16]. This device operates in the superconducting state with two junctions in series, defining a small Coulomb blockade island as in an SET, and is expected to have quantum limited energy sensitivity. The superconductivity has an effect on the nature of the back-action. The transport in this device is through coherent, Josephson tunneling. Thus, the island potential fluctuations, which are responsible for the amplifier back-action, will oscillate coherently at the Josephson frequency, which can be far from the mechanical resonance. This is in contrast to the white spectrum of potential fluctuations in a normal state SET. The low frequency island voltage noise is a result of the Johnson noise of an external shunt resistor (which is essential to provide the proper damping of the junctions). The power spectrum of the island potential fluctuations is predicted to be[15]: $S_U = \frac{1}{4\pi}\left(\frac{\Phi_0}{2\pi}\right)^2 \frac{g^2 R_s I_c^2}{eV} \coth\frac{eV}{k_B T}$, where Φ_0 is the flux quantum, $R_s = 1k\Omega$ is the resistance of the shunting resistor at temperature T, $I_c = 100nA$ is the critical current, V is the bias voltage, and g is a parameter ≈ 2. If we evaluate this expression in the limit $k_B T < eV$ where the fluctuations are minimized and are due to zero point fluctuations,

$\sqrt{S_U} \approx 400pV/\sqrt{Hz}$ which is to be compared to $1.5nV/\sqrt{Hz}$ for the optimized normal state SET.

The dissipative properties of the Josephson electrometer may have advantage compared to the normal state SET. In the normal state SET, the dissipation is located at the microscopic island and typically will hold the electron temperature $\approx 100mK$ due to the hot-electron effect[17]. This might be a problem for systems that are in close proximity to the SET which must be much colder. For the Josephson electrometer, there is no dissipation at the island since the transport is coherent Cooper-pair tunneling. The dissipation is located at a relatively remote shunting resistor where its temperature will be limited by the physical volume. The physical size of this shunting resistor is restricted to be much smaller than the wavelength of light at the Josephson frequency ω_J. For parameters taken in Ref. [15], $\omega_J = 5GHz$, thus allowing a millimeter sized shunting resistor. With dissipated power $\approx I_c^2 R_s \approx 3pW$, and resistor of volume $1mm^3$, the temperature should saturate at $\approx 5mK$.

Finally, it is interesting to speculate as to whether the back-action could be "engineered" by placing the Josephson oscillations at a frequency where the resonator has very small response (such as at the anti-nodes between each mechanical resonance[18]). Also, it may be possible to employ parametric amplification by matching the Josephson frequency to the mechanical frequency.

3 Quantum Limit for Energy Transport

Recently,[b] a nanomechanical system has demonstrated a fundamental quantum limit for energy transport[19,20]. In this experiment, the temperature ($< 1K$) and size($< 200nm$) of the sample where such that the wavelength of the thermally excited mechanical vibrations (phonons) where much larger than the cross-sectional dimensions of the waveguides. Freeze-out of the higher order mechanical vibrations was achieved, allowing heat to flow only through the remaining one-dimensional propagating modes. Figure 2 shows the nanostructure where the center body containing a heater and thermometer are connected to the environment by four phonon waveguides [21].

In this limit where the transport is through a discrete set of one-dimensional channels, the thermal conductance is expected to be bounded by a universal value $G = g_0 = \frac{\pi^2 k_B^2 T}{3h}$, independent of material characteristics and particle statistics[22] This result has been anticipated since 1983 by Pendry[23] where he calculates the maximum heat current from the maximum information rate per channel. This simple piece of physics underscores the deep connection between thermal transport and information transport[24], and further emphasizes the point of view which is becoming apparent today, that information is a primitive physical quantity[25]. This result has been identified by some as the first quantum property observed in a mechanical system[26].

4 Future prospects: Schrodinger's Whisker

An extremely intriguing possibility is the generation and detection of a superposition and ultimately, entanglement of a mechanical system with other quantum systems. One possible scheme for producing a superposition is to couple a mechanical resonator with and electronic

[b]This work was performed with Prof. M.L. Roukes at his laboratory at Caltech.

Figure 2. SEM photo on left shows a fully suspended, nanomechanical device with integrated thermal transducers which demonstrated the quantum of thermal conductance. Black regions are vacuum; central square is $4\mu m$ x $4\mu m$ and is connected to bulk by four phonon waveguides, with minimum width 200nm x 85 nm. SEM photo on right shows a set of radio-frequency, 15-380MHz, nanomechanical resonator fabricated by the Craighead group at the Cornell Nanofabrication Facility . A resonator of this type could be used for these proposed experiments.

systems that can be prepared in a quantum superposition – a mechanical cantilever coupled to a Cooper-pair box[27]. Initial estimates, show that a resonator similar to that described above would have two mean position states separated by 12 pm for two possible charge states of the Cooper-pair box, and a decoherence time of the mechanical system of 100 ns, assuming T=1 mK.[28] This proposed arrangement might be an improvement over other proposals[29] since both systems, mechanical and electronic are in a pure quantum state.

This is a mesoscopic version of Schrodinger's Cat scenario; a macroscopic system becomes entangled with the quantum state of a single microscopic object. The nanofabricated resonators described here can hardly be considered as complex as a cat, but maybe as complex as a whisker on Schrodinger's chin!

5 Conclusions

Technology is at a place which will allow for the first time, preparation, observation, and manipulation of mechanical systems at the quantum limit. Experiments of this kind are aimed at pushing the boundary between the quantum and classical world to increasingly larger length scales. These mechanical systems will be very important for demonstrating interesting and advanced detection scheme and provide important understanding of how real detectors interact with condensed matter quantum systems. An extremely exciting possibility is to harness the coherent quantum behavior now being generated in single electronic devices as a resource to generate superpositions in a mechanical system. The coupled system of quantum coherent electronic systems with mechanical systems point to a new experimental realm which might be called **Q***uantum* **E***lectro*-**M***echanics* (QEM).

Acknowlegements

I would like to acknowlege many helpful conversations with M.L. Roukes and his group at Caltech, P.C. Hammel, M. Manheimer, B.E. Kane, M. Lahaye, M.P. Blencowe, and M.N. Wybourne. This work is supported by the National Security Agency (NSA) and the

Advanced Research and Development Activity (ARDA).

References

1. A.N. Cleland and M.L. Roukes. Applied Physics Letters **69**(18) 2653 (1996).
2. Dustin W. Carr, S. Evoy, L. Sekaric, H.G. Craighead, and J.M. Parpia. Applied Physics Letters **75**(7) 920 (1999).
3. R. J. Schoelkopf, P. Wahlgren, A. A. Kozhevnikov, P. Delsing, and D. E. Prober. Science **280**(5367)1238 (1998).
4. Mark F. Bocko and Roberto Onofrio. Reviews of Modern Physics **68**(3) 755 (1996).
5. V. B. Braginsky and F. Ya. Khalili. Reviews of Modern Physics **68**(1) 1 (1996).
6. Vladimir B. Braginsky and Farid Ya. Khalili. Quantum Measurement, Cambridge University Press, 1992.
7. Kip S. Thorne, Ronald W. P. Drever, and Carlton M. Caves. Physical Review Letters **40**(11) 667 (1978).
8. M. P. Blencowe and M. N. Wybourne. Physica B **280**(1-4) 555 (1999).
9. Miles P. Blencowe and Martin N. Wybourne. Applied Physics Letters **77**(23) 3845 (2000).
10. Alexander Shnirman and Gerd Schon. Physical Review B **57**(24) 15400 (1998).
11. Y. Makhlin, G. Schon, and A. Shnirman A. Fortschritte der Physik-Progress of Physics **48**(9-11) 1043 (2000).
12. G. P. Berman, G. D. Doolen, P. C. Hammel, and V. I. Tsifrinovich. Physical Review B **61**(21) 14694 (2000).
13. Michel H. Devoret and Robert J. Schoelkopf. Nature **406** 19 (2000).
14. A. N. Korotkov. Physical Review B **49**(15) 10381 (1994).
15. A. B. Zorin. Physical Review Letters **76**(23) 4408 (1996).
16. A. B. Zorin, S. V. Lotkhov, Yu. A. Pashkin, H. Zangerle, V. A. Krupenin, T. Weimann, H. Scherer, and J. Niemeyer. Journal of Superconductivity **12**(6) 747 (1999).
17. M.L. Roukes adn M.R. Freeman, R.S. Germain, and R.C. Richards. Hot electrons and energy transport in metals at millikelvin temperatures. **55**(4) 422 (1985).
18. S. Rast, C. Wattinger, U. Gysin, and E. Meyer. Review of Scientific Instruments **71**(7) 2772 (2000).
19. K. Schwab, E. N. Henriksen, J. M Worlock, and M. L. Roukes. Nature **404**(6781) 974 (2000).
20. K. Schwab, J. L. Arlett, J. M. Worlock, and M. L. Roukes. Physica E **9**(1) 60 (2001).
21. K. Schwab, E. N. Henriksen, and M. L. Roukes. Fabrication of a nanomechanical device for quantized energy transport experiments. *Manuscript in preparation.*
22. Luis G.C. Rego and George Kirczenow. Physica B **59**(20) 13080 (1999).
23. J. Pendry. J. Phys. E **16** 2161 (1983).
24. Miles P. Blencowe and Vincenzo Vitelli. Physical Review A **62** 052104 (2000).
25. John Preskill. Quantum computation. Lecture notes for Phys 219 at Caltech.
26. Leo P. Kouwenhoven and Liesbeth C. Venema. Nature **404** 943 (2000).
27. Y. Nakamura, Yu. A. Pashkin, and J.S. Tsai. Nature **398** 786 (1999).
28. Wojciech H. Zurek, Salman Habib, and Juan Pablo Paz. Physical Review Letters **70**(9) 1187 (1993).
29. S. Bose, K. Jacobs, and P.L. Knight. Physical Review A **59**(5) 3204 (1999).

QUANTUM NOISE IN THE RADIO FREQUENCY SET

R. E. S. POLKINGHORNE, G. J. MILBURN

SRC for Quantum Computer Te chnology, University of Queensland, Brisbane QLD 4072 Australia

Reading out the state of the qubits is one of the major c hallengespresented by the Kane proposal for a solid-state quantum computer. The result of a decoherence calculation for measurement with a mechanical electroscope is presented, and a report made of progress in adapting this calculation to the promising radio-frequency SET technology.

In the quantum computer architecture proposed by Kane[1], each qubit is represented by the spin of a single phosphorus nucleus embedded in a chunk of silicon. The advantage of nuclear spins is that they are slow to decohere; by the same token, measuring their state is a formidable challenge. It is intended that the exclusion principle will allow the qubit stored in the nuclear spin to be passed to the polarisation of a donor electron. However, monitoring the motion of the electron with the necessary precision (a few hundred angstroms, the separation between adjacent nuclei), will still require a very precise electroscope.

The most promising electroscope now available is the radio-frequency single-electron transistor (RF-SET)[2], the operation of which is depicted in Figure 1. We need a model of how this instrument will act as a quantum measurement.

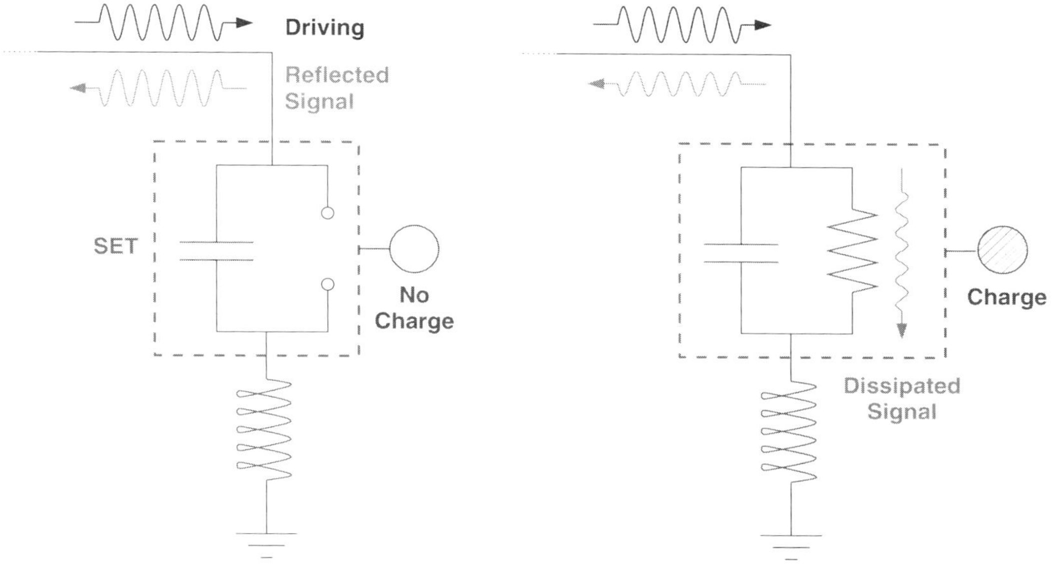

Figure 1. Operation of the radio-frequency SET. The instrument comprises an SET and an inductor, connected to a microwave source by a transmission line. If no c hargeis present to be measured, no current can tunnel through the SET. The frequency of the microwave source is chosen so that the capacitance between the leads of the SET forms a resonant circuit with the inductor, and all the signal is reflected. If c hargeis present, the SET acquires a finite tunneling resistance, which absorbs some of the signal.

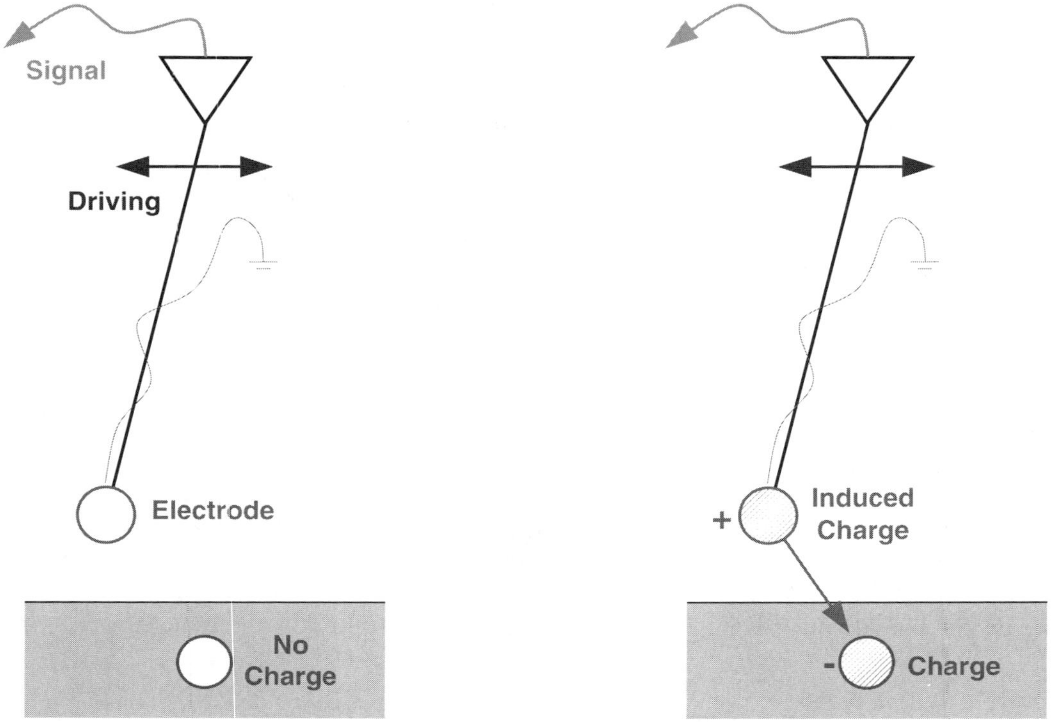

Figure 2. The mechanical electroscope. An electrode is mounted on a micro-mechanical cantilever, which is made to oscillate near a target charge. The electrode is held at a constant potential, so that charge flows between it and earth as it moves with respect to the charge being measured. This flow gives the cantilever electric potential energy as well as elastic, which changes its resonant frequency.

There has been much interest recently in electromagnetic measurements using micro-mechanical cantilevers[3,4,5,6,7]. We plan to adapt a calculation of the performance of a micro-mechanical cantilever, to be reported in an upcoming paper[8], to model the RF-SET. This type of electroscope is depicted in Figure 2. The cantilever was modeled by a simple harmonic oscillator with the standard Hamiltonian $\hbar\omega_0 a^\dagger a + 1/2$, driven coherently by a term $\hbar\gamma E(a + a^\dagger)$, coupled to the charge by a detuning term $\gamma\Delta n a^\dagger a$, where $n = 0$ if the charge is absent, or 1 if it is present. The cantilever is damped by a coupling to a readout wire at a rate γ. The current in the wire is measured to determine the vibrational state of the cantilever.

Under these conditions, the cantilever and charge are expected to obey a master equation[9]

$$\dot{\rho} = -i[(\Delta e n_1)\, a^\dagger a, \rho] - i[E\left(a + a^\dagger\right), \rho] \tag{1}$$

$$+\gamma(n+1)\mathcal{D}[a]\rho + \gamma n \mathcal{D}[a^\dagger]\rho \tag{2}$$

where

$$\mathcal{D}[a]\rho = a\rho a^\dagger - \frac{1}{2}(a^\dagger a\rho + \rho a^\dagger a) \tag{3}$$

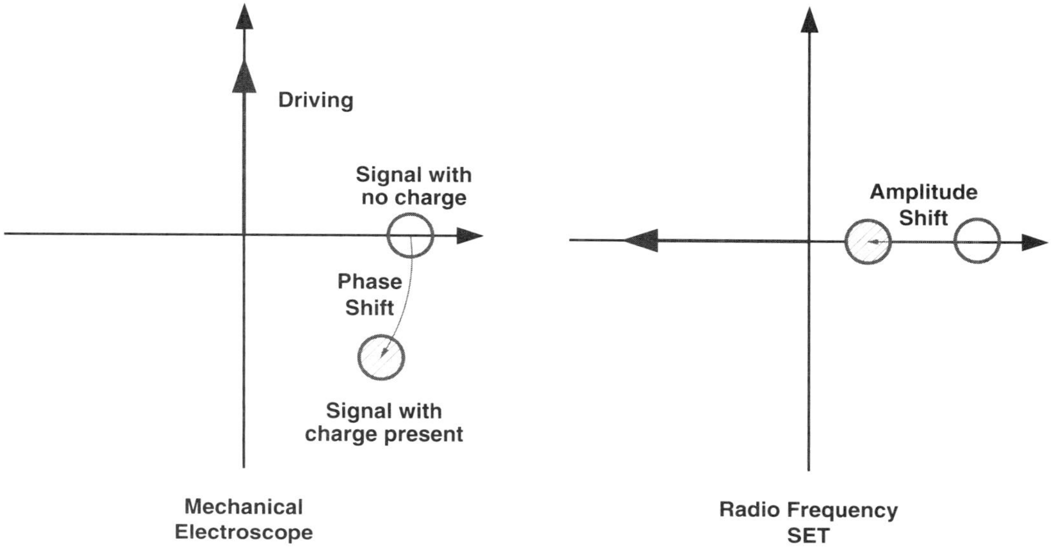

Figure 3. Modulation of the oscillation of the RF-SET circuit and the mechanical electroscope by the presence of c harge. The electroscope suffers a phase shift due to a c hange in its resonant frequency, the SET an amplitude shift due to absorption of the microwave signal.

At zero temperature, this can be solved exactly. If the charge is initially in a coherent superposition, its position will become entangled with the phase of the cantilever's vibrations (Figure 3). The coupling to the readout wire will cause this state to decohere at a rate proportional to the square of the separation of the t wo amplitudes,

$$|\alpha - \beta|^2 = \frac{16E^2\Delta^2}{1+4\Delta^2} \approx 4|\alpha|^2\Delta^2 \tag{4}$$

Cleland and Roukes give enough information about their devices for us to calculate this explicitly [3]. They state that the resonant frequency of their cantilever was 2.6 MHz, its torsional constant $\kappa = 1.1\times10^{-10}$ Nm and that the amplitude of its oscillation was 30 mrad. The information is enough for us to find α, using the definition of the annihilation operator for a torsional pendulum

$$\alpha = \langle a \rangle = \sqrt{\frac{\kappa}{2\hbar\omega}}\langle \theta_{\max} \rangle = 5.3 \times 10^6 \tag{5}$$

The normalised detuning $\Delta = \delta\omega/\gamma$ can be calculated from the frequency shift per electron $\delta\nu = 0.1$ Hz, and the quality factor $Q = \omega/\gamma = 6\,500$. This gives $\Delta = 2.4 \times 10^{-4}$, so the decay rate is $3.2 \times 10^6\,\gamma$, or $8.1 \times 10^9\,\mathrm{s}^{-1}$.

This is the decoherence rate of a qubit in the presence of a cantilever readout. It provides a lower bound on the time taken for a measurement. Any real measurement will be affected by technical noise, which is mitigated by averaging a sequence of measurements over an extended period. Hence the qubit being measured must be prevented from flipping due to the influence of the measurement for longer than the decoherence time.

Unfortunately, this calculation can not be applied directly to the RF-SET. Although the presence of charge acts on both devices by modulating the vibration of a harmonic oscillator, in the RF-SET it causes an amplitude shift rather than a phase shift. In the master equation approach taken here, this is not just a Hamiltonian term, but affects the damping terms. We are currently extending the master equation approach to treat this coupling.

References

1. B. E. Kane, Nature **393**, 133 (1999).
2. R. J. Schoelkopf et. al, Science **280**, 1238 (1998).
3. A. N. Cleland and M. L. Roukes, Nature **392**, 160 (1998).
4. J. A. Sidles et. al, *Rev. Mod. Phys.* **67**, 249 (1995)
5. Herbert Schoeller and Gerd Schön, *Phys. Rev.* B **50**, 18436 (1994)
6. A. N. Korotkov and M. A. Paalanen, *App. Phys. Lett.* **74**, 4052 (1999)
7. G. P. Berman et al, Formation and Dynamics of a Schrödinger-Cat State in Continuous Quantum Measurement, arXiv preprint `quant-ph/0101035`.
8. R. E. S. Polkinghorne and G. J. Milburn, Single electron measurements with a micro-mechanical resonator, Submitted to *Phys. Rev.* A.
9. G. J. Milburn, K. Jacobs and D. F. Walls, *Phys. Rev.* A **50**, 5256 (1994)

NMR QUANTUM COMPUTING – LESSONS FOR THE FUTURE

LIEVEN VANDERSYPEN* AND ISAAC CHUANG†
*† *Solid-State and Photonics Laboratory, Stanford University, Stanford, CA 94305*
*† *IBM Almaden Research Center, San Jose, CA 95120*
† *MIT Media Laboratory, Cambridge, MA 02139*

Future physical implementations of large-scale quantum computers will face significant practical c hallenges. Many useful lessons can be drawn from present results with Nuclear Magnetic Resonance realizations of controllable t wo, three, five, and seven qubit quantum systems. We summarize various experimental methods and theoretical procedures learned in this work which will be of considerable value in building and testing quantum processors with a wide variety of physical systems.

1 Requirements for quantum computers

DiVincenzo's five criteria for physical realization of quantum computers[1,2] provide an excellent starting point for enumerating insight gained from present NMRimplementations of small quantum processors. These criteria are:

1. Quantum bits: finite Hilbert space quantum systems which can be composed in direct product form.
2. Long coherence: high isolation of the system from the environment, such that quantum superposition states have long lifetimes.
3. Universal control: ability to subject the quantum system to a controllable sequence of unitary transforms which can approximate an arbitrary operator.
4. Initialization: ability to prepare the quantum systems in a fiducial input state, such as the ground state.
5. Measurement: readout capability to projectively measure the final quantum state.

We go through each of these in turn in four sections below, and describe how each criterion is satisfied in NMR quantum computation, starting with a combination of 1 and 2. In Section 6, we conclude by briefly describing our recent experiment demonstrating a five-qubit quantum computation.

2 Robust qubits

Some perspective on the wide range of possible qubit realizations can be obtained by comparing a few choices, as shown in Fig. 2. From this quick survey, one of the most striking realizations is that only a few fundamental physical degrees of freedom are particularly good for qubits, and among these, spin has many advantages.

NMR quantum computation uses the spin of atomic nuclei as qubits. The atoms are usually chosen to have spin-1/2 nuclei, and are elements of molecules t ypically dissolved in a liquid. Hydrogen (protons), Fluorine, Carbon-13, and Nitrogen-15 are common choices, since they are relatively easy to place in artificially synthesized molecules. Neighboring nuclei couple with each other via a magnetic dipole-dipole interaction and via an electron mediated through-bond coupling arising from the Fermi contact interaction between electrons and the nuclei.

System	τ_Q	τ_{op}	$n_{op} = \lambda^{-1}$
Nuclear spin	$10^{-2} - 10^{8}$	$10^{-3} - 10^{-6}$	$10^{5} - 10^{14}$
Electron spin	10^{-3}	10^{-7}	10^{4}
Ion trap (In^+)	10^{-1}	10^{-14}	10^{13}
Electron – Au	10^{-8}	10^{-14}	10^{6}
Electron – GaAs	10^{-10}	10^{-13}	10^{3}
Quantum dot	10^{-6}	10^{-9}	10^{3}
Optical cavity	10^{-5}	10^{-14}	10^{9}
Microwave cavity	10^{0}	10^{-4}	10^{4}

Figure 1. Crude estimates of decoherence and operation times for various qubit realizations. τ_Q gives decoherence times in seconds, τ_{op} operation time in seconds, and $n_{op} = \lambda^{-1} = \tau_Q/\tau_{op}$, the maximum number of operations[3,2].

Nuclear spins can have amazingly long coherence times. The spin-spin or phase coherence time, T_2, is often on the order of seconds (for example, the protons in chloroform have $T_2 \approx 5$ seconds when properly prepared). This is especially true in liquids, where the rapid tumbling of dilute solute molecules averages away most couplings to other molecules. In such systems, the spin-lattice or energy relaxation time T_1 is typically equal to T_2 (there may be a factor of 2 difference, depending on how they are defined). And in solids, such as frozen ^{129}Xe, T_1's on the order of several hours is easily achieved. Solid state NMR usually faces very short T_2 times, on the order of microseconds or less, but this is not fundamental; in dilute spin systems, it should be possible to reduce and control long range spin-spin interactions so that long T_2's are recovered[4,5,6]. Part of the reason nuclear spins have such long coherence times is because their Larmor frequencies in typical magnetic fields, of 1-10 Tesla, place them in a regime of frequencies of 10-1000 MHz, which is nicely separated from other timescales (such as spin-spin couplings, spin-lattice couplings, tumbling, and other decoherence mechanisms) in the system. Electrons are also spin-1/2 particles and can also have relatively long coherence times, but their spin precession frequencies are about a thousand times higher than that of nuclear spins, making them harder to work with, although there are promising possibilities using semiconductor heterostructures[7] and quantum dots[8].

Another notable characteristic of quantum computation with spin resonance has been that the spins on the molecule stay where they are; they are not moved through logic gates as we usually think of when performing computations in modern digital computers. Instead, in NMR quantum computation, as we shall see below, quantum gates are brought to the spins. No physical reorganization of the spins is performed.

Lesson 1: a. Spins make good qubits
b. Bring gates to spins

3 Coherent universal control techniques

The NMR Hamiltonian for a typical molecule (of spin-1/2 nuclei) in liquid solution is, to an excellent approximation,

$$H = \sum_k \hbar\omega_k Z_k + \sum_{j,k} \hbar 2\pi J_{jk} Z_j \otimes Z_k + H^{\mathrm{RF}} + H^{\mathrm{env}}\,, \tag{1}$$

where Z_k is the Pauli spin angular momentum operator for spin k, ω_k is its Larmour precession frequency, J_{jk} represents the strength of the coupling between spins j and k (usually $1-300$ Hz), $H^{\rm RF}$ is due to externally applied radiofrequency (RF) fields, and $H^{\rm env}$ is the coupling to the environment – systems other than the RF, such as higher order spins on the same molecule, electrons, vibrational degrees of freedom, and other molecules. This expression also assumes that only through-bond couplings are left after motional averaging has removed the effect of dipole-dipole couplings.

The system dynamics is controlled by $H^{\rm RF}$, which in the rotating frame is, to good approximation,

$$H^{\rm RF} = \sum_k P_k(t) \left[X_k \cos\phi_k(t) + Y_k \sin\phi_k(t)\right] , \qquad (2)$$

where X_k and Y_k are Pauli spin operators, and $P_k(t)$, $\phi_k(t)$ are classical time-varying variables which correspond to the amplitude and phase of RF pulses applied at frequency ω_k. Theoretically, control over P and ϕ is sufficient to enable arbitrary unitary operations to be approximated, so long as a pathway of nonzero J_{jk} couplings between any pair of spins.

Single qubit operations are performed using RF pulses. Typically, pulses corresponding to $\pi/2$ rotations of a spin have lengths of ≈ 10 microseconds when applying 50 to 300 Watts of power. Less power can be used to allow longer pulses, which can then be tailored in shape to provide more frequency selectivity. The phase determines the axis of rotation and must be set with respect to a reference frame which rotates along with the precession motion of the spin. Most conveniently, this is done via a separate stable and accurate reference oscillator for each spin. In practice, fewer reference oscillators than qubits may be available, in which case a single oscillator can be used to keep track of the evolution of several spins at nearby frequencies via a time counter. The same reference oscillator is also used as the source for the RF pulses. In case one frequency source must be used to address several qubits, phase ramping techniques (the use of a phase shifter which increments the phase linearly throughout the pulse) can be used to shift the signal produced by the source to a different frequency, or even to multiple other frequencies. Alternatively, the frequency of the oscillators could be jumped back and forth to the respective spin Larmour frequencies. These techniques require accurate control over the amplitude and phase of the RF pulses, on a time scale several hundred times shorter than the total pulse duration.

Two-qubit gates require a qubit-qubit interaction term in the Hamiltonian, which in NMR is provided by the natural J coupling. On the one hand, this is convenient since otherwise couplings would have to be engineered, but on the other hand it is not possible to turn off the coupling Hamiltonian. Fortunately, NMR spectroscopy has developed a technique known as *r efœusing* to undo the evolution of undesired terms in the Hamiltonian, even if they cannot be actually turned off. During time intervals of free evolution, this is accomplished by applying π pulses (mathematically, $R_{x1}^2 = e^{-i\pi X_1/2}$) which reverse the evolution under select terms of the Hamiltonian, by virtue of the property

$$R_{x1}^2 e^{-iaZ_1 t} R_{x1}^2 = e^{iaZ_1 t} . \qquad (3)$$

By sending π pulses at carefully chosen instances throughout the time interval, it is possible to effectively annihilate the effect of undesired terms in the Hamiltonian while preserving the effect of the desired terms. Starting from widely used traditional techniques, refocusing schemes have been designed specifically for NMR quantum computing[9,10,11], and these can be used directly in many future implementations schemes.

The natural Hamiltonian is active not only in between but also during the RF pulses. In many cases this built-in Hamiltonian is not at all negligible compared to the RF Hamiltonian, especially if the RF pulses are long (on the order of a few ms) and the RF power is correspondingly low. Fortunately, RF pulses with shaped amplitude and phase envelopes can be designed to exhibit self-refocusing behavior, thereby reducing the effect of certain couplings. The couplings whose effect cannot be neutralized in this way must be taken care of in a similar manner as is done for the coupled evolution during delay times.

Lesson 2:	a. Phase coherent single-qubit control is required
	b. Refocusing can turn off interactions
	c. Take advantage of shaped pulses

4 Input state preparation

Quantum algorithms generally assume or demand that qubits can be prepared in a pure state, usually the ground state $|0\rangle$. However, a nuclear spin in thermal equilibrium at room temperature and subject to a reasonable static magnetic field is in a highly mixed state, given by the density matrix

$$\frac{1+\epsilon}{2}|0\rangle\langle 0| + \frac{1-\epsilon}{2}|1\rangle\langle 1|\,, \tag{4}$$

where the polarization $\epsilon = \hbar\omega/2k_BT$ (k_BT is the thermal energy) is typically about 10^{-6} to 10^{-5}. In an n spin system in thermal equilibrium, the populations of each of the 2^n basis states are thus almost, albeit not completely, equal, whereas ideally only the ground state should be populated at the start of a computation.

Cooling down the sample to absolute zero temperature would put the spins in a pure state, but this is experimentally demanding and would freeze liquid samples. NMR quantum computation with liquid-state molecules was made possible by an invention by Cory, Fahmy and Havel[12] and Gershenfeld and Chuang[13] to prepare from a thermal ensemble a state which produces measurement results and behaves dynamically as if it were at zero temperature.

This procedure requires the creation of "effective pure" states, with a density matrix of the form

$$\rho = \frac{1-\alpha}{2^n}\mathbf{1} + \alpha|\psi\rangle\,\langle\psi|\,. \tag{5}$$

Because states with a density matrix proportional to the identity matrix are not observerable in NMR (only population *differences* produce observable signals) and furthermore do not evolve during unitary operations, an effective pure state gives the signal and has the dynamical behavior of just the second term of Eq. 5, which represents a pure state. Thus, an ensemble of nuclear spins at room temperature can be made to look like an ensemble of spins at zero temperature, up to a decrease in signal strength by α.

Three methods are known to prepare effective pure states: logical labeling[13], spatial labeling[12] and temporal labeling[14]. The first and third method only require unitary operations such as those described in Sec. 3; the second also requires non-unitary operations,

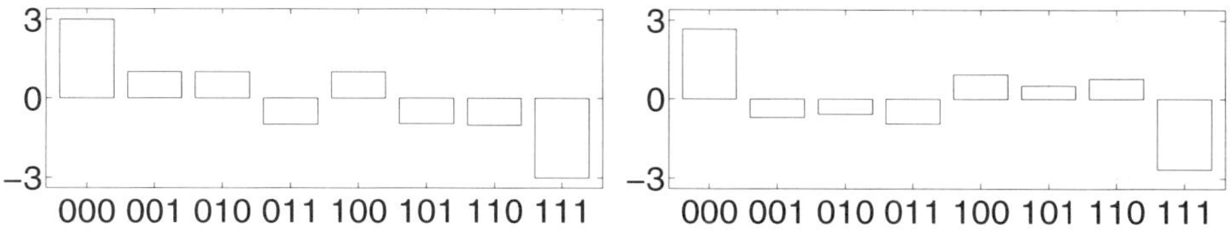

Figure 2. Experimentally determined populations (in units of $\epsilon/8$, and relative to the average) of the states $|000\rangle, \ldots, |111\rangle$ (Left) in thermal equilibrium and (Right) after logical labeling. The populations were determined by partial state tomography [19]. (©1999 The American Physical Society)

which in NMR can elegantly be approximated by a pulsed magnetic field gradient on a homogeneous sample[12].

Logical labeling consists of selecting a submanifold spanned by 2^k states from the Hilbert space of n coupled spins, by conditioning upon the state of $n-k$ ancilla spins, in such a way that the state of the submanifold has the characteristics of an effective pure state, i.e. all populations but one are equal. For example, the eight populations of a three-spin homonuclear system deviate from the average by approximately $\frac{1}{8}\{3\epsilon, \epsilon, \epsilon, -\epsilon, \epsilon, -\epsilon, -\epsilon, -3\epsilon\}$ for the states $|000\rangle$ through $|111\rangle$ respectively. By applying a short pulse sequence[15], these populations can be permuted to obtain $\frac{1}{8}\{3\epsilon, -\epsilon, -\epsilon, -\epsilon, \epsilon, \epsilon, \epsilon, -3\epsilon\}$. Note that at this point qubits 2 and 3 are in an effective pure ground state conditioned upon qubit 1 being in the state $|0\rangle$. We implemented this procedure experimentally on the three ^{19}F nuclei of CF_2CFBr (see Fig. 2) and subsequently executed the t wo-qubit Grover algorithm within the logically labeled subspace[15].

Temporal labeling consists of adding up the spectra of multiple experiments, where each experiment starts off with a different state preparation pulse sequence consisting of CNOT and NOT operations. These operations are designed to transform the thermal equilibrium deviation density matrix, for this purpose best written as $\rho_{th} \propto ZII + IZI + IIZ$ ($n = 3$ terms, where ZII is shorthand for $Z \otimes I \otimes I$ and I is the 2×2 identity matrix), into a different set of n terms such that the combined sum of the multiple experiments corresponds to the desired effective pure state density matrix $\rho_e \propto ZII + IZI + IIZ + ZZI + ZIZ + IZZ + ZZZ$ ($2^n - 1 = 7$ terms). In principle, the summation of $\lceil (2^n - 1)/n \rceil$ experiments thus suffices to create all desired terms[16] (the original temporal labeling scheme involved $2^n - 1$ experiments[14]), although for heteronuclear systems a slightly larger number of experiments is generally needed. We used this technique to prepare an effective pure state on a five-spin system, as shown in Fig. 3.

These schemes for producing effective pure states come at a cost. Either the signal strength decreases exponentially with the number of qubits ($\alpha \propto n/2^n$ in logical and spatial labeling), or the number of experiments increases exponentially (as $2^n/n$ in temporal labeling).

However, this cost is not fundamental: Schulman and Vazirani developed an improved scheme[17], whose cost is only polynomial. The essential building block of this cooling algorithm has been demonstrated[18] on the three fluorine spins of CF_2CFBr. This algorithm requires only a linear overhead in the number of spins needed to extract n pure qubits and a quasi-linear overhead in the number of operations; this algorithm is thus efficient. The prefactor in the overhead in qubits is $\approx 1/\epsilon^2$, which is unreasonably high (about 10^9) for

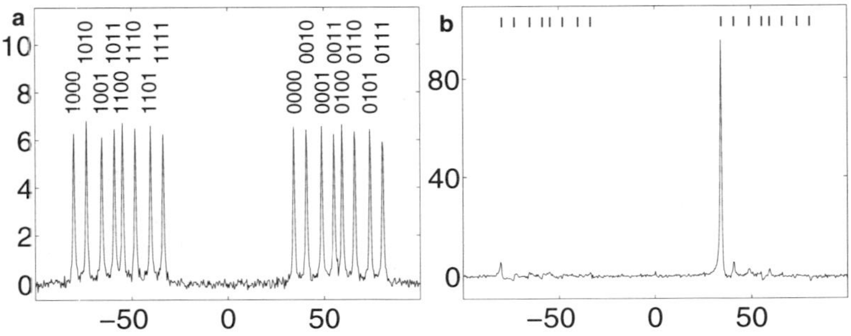

Figure 3. **a** Experimentally measured spectrum (real part, in arbitrary units; relative frequencies are in units of Hz) of one spin of a five-spin molecule in thermal equilibrium. The 16 lines arise from the J couplings; the line of spin i is shifted by $\pm J_{ij}/2$, depending on whether spin j is in $|0\rangle$ or $|1\rangle$. In equilibrium, all 16 lines in each spectrum have virtually the same intensity, and they can be labelled as shown, taking into account the sign and magnitude of the J couplings. **b** The same spectrum when the five spins are in an effective pure state. Only the line labelled 0000 is present. (©2000 The American Physical Society)

thermally equilibrated nuclear spins at room temperature. However, for systems with much higher initial polarizations, this number is quite reasonable and Schulman-Vazirani cooling can be very useful.

Lesson 3: Zero temperature dynamics can be observed in high temperature systems

5 Measurement

Readout of the nuclear spin states is accomplished by recording the oscillating voltage induced by the precession of the spin magnetic moments in a pick-up coil surrounding the sample. This RF signal is mixed down with and phase-referenced against the signal of a local reference oscillator, and recorded with a phase sensitive detector. The resulting time-domain signal is Fourier transformed to obtain spectra such as the ones shown in Fig. 3a. The amplitude and phase of the spectrum give information about the state of the qubits. With properly set receiver phase settings, a positive or negative line indicates that the spin was along $+z$ ($|0\rangle$) or $-z$ ($|1\rangle$) respectively. The information contained in the line-splittings can be used to determine the state of several qubits from inspection of the multiplet of just one spin.

Several problems arise in such a measurement scheme. Most importantly, because the magnetic moment of a single nuclear spin is extremely small, an ensemble of about 10^{18} molecules are used. These are all subject to the same static and RF magnetic fields and thus undergo (ideally) identical operations. However, the signal of all the spins are averaged together, and in practice it is not possible to distinguish one spin's signal from another.

The use of an ensemble demands a modification or extension of certain quantum algorithms, such as Shor's algorithm, which in their original description assume projective measurements on single quantum systems, as opposed to weak ensemble-averaged measurements. Fortunately, such modifications can be made for all the known quantum algorithms[13]. The use of ensembles of quantum systems, which is built-in in current NMR experiments, may thus also prove a valuable and time-saving approach for other quantum computer implementations.

Lesson 4:	a. Spin magnetic moments are small
	b. Ensemble measurements are acceptable

6 Experiments

The simplest possible non-trivial quantum computations are the two-qubit Deutsch-Jozsa and Grover algorithm, and those were also the first ones implemented with NMR[19,20,21,22]. Both provide an excellent test bed for future quantum computer implementations: despite their simplicity, they require that all of the ingredients of a quantum computer be put together in a working device. Three-qubit experiments represent the next interesting step because now a network of interactions between three pairs of qubits is involved[23,24]. Evidently, only the realization of more and more quantum bits really gives a feel for what it would take to build a quantum computer large enough to outperform classical machines.

We recently implemented a quantum algorithm for order-finding on a 5-spin molecule[16]. This algorithm is a generalization of Shor's factoring algorithm[25] and represents a very demanding but also realistic and complete test of a 5-qubit computer: all spins must be prepared in an effective pure state, the quantum circuits (Fig. 4) include two-qubit gates between each pair of qubits and the implementation involves tens of two-qubit gates and hundreds of RF pulses. This puts very high demands on the NMR properties of the molecule (J-couplings, chemical shifts, coherence times) and on the quantum control techniques for steering the spin dynamics. Despite these difficulties, we successfully ran various instances of this algorithm, employing many of the techniques reviewed in the preceding sections. The output spectra immediately reveal the correct answer to the computation (the initially unknown order of a permutation available as a black box), as shown in Fig. 5.

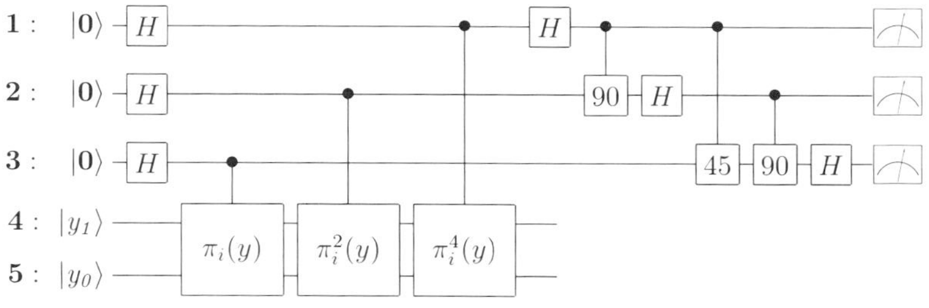

Figure 4. Outline of the quantum circuit for an algorithm to find the order r of a permutation π on four elements. The starting element y is stored in the second register (qubits 4 and 5) and the first register (qubits 1, 2 and 3) represents the number of times the permutation is executed. The transformation $|x\rangle|y\rangle \mapsto |x\rangle|\pi^x(y)\rangle$ corresponds to permuting x times starting from y and was implemented in three steps, using the fact that $\pi^x = \pi^{x_0}\pi^{2x_1}\pi^{4x_2}$, where $x_2x_1x_0$ is the binary representation of x. Each of these three operations is a permutation on qubits 4 and 5, controlled by qubits 3, 2 and 1 respectively. The details of the exponentiated permutations depend on π. The controlled permutation is followed by the quantum Fourier transform (QFT) on the first register, where the boxed 90 and 45 represent rotations about $\hat{z}$ over those angles. (©2000 The American Physical Society)

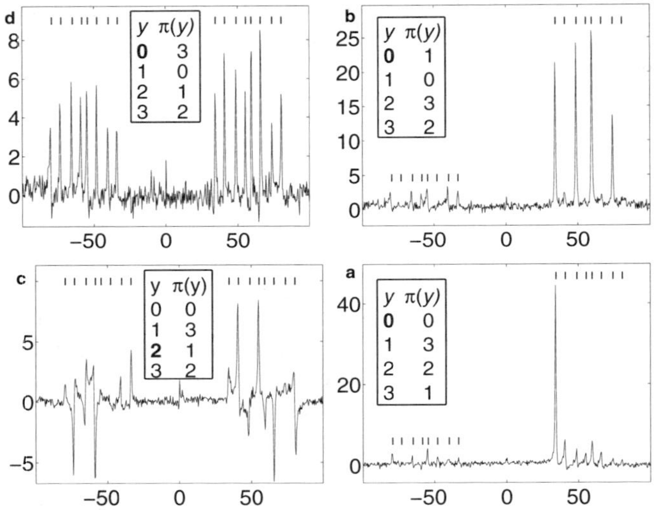

Figure 5. The same spectrum as in Fig. 3, after execution of the order-finding algorithm. The 16 marks on top of each spectrum indicate the position of the 16 lines in Fig. 3 a. The respective permutations are shown in inset, with the input element set in bold. The prediction is that one line should be visible (and up) if $r = 1$, four lines when $r = 2$, sixteen lines for $r = 4$ and finally the *net* area under the lines should be zero when $r = 3$. (©2000 The American Physical Society)

The results of Fig. 5, as any experimental results, slightly deviate from the ideally expected results. This is in part due to decoherence, but also in large part due to erroneous unitary evolutions. While most such systematic errors could in principle be corrected using quantum error correction, this would entail a very large overhead. Instead, it is in principle also possible to characterize the actual (as opposed to ideal) evolution of the quantum system during each of the elementary logic operations, and then to design the *sequence* of operations such that erroneous evolutions which take place at one point in time, unwind at a later point in time. This is commonly done in many standard NMR pulse sequences, but is much more complex and less transparent in the course of a quantum computation. It is therefore encouraging that a very large degree of cancellation of erroneous evolution has been observed in NMR quantum computing experiments, allowing the successful concatenation of hundreds of two-qubit gates involving about 1350 RF pulses[24]. A study of the potential for similar cancellation behavior will be important for any proposed quantum computer realization.

Acknowledgments

LV gratefully acknowledges the support of a Yansouni Family Stanford Graduate Fellowship.

References

1. D. P. DiVincenzo, *Fortschr. Phys.* **48**, 771 (2000).

2. M. A. Nielsen and I. L. Chuang, *Quantum computation and quantum information* (Cambridge University Press, Cambridge, UK, 2000).
3. D. P. DiVincenzo, *Phys. Rev. A* **51**, 1015 (1995).
4. Bruce Kane, *Nature* **393**, 133 (1998).
5. D. G. Cory et al., *Fortschr. Phys.* **48**, 875 (2000).
6. T. D. Ladd, J. R. Goldman, A. Dana, F. Yamaguchi, and Y. Yamamoto, *arXive e-print quant-ph/0009122* (1998).
7. Rutger Vrijen et al., *Phys. Rev. A* **62**, 012306 (2000).
8. Daniel Loss and David P. DiVincenzo, *Phys. Rev. A* **57**, 120 (1998).
9. D.W. Leung, I.L. Chuang, F. Yamaguchi, and Y. Yamamoto, *Phys. Rev. A* **61**, 042310 (2000).
10. J.A. Jones and E. Knill. *J. Magn. Reson.* **141**, 322 (1999).
11. N. Linden, E. Kupcce, and R. Freeman, *Chem. Phys. Lett.* **311**, 321 (1999).
12. D. G. Cory, A. F. Fahmy, and T. F. Havel, *Proc. Nat. Acad. Sci. USA* **94**, 1634 (1997).
13. N. Gershenfeld and I. L. Chuang, *Science* **275**, 350 (1997).
14. E. Knill, I. Chuang, and R. Laflamme, *Phys. Rev. A* **57**, 3348 (1998).
15. L. M. K. Vandersypen, C. S. Yannoni, M. H. Sherwood, and I. L. Chuang, *Phys. Rev. Lett.* **83**, 3085 (1999).
16. L. M. K. Vandersypen, M. Steffen, G. Breyta, C. S. Yannoni, R. Cleve, and I. L. Chuang, *Phys. Rev. Lett.* **85**, 5452 (2000).
17. L. J. Schulman and U. Vazirani, *Proc. 31st Ann. ACM Symp. on Theory of Computing (STOC '99)*, p. 322 (1999).
18. D. Chang, L.M.K. Vanderspen, and M. Steffen, to appear in *Chem. Phys. Lett.*, see *arXive e-print quant-ph/001055* (2000).
19. I. L. Chuang, N. Gershenfeld, and M. Kubinec, *Phys. Rev. Lett.* **18**, 3408 (1998).
20. I. L. Chuang, L. M. K. Vandersypen, X. L. Zhou, D. W. Leung, and S. Lloyd, *Nature* **393**, 143 (1998).
21. J. A. Jones and M. Mosca, *J. Chem. Phys.* **109**, 1648 (1998).
22. J. A. Jones, M. Mosca, and R. H. Hansen, *Nature* **393**, 344 (1998).
23. N. Linden, H. Barjat, and R. Freeman, *Chem. Phys. Lett.* **296**, 61 (1998).
24. L. M. K. Vandersypen, M. Steffen, G. Breyta, C. S. Yannoni, and I. L. Chuang, *Appl. Phys. Lett.* **76**, 646 (2000).
25. A. Yu. Kitaev, *arXive e-print quant-ph/9511026,* (1995).

QUANTUM NETWORKS BASED ON CAVITY QED

H. MABUCHI, M. ARMEN, B. LEV, M. LONCAR, J. VUCKOVIC, H. J. KIMBLE,
J. PRESKILL, M. ROUKES, A. SCHERER

California Institute of Te chnolgy, Mail Code 12-33, Pasadena, CA, 91125, U.S.A.

S. J. VAN ENK

Lucent Technologies Bell Labs, 600-700 Mountain Ave., Murray Hill, NJ, 07974, U.S.A.

We review an ongoing program of interdisciplinary research aimed at developing hardware and protocols for quantum communication networks. Our primary experimental goals are to demonstrate *quantum state mapping* from storage/processing media (internal states of trapped atoms) to transmission media (optical photons), and to investigate a nanotechnology paradigm for cavity QED that would involve the integration of magnetic microtraps with photonic bandgap structures.

1 Introduction

No one can dispute that we live in an age of information; living in such an age, we experience an ever-increasing need for technology that allows us to share information in a secure, efficient, and reliable way. Quantum technologies show great potential for revolutionizing the methods by which we collect and distribute information, in diverse engineering contexts that range from computer architecture to the Internet. Seminal theoretical research in the field has identified a portfolio of tasks for which quantum methods are provably superior to their classical counterparts. For example, it has recently been shown that the communication complexity of a fundamental task such as appointment-scheduling can be drastically reduced through the use of quantum resources [1]. It is also known that the elementary protocol of quantum state teleportation [2] could provide a means for distributing cryptographic key with absolute and verifiable security [3], and that related methods could be used

Figure 1. Transmission of quantum information between nodes in a quantum network. A qubit is initially stored in the internal state of an atom trapped within a high-finesse optical cavity , at the sending node (each node contains multiple atoms). 1: the atomic state is mapped to that of an optical photon via cavity QED, 2: the photon travels through a fiber to the receiving node, 3: the photonic state is mapped into that of a trapped atom at the receiving node. Multiple nodes connected in this fashion form a *quantum network*.

to enhance the capacity of noisy communication channels [4].

An ongoing program of research at Caltech focuses on development of the hardware and error correction protocols required to construct a *quantum network* (see Fig. 1). Here quantum nodes with memory and local processing capabilities will be connected by quantum communication channels that allow robust transmission of coherent quantum information among nodes of the network. Such a network would be capable of performing distributed quantum computations [5]. In our proposed implementation, trapped neutral atoms will provide quantum memory at each node of the network, and optical cavities will be utilized both to perform quantum gates and to transfer quantum information between nodes [6]. With a long-term view towards developing integrated and robust hardware for quantum nodes, much of our work is aimed at exploring a revolutionary new technical paradigm for cavity QED based on magnetic microtraps and optical photonic bandgap structures [7]. In the following sections, we first provide an overview of this new technical paradigm and then describe the quantum repeater protocol that will enable robust quantum communication in the presence of channel decoherence.

2 Nanotechnology for optical cavity QED with neutral atoms

During the past decade, advances in semiconductor crystal growth technique have led to the rapid development of optical microcavity devices. With the ultrahigh precision that can now be obtained over layer thicknesses, high reflectivity mirrors can easily be grown by depositing alternating layers of high- and low-index materials to define high-Q cavities in the vertical dimension. Recently, it has also become possible to microfabricate high reflectivity mirrors with *horizontal* orientation by etching periodic arrays of holes into a semiconductor substrate. These "photonic crystals" can be designed to fully suppress horizontal light propagation within certain frequency bands, called photonic bandgaps (see Fig. 2) [8,9].

When combined with high index-contrast slabs, in which the propagation of light is already confined to a plane, photonic bandgap mirrors can be used to confine light within "nanocavities" of extremely small volume. Such cavities can be produced simply by introducing defects into the photonic bandgap structure, for example by omitting one hole from the periodic array. The resonance wavelength of such a cavity can be lithographically defined by adjusting the precise geometry of the holes immediately surrounding the defect. In previous work this strategy has been employed to construct microcavity semiconductor lasers [10] with mode volumes as small as $2.5\,(\lambda/2n_{\text{slab}})^3$.

Using the same fabrication techniques, it is also possible to guide, bend, filter, and sort light in 2-D photonic crystals. Such photonic waveguides could clearly be used as optical channels between nanocavities, and could also be used to define simple interferometers to combine cavity inputs/outputs. Tapered photonic waveguides could be pigtailed to standard optical fibers, as a means of bringing light into or out of the semiconductor slab. Using the techniques described above, we can easily envision fabricating semiconductor wafers in which multiple interconnected optical cavities are embedded at an areal density $\sim 10^6$ cm^{-2}. Such arrays of PBG nanocavities could replace the Fabry-Perot cavities that have been used in previous experiments on optical cavity QED with neutral atoms.

The basic parameter that measures coupling strength between an atom and an optical cavity is the so-called "vacuum Rabi frequency" $g \equiv \frac{d \cdot E_1}{2\hbar}$, where d is the atomic dipole-transition matrix element and E_1 is the electric field per photon in the cavity. To achieve

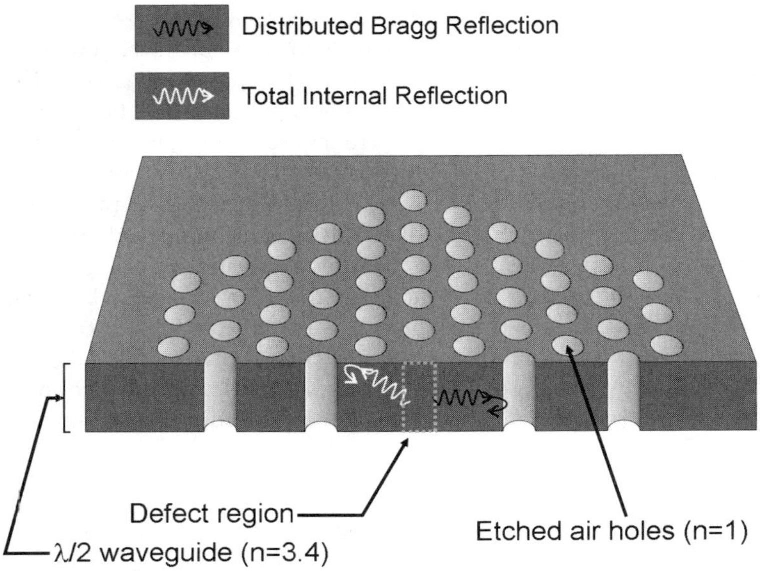

Figure 2. Schematic cross-section of the photonic crystal structure.

strong coupling we must have g sufficiently large compared to both the atomic dipole decay rate $\gamma_\perp$ and the cavity field decay rate κ. For the purpose of quantum information processing, we require that both the critical photon number m_0 and the critical atom number N_0 be less than one, where $m_0 = \frac{\gamma_\perp^2}{2g^2}$ and $N_0 = \frac{2\kappa\gamma_\perp}{g^2}$. As the dipole matrix element d is fixed by atomic structure, the only way to increase $g/\gamma_\perp$ is to construct cavities in which E_1 is as large as possible.

We have performed numerical calculations (via finite-difference time-domain modeling [12]) of the mode properties for several possible PBG defect cavities. In one case we have obtained $V_{\text{eff}} = 0.09\,(\lambda/2)^2$ together with $Q \approx 1.6 \times 10^4$. This particular geometry incorporates a central "defect" hole of reduced diameter $\sim$ 108 nm. To couple most strongly with the cavity field, an atom should be trapped at the center of this hole. While further optimization of the geometry may still be possible, these parameters already lead to $g \approx 5.6$ GHz with $\kappa = 11$ GHz, or $m_0 \approx 5.4 \times 10^{-8}$ photons and $N_0 \approx 1.8 \times 10^{-3}$ atoms.

In addition to the creation of semiconductor nanostructures, modern microfabrication techniques also allow the patterning of conducting wires and ferromagnetic materials at the $1 - 10\ \mu$m scale. As a result, it should be possible to construct magnetic microtraps with field gradients on the order of 10^8 G/cm and field curvatures $\sim 10^8$ G/cm^2 [14,15]. With this magnitude of field curvature, *e.g.*, an Ioffe configuration trap would hold a Cs atom with $\eta \sim 0.35$ in the axial coordinate and $\eta \sim 0.035$ in the radial coordinates [14], and provide a well-defined bias field ~ 1 G at the trap center.

Our initial integration scheme will be to deposit micron-scale wires on the surface of a photonic crystal membrane, such that each PBG cavity is surrounded by a concentric microtrap. Fig. 3 shows one possible configuration. Current-carrying wires arranged in a

(a)

(b)

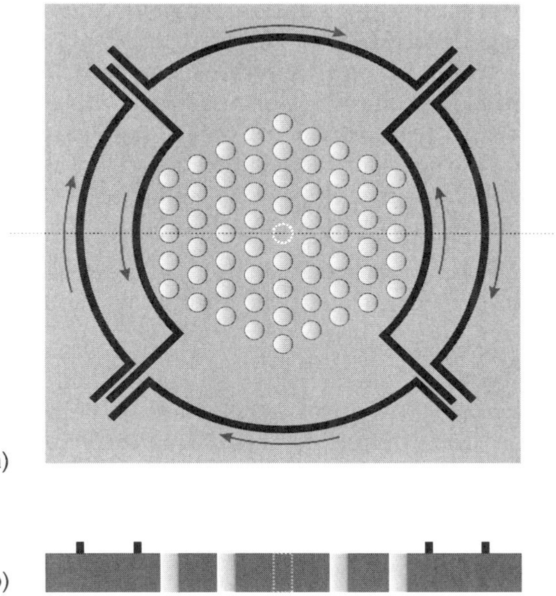

Figure 3. Schematic of an integrated PBG cavity and planar magnetic microtrap. (a) Top view showing arrangement of the current-carrying wires. Arrows indicate the direction of current flow. (b) Cross-section through the defect cavity (as indicated by the dashed line in (a)), showing wire deposition on the surface of the semiconductor slab.

suitable modification of Weinstein and Libbrecht's 'Ioffe *c*' configuration [14] can project a magnetic field with a stable minimum at the geometric center of the PBG cavity, such that one or more atoms could be confined within the defect hole and (as discussed above) would therefore experience strong coupling. A similar geometry can be envisioned for a microtrap based on permanent magnets, rather than current-carrying wires. This would involve the use of techniques for electroplating ferromagnetic material (such as nickel [16]) into selected holes around the perimeter of each PBG cavity.

3 Quantum repeater architecture

Any quantum communication scheme will have to deal with losses and other noise (such as phase shifts) in the quantum channel connecting the sending and receiving nodes. Unfortunately, standard quantum error correction schemes as developed for quantum computing will not be practical for two reasons. First, such schemes typically work only if the error probability is sufficiently small, and second, they would require a large overhead in qubits to protect a single qubit against one error. Here we review an error correction protocol that solves for photon absorption errors to all orders (i.e., it works irrespective of the error rate) and that nevertheless requires only a moderate overhead in resources. Subsequently we describe a quantum repeater scheme that allows one to reinforce the quantum signal at intermediate nodes, thus allowing communication over long distances.

Quantum error correction schemes require entangled states of at least 5 qubits to protect one qubit against a single general error. In the specific physical setup described in the present proposal, however, not all possible errors are equally likely to occur. In particular, whereas photon absorption is a common error, the inverse process, the spontaneous creation of a photon with a particular polarization and optical frequency is extremely unlikely. This fact allows one to encode the quantum information in such a way that the photon absorption error can be corrected to all orders with just two entangled atoms per node. Without going into the technical details (see [17,18]), we present here just the basic ideas.

1. The logical $|0\rangle$ and $|1\rangle$ are encoded in the absence and presence of a photon in a particular mode, respectively, so that the error $|0\rangle \mapsto |1\rangle$ has been eliminated.

2. We use an *auxiliary* qubit in the sending node that contains no information about the actual qubit we wish to communicate, to send a photon. This way, if that photon is absorbed, no information will be lost. This part of the protocol is repeated until we know that the photon arrived. Only then do we involve the qubit that contains the actual information.

3. In order to detect photon absorption, we do not have to monitor the environment. The auxiliary atom in the sending node is used twice: its state is communicated in two transmissions to two atoms in the receiving node. Between the two transmissions a NOT operation is applied (i.e., we interchange $|0\rangle \leftrightarrow |1\rangle$) to the sending atom. The two atoms receiving the information should, therefore, be in different states. Photon absorption can thus be unambiguously detected by measuring whether the two atoms are in the state $|0\rangle|0\rangle$.

4. As a bonus, the application of the NOT operation symmetrizes and thereby corrects for all systematic phase errors.

Two atoms per node are sufficient to achieve a high communication fidelity F (defined as the overlap between the actual and desired final state). If there is need for further correction of the remaining phase errors, one extra atom per node is needed [18].

The probability of photon absorption increases exponentially with the length of the communication channel. Thus, while the above error correction scheme in principle works just as well for large-distance communication, it would require an exponentially large number of transmissions. In order to reduce the number of transmissions, we need, just as for classical communication, intermediate nodes—quantum repeaters—where the signal is "amplified." Of course, since true amplification of quantum signals is not possible, the procedure used is more subtle than in the classical case, and is in fact a variant of standard entanglement purification schemes [19], modified to encompass the restrictions on the amount of physical resources.

Several quantum repeater schemes have been discussed in Ref. [20]. The most efficient scheme takes the form of a concatenated error correction code. It makes use of two simple sub-procedures.

1. Two pairs of imperfect EPR pairs of entangled qubits can be used to distill a single pair of *higher* fidelity between nodes A and B.

2. Two pairs of imperfect entangled states connecting nodes A with B, and B with C, resp., can be used to create a single EPR pair of *lower* fidelity between nodes A and C.

When one has more than one intermediate mode, repeating procedure 2 would lead to a fidelity that decreases exponentially with the total number of nodes. In order to achieve a certain minimum final fidelity, one has to periodically purify the entangled states obtained thus far with procedure 1. The whole protocol is then a concatenated procedure in that the purification has to be repeated at different levels: starting at the lowest level one purifies all EPR pairs between adjacent modes, at the next level one purifies the lower-fidelity EPR pairs between more distant nodes as gotten by procedure 2, and so on.

Acknowledgments

This research has been supported by the Caltech MURI Center for Quantum Networks.

References

1. H. Buhrman, R. Cleve, and A. Wigderson, LANL e-print quant-ph/9802040, http://xxx.lanl.gov.
2. C. H. Bennett, G. Brassard, C. Crepeau, R. Jozsa, A. Peres, and W. K. Wootters, Phys. Rev. Lett. **70**, 1895 (1993).
3. H.-K. Lo and H. F. Chau, Science **283**, 2050 (1999).
4. C. H. Bennett, P. W. Shor, J. A. Smolin, and A. V. Thapliyal, Phys. Rev. Lett. 85, 3081 (1999).
5. R. de Wolf, in press (2000).
6. J. I. Cirac, P. Zoller, H. J. Kimble, and H. Mabuchi, Phys. Rev. Lett. **78**, 3221 (1997).
7. J. Vuckovic, M. Loncar, H. Mabuchi, and A. Scherer, submitted to Phys. Rev. A (2000).
8. E. Yablonovitch, Phys. Rev. Lett. **58**, 2059 (1987).
9. S. John, Phys. Rev. Lett. **58**, 2486 (1987).
10. O. Painter, R. K. Lee, A. Scherer, A. Yariv, J. D. O'Brien, P. D. Dapkus, and I. Kim, Science **284**, 1819 (1999).
11. J. D. Joannopoulos, R. D. Meade, and J. N. Winn, *Photonic Crystals* (Princeton University Press, 1995).
12. J. Vuckovich, O. Painter, Y. Xu, A. Yariv, and A. Scherer, IEEE J. of Quantum Electronics **35**, 1168 (1999).
13. R. K. Lee, O. J. Painter, B. Kitzke, A. Scherer, and A. Yariv, Electronics Lett. **35**, 569 (1999).
14. J. D. Weinstein and K. G. Libbrecht, Phys. Rev. A **52**, 4004-4009 (1995).
15. M. Drndic, K. S. Johnson, J. H. Thywissen, M. Prentiss, and R. M. Westervelt, Appl. Phys. Lett. **72**, 2906-2908 (1998).
16. M. Todorovic, S. Schultz, J. Wong, and A. Scherer, Appl. Phys. Lett. **74**, 2516 (1999).
17. S.J. van Enk, J.I. Cirac and P. Zoller, Phys. Rev. Lett. **78**, 4293 (1997).
18. S.J. van Enk, J.I. Cirac and P. Zoller, Science **279**, 205 (1998).
19. C.H. Bennett, G. Brassard, S. Popescu, B. Schumacher, J.A. Smolin and W.K. Wootters, Phys. Rev. Lett. **76**, 722 (1996); D. Deutsch, A. Ekert, C. Macchiavello, S. Popescu and A. Sanpera, Phys. Rev. Lett. **77**, 2818 (1996).
20. W. Dür, H.J. Briegel, J.I. Cirac and P. Zoller, Phys. Rev. A **59**, 169 (1999).

EFFICIENT LINEAR OPTICS QUANTUM COMPUTATION

E. KNILL, R. LAFLAMME
Los Alamos National Laboratory, MS B265, Los Alamos, NM 87545

G. J. MILBURN
Center for Quantum Computer Te chnology, Univ. of Queensland, St. Lucia, Australia

Two qubit gates for photons are generally thought to require exotic materials with huge optical nonlinearities. We show here that, if we accept t wo qubit gates that only work conditionally, single photon sources, passive linear optics and particle detectors are sufficient for implementing reliable quantum algorithms. The conditional nature of the gates requires feed-forward from the detectors to the optical elements. Without feed forward, non-deterministic quantum computation is possible. We discuss one proposed single photon source based on the surface acoustic wave guiding of single electrons.

1 Quantum gates for photons.

One of the earliest proposals [1] for implementing quantum computation was based on encoding each qubit in t wo optical modes together containing exactly one photon. However it is extremely difficult to unitarily couple t wo optical modes containing few photons. Here we consider the question of what can be achieved in principle using combinations of only the simplest optical elements: passive linear optics, photodetectors, and single photon sources.

The dynamics implemented by passive linear optics conserves the total number of bosons in the modes. It is therefore convenient to describe them by their effect on the creation operators. Specifically, if U is the unitary operator associated with the evolution, then U takes the state ${\mathbf{a}^{(l)}}^\dagger|\mathbf{0}\rangle$ to

$$U{\mathbf{a}^{(l)}}^\dagger|\mathbf{0}\rangle = U{\mathbf{a}^{(l)}}^\dagger U^\dagger|\mathbf{0}\rangle \tag{1}$$

$$= \sum_m U_{ml}{\mathbf{a}^{(m)}}^\dagger|\mathbf{0}\rangle, \tag{2}$$

using the fact that $U^\dagger|\mathbf{0}\rangle = |\mathbf{0}\rangle$. The matrix defined by the coefficients U_{ml} must be unitary, and furthermore, for all unitary U_{ml}, there is a sequence of phase shifter and beam splitter evolutions which implements the corresponding operation [2]. For a named optical element X, let $U(X)$ be the unitary matrix associated with X according to the above rules. The unitary matrices associated with phase shifters ${P_\theta}^{(l)}$ and beam splitters ${B_\theta}^{(lm)}$ are:

$$U({P_\theta}^{(1)}) = e^{i\theta}$$

$$U({B_\theta}^{(12)}) = \begin{pmatrix} \cos(\theta) & -\sin(\theta) \\ \sin(\theta) & \cos(\theta) \end{pmatrix}.$$

General linear optical elements have Hamiltonians which consist of terms at most quadratic in the annihilation and creation operators.

We choose the traditional encoding of photonic qubits based on a "dual rail" logic, using t wo modes and one photon;

$$|0\rangle_L = |1\rangle_1 \otimes |0\rangle_2 \tag{3}$$

$$|1\rangle_L = |0\rangle_1 \otimes |1\rangle_2 \quad . \tag{4}$$

The modes could be two input modes to a beam splitter distinguished by the different directions of the wave vector, or they could be distinguished by polarisation. In the case of a beam splitter a single qubit gate is easily implemented by the linear transformation

$$a_i(\theta) = U(\theta)^\dagger a_i U(\theta) \tag{5}$$

with $U_{12}(\theta) = \exp\left[i\theta(a_1^\dagger a_2 + a_1 a_2^\dagger)\right]$. Thus

$$\begin{aligned} a_1(\theta) &= \cos\theta a_1 - i\sin\theta a_2 \\ a_2(\theta) &= \cos\theta a_2 - i\sin\theta a_1 \end{aligned}$$

The description in the logical basis becomes,

$$|0\rangle_L \to \cos\theta_1 |0\rangle_L - i\sin\theta_1 |1\rangle_L \tag{6}$$

In addition to a single qubit gate, universal quantum computation requires at least a two qubit gate. A simple choice is the conditional sign shift gate,

$$|x\rangle_L |y\rangle_L \to e^{i\pi x.y} |x\rangle_L |y\rangle_L \tag{7}$$

Unfortunately this is difficult to implement with the optical qubits as photons do not interact except via matter. The mutual Kerr nonlinear phase shift is one such interaction,

$$U_{NL} = \exp[i\pi a_1^\dagger a_1 a_2^\dagger a_2] \tag{8}$$

In practice it is not possible to get a single photon phase shift of π, which this transformation implies, without adding a considerable amount of noise from the electronic systems that mediate the interaction. However, as we now show, we can get an effective nonlinear sign shift with linear optical transformations, such as a beam splitter, by coupling the modes of interest to ancilla modes and performing single photon detection on these ancilla modes. Under the right circumstances the conditional states corresponding to observed counts at the ancilla modes will effect the required conditional phase shift.

We first give a simple illustration of the principle. Suppose we mix two modes at a beam splitter and count photons only on one output mode, say mode a_2. We will assume that mode a_2 is initially in the vacuum state. The conditional state of mode a_1, given a count n is given by

$$|\psi^{(n)}\rangle_1 = \hat{E}_n |\psi\rangle_1 \tag{9}$$

where $|\psi\rangle_1$ is the input state for mode a_1 and the measurement operator is

$$\hat{E}_n = {}_2\langle n|U(\theta)|0\rangle_2 . \tag{10}$$

The case of interest here, where $n = 0, 1$, is given by

$$\hat{E}_0 = \sum_{n=0}^{\infty} \frac{(\cos\theta - 1)^n}{n!} (a_1^\dagger)^n a_1^n \tag{11}$$

$$\hat{E}_1 = \cos\theta \hat{E}_0 - \sin^2\theta a_1^\dagger \hat{E}_0 a_1 \tag{12}$$

In this way a simple beam splitter can implement a conditional phase shift for a particular number state in mode a_1.

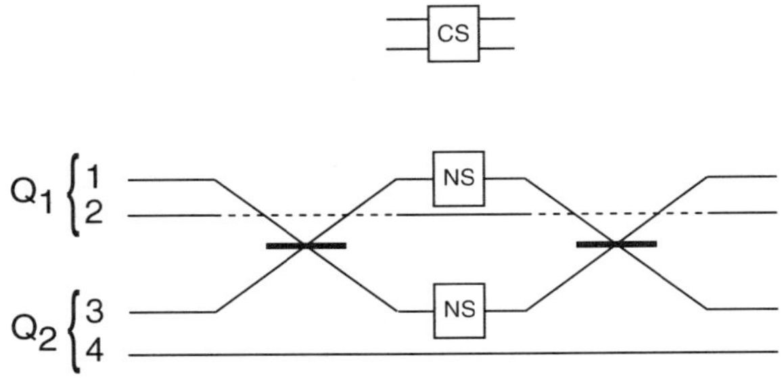

Figure 1. A c-sign two qubit gate implemented using two non deterministic nonlinear sign shift gates.

We seek to implement the conditional nonlinear sign shift for the state of mode a_1 defined by,

$$|\phi\rangle = \alpha_0|0\rangle_0 + \alpha_1|1\rangle_1 + \alpha_2|2\rangle_1 \rightarrow \alpha_0|0\rangle_1 + \alpha_1|1\rangle_1 - \alpha_2|2\rangle_1 \tag{13}$$

We use two ancilla modes, a_2, a_3 which can be regarded as a representing a single qubit in dual rail logic: $|0\rangle_L = |1\rangle_2|0\rangle_3 \quad |1\rangle_L = |0\rangle_2|1\rangle_3$ In the *first step*, single qubit rotation is performed on modes $2-3$, prepared in state $|0\rangle_L$, with the unitary operator $U_{23}(\theta_1)$,

$$|0\rangle_L \rightarrow \cos\theta_1|0\rangle_L - i\sin\theta_1|1\rangle_L \tag{14}$$

In the *second step*, mode a_2 of this qubit is coupled to the signal mode,

$$U_{12}(\theta_2)\left[\cos\theta_1|0\rangle_L - i\sin\theta_1|1\rangle_L\right]|\psi\rangle_1 \tag{15}$$

where $|\psi\rangle_1$ is the signal input state. In the *third step* a final qubit rotation through an angle θ_3 is again performed on modes a_2, a_3 with the unitary $U_{23}(\theta_3)$, and a photon counter records a count on each of the qubit modes. We will show that the conditional sign shift is implemented when the count is $n_2 = 1,\ n_3 = 0$. This corresponds to projecting out the logical state $|0\rangle_L$ on the qubit in modes a_2, a_3.

The corresponding conditional state is then seen to be,

$$|\psi^{(0)}\rangle_1 = \cos\theta_1\cos\theta_3\hat{E}^{(1)}(\theta_2)|\psi\rangle_1 - \sin\theta_1\sin\theta_3\hat{E}^{(0)}(\theta_2)|\psi\rangle_1 \tag{16}$$

For the desired conditional nonlinear sign shift gate (NS gate) we require,

$$\begin{aligned}
\cos\theta_1\cos\theta_3\cos\theta_2 - \sin\theta_1\sin\theta_3 &= \lambda \\
\cos\theta_1\cos\theta_3\cos 2\theta_2 - \sin\theta_1\sin\theta_3\cos\theta_2 &= \lambda \\
\cos\theta_1\cos\theta_3\cos\theta_2(\cos^2\theta_2 - 2\sin^2\theta_2) - \sin\theta_1\sin\theta_3\cos^2\theta_2 &= -\lambda
\end{aligned}$$

The solution is easily seen to be $\theta_1 = \theta_2 = 22.5^o \quad \theta_2 = 114.47^o$ with $\lambda = -1/2$. The success probability is $\lambda^2 = 1/4$. In figure 1 we show how two non deterministic NS gates can be used to implement a two qubit conditional phase shift gate, a c-sign gate. As success requires the two NS gates to work the overall probability of success is 1/16.

A cascaded sequence of non deterministic gates is useless for quantum computation as the probability of many gates working in sequence would decrease exponentially. We now show how to avoid this by using a teleportation protocol to only implement a gate in a quantum circuit if it works. In essence we hold back the gate until we are sure it works and then teleport it onto the required stage of the computation. The idea that teleportation can be used for universal quantum computation was first proposed by Gottesman and Chuang [3]. The idea is to prepare a suitable entangled state for a teleportation protocol with the required gate already applied. We use a non deterministic NS gate to prepare the required entangled state, and only complete the teleportation when the this stage is known to work. The teleportation step itself is non deterministic, but we will show that by using the appropriate entangled resource the teleportation step can be made near deterministic. The near deterministic teleportation protocol requires only photon counting and fast feedforward. We do not need to make measurements in a Bell basis.

A basic quantum teleportation protocol for transferring the state $\alpha_0|0\rangle_0 + \alpha_1|1\rangle_0$ to mode 3 begins with adjoining the "entangled" ancilla state $|\mathrm{t}_1\rangle_{12} = |01\rangle_{12} + |10\rangle_{12}$ to mode 1. (We omit normalisation constants wherever possible.) Note that in this case, the ancilla state is easily generated from $|10\rangle_{12}$ by means of a beam splitter.

The second step is to measure modes 0 and 1 in the basis $|01\rangle_{01} \pm |10\rangle_{01}, |00\rangle_{01} \pm |11\rangle_{01}$ (the "Bell basis"). We decompose the measurement into two steps. The first step determines the parity p of the number of bosons in modes 0 and 1 ("parity measurement"). The second determines the sign s in the superposition. Consider the case where $p = 1$. Then if $s =$ '+', the state of mode 2 is $\alpha_0|0\rangle_2 + \alpha_1|1\rangle_2$. If $s =$ '-', the state is $\alpha_0|0\rangle_2 - \alpha_1|1\rangle_2$, which can be restored to the starting state by using a phase shifter. For $p = 0$, the situation is similar except that $|0\rangle_2$ and $|1\rangle_2$ are flipped (and cannot be easily un-flipped using linear optics). The key property of quantum teleportation is that the input state appears in mode 2 up to a simple transformation without having interacted with mode 2 after the preparation of the initial ancilla state.

Consider the parity measurement. Applying a balanced beam splitter to modes 0 and 1 and then measuring the number of photons in the two modes successfully determines the parity, and if it is odd, the sign. As a result, it can be used to perform the teleportation with success probability 1/2. We refer to the partial Bell-state measurement as $\mathbf{BM}_1$ and to the corresponding teleportation protocol as $\mathbf{T}_{1/2}$.

The next step is to use $\mathbf{T}_{1/2}$ to design a conditional sign flip c-$\sigma_{z1/4}$ which succeeds with probability 1/4. To see how to do this, observe that to implement c-σ_z on two bosonic qubits in modes 1, 2 and 3, 4 respectively, we could try to first teleport the first modes of each qubit to two new modes (labelled 6 and 8) and then apply c-σ_z to the new modes. When using $\mathbf{T}_{1/2}$, we may need to apply a sign correction. Since this commutes with c-σ_z, there is nothing preventing us from applying c-σ_z to the prepared state before performing the measurements. The implementation is shown in Fig. 2 and now consists of first trying to prepare two copies of $|\mathrm{t}_1\rangle$ with c-σ_z already applied, and then performing two $\mathbf{BM}_1$ measurements. Given the prepared state, the probability of success is $(1/2)^2$. The state can be prepared using c-$\sigma_{z1/16}$, which means that the preparation has to be retried an average of 16 times before it is possible to proceed.

To improve the probability of successful teleportation to $1 - 1/(n+1)$, we generalise the initial entanglement by defining

$$|\mathrm{t}_n\rangle_{1\ldots(2n)} = \sum_{j=0}^{n} |1\rangle^j |0\rangle^{n-j} |0\rangle^j |1\rangle^{n-j}. \tag{17}$$

The notation $|a\rangle^j$ means $|a\rangle|a\rangle\ldots$, j times. The modes are labelled from 1 to $2n$, left to

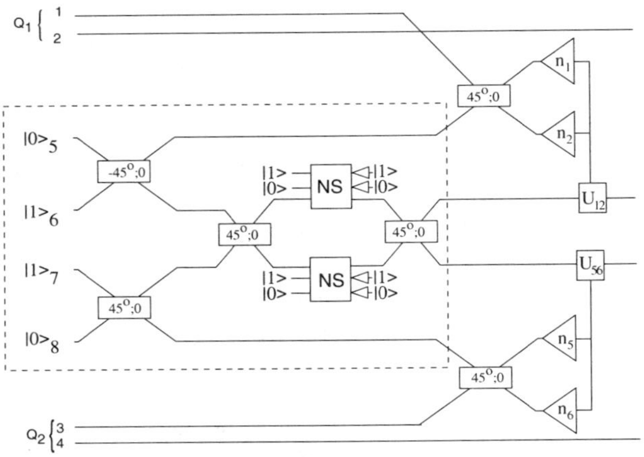

Figure 2. A c-sign two qubit gate with teleportation to increase success probability to 1/4.

right. Note that the state exists in the space of n bosonic qubits, where the k'th qubit is encoded in modes $n+k$ and k (in this order).

To teleport the state $\alpha_0|0\rangle_b + \alpha_1|1\rangle_b$ using $|t_n\rangle$ use the measurement $\mathbf{BM}_n$ implemented by first applying an $n+1$ point Fourier transform $\hat{\mathbf{F}}_{n+1}$ on modes 0 to n. $\hat{\mathbf{F}}_{n+1}$ is defined by

$$u(\hat{\mathbf{F}}_{n+1})_{kl} = \omega^{kl}/\sqrt{n+1}, \tag{18}$$

where $\omega = e^{i2\pi/(n+1)}$ and $k, l \in 0 \ldots n$. It is by definition implementable with passive linear optics. Using the parallel fast Fourier transform (see page 795 of [4]), it can be implemented with $O(n\log(n))$ elements and depth $O(\log(n))$, for n a power of 2. Alternatively, a multiport generalisation of the Mach-Zehnder interferometer can be used [5]. After applying $\hat{\mathbf{F}}_{n+1}$, $\mathbf{BM}_n$ measures the number of photons in each of the modes 0 to n.

Suppose $\mathbf{BM}_n$ detects k bosons altogether. It is possible to show [6] that if $0 < k < n+1$, then the teleported state appears in mode $n+k$ and only needs to be corrected by applying a phase shift. The modes $2n-l$ are in state 1 for $0 \leq l < (n-k)$ and can be reused in future preparations requiring single bosons. The modes are in state 0 for $n-k < l < n$. If $k=0$ we learn that the input state is measured and projected to $|0\rangle_b$ and if $k = n+1$, it is projected to $|1\rangle_b$. The probability of these two events is $1/(n+1)$, regardless of the input. We will make use of the fact that failure is detected and corresponds to measurements in the basis $|0\rangle, |1\rangle$ with the outcome known. Note that both the necessary correction and which mode we teleported to are unknown until after the measurement.

There are two problems with the methods shown so far. The first is that for large n, the obvious networks for preparing the required states have very low probabilities of success. It is therefore desirable to avoid using the methods with large n. The second problem is that to get decent success probabilities for coupling gates does large n, particularly if it is necessary to meet the accuracy requirements of reliable quantum computing.

To achieve exponential improvements in the probability of success for gates and state production with small n, we use quantum codes and exploit the properties of the failure behaviour of c-$\sigma_{zn^2/(n+1)^2}$. For details see Knill et al. [6]. As a result it is possible to iterate the methods to efficiently achieve essentially perfect quantum computation. This iteration is known as *concatenation* and underlies the accuracy threshold theorems of fault tolerant quantum computation [7,8,9,10].

To conclude we summarise the physical requirements for linear optics quantum computation. The Scheme requires: (i) single photon sources (ii) fast efficient single photon detectors, (iii) low photon absorption and (iv) fast electro-optics feed forward. There is currently a great deal of interest in schemes for single photon sources[11] and recently a particularly useful scheme was demonstrated[?]. We expect that reliable and fast single photon sources are not far away. In the mean time however the basic elements of linear optics QC could be demonstrated with conditional non deterministic single photon sources, such as are used in parametric down conversion experiments[12]. The requirement for fast single photon detectors that can distinguish zero, one and t wo photons is a difficult one to meet with current technology, but is achievable by a variety of means. Low photon absorption is not in principle a problem. As was demonstrated by Knill et al.[6], there are linear optics protocols that can detect for photon loss. The requirement of fast electro optic feed forward is perhaps the biggest hurdle to overcome for demonstrating a simple quantum circuit. The near deterministic teleportation steps require some kind of delay line before measurement results are fed forward. This places strong bandwidth requirements on any electro-optical processes. Despite these difficulties we expect that the basic elements of linear optics quantum computation will be demonstrated in the next few years and in the long run reliable linear optics quantum computation may be no more difficult than other schemes for large scale quantum computation.

heading for

References

1. G.J.Milburn, Phys. Rev. Lett. **62**, 2124–2127 (1988).
2. M. Reck, A.Zeilinger, H. J.Bernstein, and P.Bertani, Phys. Rev. Lett., **73**, 58–61 (1994).
3. D.Gottesman, and I.L.Chuang, Nature, **402**, 390–393 (1999).
4. T.H.Cormen, C.E.Leiserson, and R.L.Rivest, *Introduction to Algorithms*, MIT Press, Cambridge, Mass, (1990).
5. G.Weihs, M.Reck, H.Weinfurter, and A.Zeilinger, Opt. Lett. **21**, 302–304 (1996).
6. E.Knill,R.Laflamme and G.J.Milburn, Nature, **409**, 46, (2001).
7. D.Aharonov, and M. Ben-Or, In *Proceedings of the 29th Annual ACM Symposium on the Theory of Computation (STOC)*, 176–188 (ACM Press, New York, New York, 1996).
8. A.Y.Kitaev, Quantum computations: algorithms and error correction. Uspekhi Mat. Nauk. **52**, 53–112 (1997).
9. E.Knill, R.Laflamme, and W.H.Zurek, Science, **279**, 342–345 (1998).
10. J.Preskill, Proc. R. Soc. Lond. A, **454**, 385–410 (1998).
11. C.Foden,V.I.Talyanskii,G.J.Milburn,M.L.Leadbeater, and M. Pepper, Phys. Rev. A, **62**, 011803(R), 1-4(2000).
12. D.Bouwmeester, J.-W.Pan, M.Daniell,H.Weinfurter,A.Zeilinger, Phys. Rev. Lett. **82**, 1345, (1999).

OPTICAL QUANTUM INFORMATION PROCESSING: EXPERIMENTS WITH CW LASER LIGHT

H-A.BACHOR, PING KOY LAM, W.BOWEN, R.SCHNABEL, S.BENNETTS, B.BUCHLER, U.L.ANDERSEN

Department of Physics, The Faculties, The Australian National University Canberra ACT 0200, Australia

T.C.RALPH

Department of Physics, University of Queensland, QLD 4072, Australia

Experiments and devices based on continuous waves (CW) optical beams can provide solutions to optical quantum information processing which are analogous to those carried out with single photon entangled states. We report on our experimental achievements and goals for using squeezed light and entangled beams for quantum nondemolition measurements, dense coding, teleportation and quantum cryptography .

1 Introduction

Optical beams can carry quantum information (QI) and recently many techniques for QI communication and QI processing which rely on exploiting the properties of optical entanglement have evolved and extend well beyond the limits of conventional classical optics. These techniques include quantum nondemolition measurements (QND), quantum cryptography, dense coding and teleportation. The most established technology is the use of correlated, or entangled pairs of single photons [1]. Here each individual photon carries the information, for example, in the form of polarisation. An alternative is to represent the information in the form of continuous variables, for example the quadrature amplitudes at a fixed modulation frequency [2] . This can be implemented by using continuous wave (CW) laser beams, as reported here and by other groups [3,4,5], or by continuous trains of soliton pulses as demonstrated in Erlangen [6]. In the following sections, we present our latest experimental results.

2 Description of QI on CW beams and terminology

We are using laser beams with one single spatial mode. These beams carry information in the form of temporal modulation. We describe the modulation and the noise in terms of the amplitude quadrature X1 and phase quadrature X2. A multitude of modulation frequencies Ω, each forming one communication channel, can be used simultaneously. We quantify the signals and the noise through the corresponding variances $V1(\Omega)$ and $V2(\Omega)$ which are both normalized to the standard quantum limit. In this description all optical beams obey the Heisenberg uncertainty principle $V1 \cdot V2 \geq 1$ and those which obey it in the equality are called minimum uncertainty optical states. A classical beam has $V1 \geq 1$ and $V2 \geq 1$.

The detection of the information in one frequency channel is achieved with homodyne detection. This is essentially the beating of the DC component of the beam itself, or of a local oscillator beam at the optical frequency ν_L, with the sidebands at $\nu_L \pm \Omega$ (see Fig.1). A minimum uncertainty classical beam, such as a perfect laser beam, has no extra photons

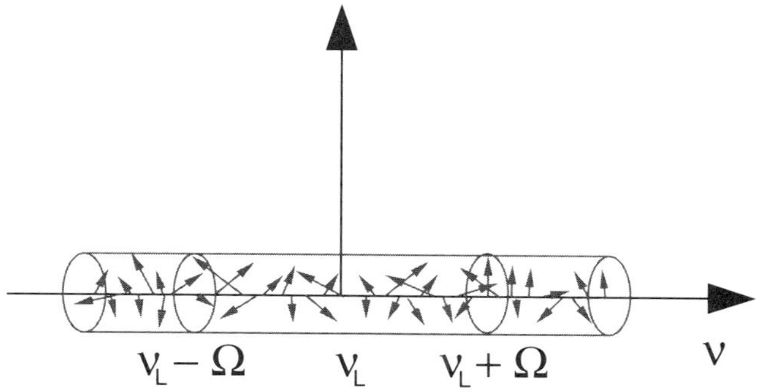

Figure 1. Representation of the modulation and noise in a continuous wave laser beam. Each pair of small frequency intervals at $\nu_L \pm \Omega$ represents one distinct information channel. For classical light the noise in the two sidebands is independent, not correlated. For squeezed light the noise is correlated. The cylinder represents the vacuum noise.

in the sidebands ($n_\Omega = 0$). Any information, or noise, requires extra energy and the photon number, which is usually small, is given by

$$n_\Omega = \frac{V1(\Omega) + V2(\Omega)}{4} - \frac{1}{2} \quad , \tag{1}$$

where n_Ω times the detection bandwidth is in the units of photons per second.

In general there is a distinct analogy between single photon experiments and CW beam sideband experiments since the photon number in these sidebands can be extremely low. Thus quantum information can be encoded onto the beams by modulations and superposition states in the X1, X2 phase space.

3 Building blocks for QI processing

For the communication and processing of quantum information we require non-classical forms of light. These are the so called squeezed states where a nonlinear optical process is used to suppress selectively the noise in one quadrature, for example $V1 < 1$, with the consequence that the other quadrature is increased , $V2 \geq 1/V1$. A number of nonlinear processes can be used and after years of development the optical parametric oscillator (OPO) and the nonlinear soliton interaction in fibers have emerged as some of the best and most reliable techniques [2]. At the ANU we can generate noise suppression by 5.5 dB (V1=0.28) routinely [7]. Similarly intensity squeezing with solitons of 5 dB (V1 = 0.32) is state of the art with optical fibres [6]. The degree of squeezing, and even the quadrature in which squeezing is observable, can vary with the detection frequency.

Our second experimental component is the noiseless transfer of information in *one* quadrature using electro-optic modulators in a control loop. We have shown that this can be done without introducing any extra noise, even for non-classical input beams [8]. For example we have shown that we can amplify a squeezed laser beam with a input signal to noise ratio (SNR) of 2.8 dB at 10 MHZ and send it along a communication line with more than 98 % attenuation and still achieve an SNR of 2.6 dB at the output. This is vastly

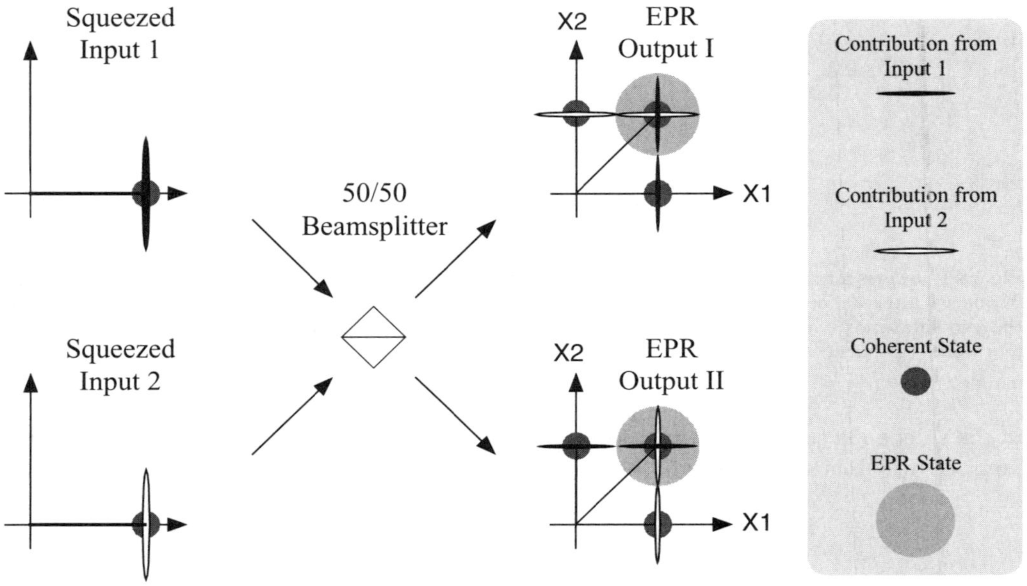

Figure 2. Diagrammatic representation of entangled beams in a phase space diagram. Two squeezed beam are combined by a 50/50 beamsplitter, the outgoing beams (I) and (II) have large uncertainties (shaded area), but measurements between (I) and (II) show correlations within the uncertainty circle. The error of inferance is given by the overlap of the squeezed beams.

superior to a conventional, phase independent laser amplifier which would have achieved an output SNR degraded by almost 3 dB (the "noise penalty" in the limit of large gain). Furthermore, we have shown that the electro-optic feedforward loop when used in combination with a squeezed vacuum meter input is a quantum nondemolition (QND) device with results given in Figure 4.

Finally, we require the CW equivalent of entangled photon pairs in order to transport QI of *two* noncommuting quadratures via teleportation. It was shown by Ralph and Lam that the combination of two phase locked squeezed beams (SQ1) and (SQ2) with orthogonal squeezing quadratures, combined on a 50/50 beam splitter, is fully analogous to an EPR-entangled photon pair. Fig. 2 shows schematically this arrangement for producing entangled beams (I) and (II). The individual output beams, either (I) or (II) are very noisy with $V1_I = V2_{II}$ being determined by the size of the anti-squeezing quadrature of the input beams. However, within this noise strong correlations exist between the measurements carried out with the two output beams. Measuring both quadratures of one beam allows us to predict the quadratures of the other. The uncertainty of this is given by the degree of squeezing , that is $V2_I = V1_{II}$. The degree of entanglement can be quantified by the conditional variance $V_{I|II}$ between the output beams which can be deduced by measuring the sum and differences of the photocurrents proportional to $V1_I, V2_I, V1_{II}, V2_{II}$.

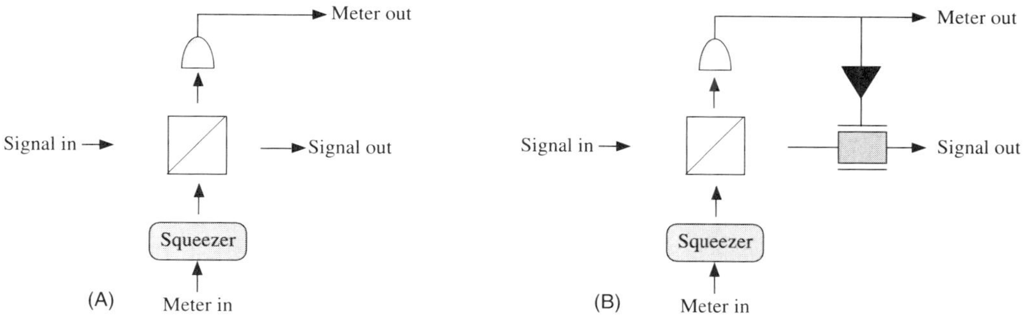

Figure 3. (A) Schematic diagram for a QND measurement, and (B) pre- and post-enhanced QND measurement with electro-optic feed-forward.

It is interesting that for a squeezed beam the sidebands at $\nu_L \pm \Omega$ are correlated to each other, for an entangled pair there are no such correlation for a single beam on its own; but in contrast there is now a correlation between the sidebands of the two beams within the entangled pair. The photon number in these sidebands (Eqn.(1)) is not much greater than 1 for the degree of squeezing that is achievable. Thus the number of photons measured in a detection interval in our entanglement experiments is comparable to that of single photon experiments.

4 Results of QND measurements

We have reported excellent noise suppression of the vacuum state with a single OPO [7]. In a first step this light is now applied to demonstrate the quantum nondemolition measurement of a signal input beam (S_{in}). The input beam is combined on a beam splitter with the squeezed beam (M_{in}), which is regarded as the meter beam, as shown in Fig.3A. The quality of this QND device can be determined by measuring two independent quantities: the signal and meter transfer coefficient $T_{S+M} = T_S + T_M$ where T_S and T_M are the ratios of the SNRs from the signal output (S_{out}) and the meter output (M_{out}), respectively, to the sigal input beam (S_{in}) and, secondly, the conditional variance $V_{S|M}$ between the signal (S_{out}) and the meter (M_{out}) output beams. A true QND scheme satisfies $T_{S+M} > 1$ and $V_{S|M} < 1$. A perfect QND scheme achieves $T_{S+M} = 2$ and $V_{S|M} = 0$. In Fig.4 we show the results achieved to date by various groups and various experiments [8] and our result (labeled 1) obtained with a 50/50 beamsplitter and squeezing of 4.5 dB. We can improve the SNR, but not the conditional variance, by feeding the photocurrent from the meter beam (M_{out}) forward to the signal beam (S_{out}), as described by B.C.Buchler et al.[9], see Fig. 3B. The results are shown in Fig. 4 where (2) is the experimental result with aditional feedforward. They are the best QND results to date. (3) represents our prediction for a 90/10 beamsplitter. This QND experiment is the single quadrature equivalent of the teleportation experiment planned and confirms that the technology is sufficiently advanced to contemplate QI processing.

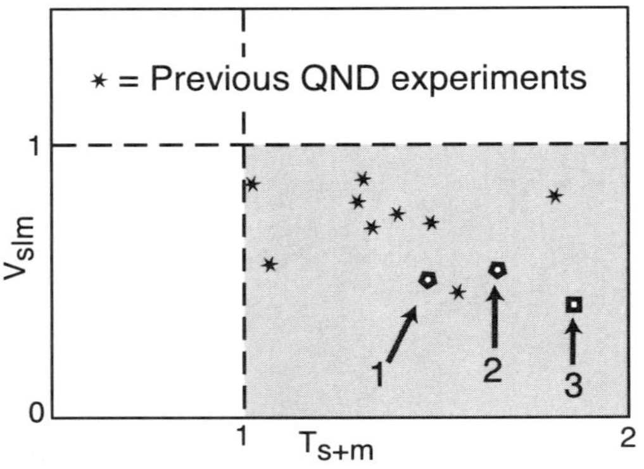

Figure 4. New experimental results for quantum nondemolition measurements (QND). The quality of a QND is given by the two parameters, T_{S+M} which describes the reduction in SNR and $V_{S|M}$ which describes the correlation between the signal and the meter. $T_{S+M} = 2$ and $V_{S|M} = 0$ is the perfect QND result. All present results are plotted (stars) together with our new result achieved with OPO squeezed light (1) and electro-optic feedforward using a 50/50 beamsplitter (2). (3) represents our prediction for a 90/10 beamsplitter.

5 Experiments in teleportation

We are presently assembling a source of entangled beams based on our well tested OPO technology. It contains a low noise laser at 1064 nm, converted into about 500 mW of green light at 532 nm which pumps two identical OPOs, operating in the degenerate mode at 1064 nm. Each will produce a beam with at least 4 dB noise reduction. The beams are phaselocked and produce a pair of entangled beams. The major difficulty is the complexity of operating simultaneously all the multitude of electronic feedback loops which are required to control the low bandwidth of the laser and the phase of all the seed beams, the local oscillator beams and the squeezed beams. Once the entanglement has been measured we can proceed in a series of QI experiments: Teleportation is at the heart of our program. Quantum information which is placed as phase and amplitude modulation, first on classical, coherent beams and later on squeezed beams is noiselessly transferred. The pair of entangled light beams is used for the detection and the reconstruction of the information. We measure the information transfer in both quadratures ($T1+T2$) and the conditional variance ($V_{in|out}$) of the input and the output beams to quantify the properties of the teleporter in a similar way as shown for the QND experiment above. According to our theoretical predictions [10] this experiments should extend the results already observed by Furusawa et al.[3].

6 Conclusion

Continuous wave beams can be used for QI processing. The ideas are well established and the current concern is to overcome the technological challenge of operating two sources of squeezed light simultaneously with high efficiency. In the near future entangled beams will be available routinely and our first three demonstrations of this technology will be dense coding, quantum teleportation and quantum cryptography all using continuous variables.

After these fundamental studies we expect applications in the area of fast, high bandwidth cryptography and the transmission of quantum information in optical networks and potentially within a quantum computer.

Acknowledgments

We would like to thank Gerd Leuchs, Christine Silberhorn, Natalia Korolkova and Friedrich König for their stimulating collaboration. This work is financially supported by the Australian Research Council.

References

1. C.H.Bennett and D.P.DiVincenzo, Nature **404**, 247 (2000)
2. H.-A.Bachor *A guide to experiments in Quantum Optics* (Wiley-VCH, Berlin, 1998)
3. A.Furusawa, J.L.Sørensen, S.L.Braunstein, C.A.Fuchs, H.J.Kimble and E.S.Polzik, Science **282**, 706 (1998)
4. J.Hald, J.L.Sørensen, C.Schori and E.S. Polzik, Phys.Rev.Lett. **83**, 1319 (1999)
5. H.Wang, Y.Zhang, Q.Pan, H.Su, A.Porzio, C.Xie and K.Peng, Phys.Rev.Lett. **82**, 1414 (1999)
6. G.Leuchs, T.C.Ralph, Ch.Silberhorn and N.Korolkova, J.Mod.Opt. **46**, 1927 (1999)
7. P.K.Lam, T.C.Ralph, B.C.Buchler, D.E.McClelland, H-A.Bachor and J.Gao, J.Opt.B: Semiclass.Opt. **1**, 469 (1999)
8. P.Grangier, J.A.Levenson and J.-P.Poizat, Nature **396**, 537 (1998)
9. P.K.Lam, E.H.Huntington, T.C.Ralph and H-A.Bachor, Phys.Rev.Lett. **79**, 1471 (1997)
10. T.C.Ralph and P.K.Lam, Phys.Rev.Lett. **81**, 5668 (1998)

LIGHT-MEDIATED TRANSFER AND ENTANGLEMENT OF MOTIONAL QUANTUM STATES IN A DISTRIBUTED NETWORK

A.S. PARKINS

Department of Physics, University of Auckland, Private Bag 92019, Auckland, New Zealand
E-mail: s.parkins@auckland.ac.nz

H.J. KIMBLE

Norman Bridge Laboratory of Physics 12-33, California Institute of Technology, Pasadena, CA 91125, U.S.A.
E-mail: hjkimble@its.caltech.edu

We describe a scheme for distributing entanglement of orthogonal motional modes between distantly separated harmonically-trapped atoms using a light-mediated quantum state transfer procedure based on interactions in cavity QED.

1 Introduction

In addition to being of fundamental interest in physics, quantum entanglement is an essential resource for quantum information processing and distribution. Popular candidates for experimental investigation in this context include trapped atoms, cavity quantum electrodynamics (cavity QED), and nonclassical light fields, using which a variety of impressive quantum state manipulations have now been demonstrated, including elementary quantum logic operations, preparation of quantum-mechanically entangled pairs and quadruplets of atoms, and quantum teleportation.

In recent work we have proposed a scheme that offers the possibility of combining the features of trapped atoms and nonclassical light fields in a distributed quantum network;[1] in particular, a scheme that enables motional quantum states to be coupled in a simple manner to propagating (nonclassical) light fields via interactions in cavity QED. This coupling enables the *deterministic* generation and distribution of quantum entanglement between different (and distant) atoms and/or light fields. One particular example we have considered is the preparation of an Einstein-Podolsky-Rosen (EPR) state in the positions and momenta of a pair of distantly-separated trapped atoms,[2] which in turn leads to a scheme for the teleportation of motional quantum states.[3]

For these proposals, entanglement of the atoms' motional states is achieved through the transfer of entanglement from the quantum-correlated output light fields from a nondegenerate parametric amplifier. However, the trapped-atom cavity QED setup can itself act as a source of nonclassical light fields; nonclassical motional states (prepared by some means independent of the atom-cavity coupling) can be transferred to the propagating output field from the cavity, which can in turn be coupled to the motion of a second atom confined inside another, distant cavity.[4] It follows that "local" entanglement of orthogonal motional modes of a single atom trapped inside an optical cavity can be transformed, via the motion-light coupling, into "nonlocal" entanglement of the motional modes of distantly-separated atoms. In this way, the need for nonlinear (quantum) optical devices, such as nondegenerate parametric amplifiers, is eliminated, and all of the desired operations are achieved using only trapped-atom cavity QED configuratons.

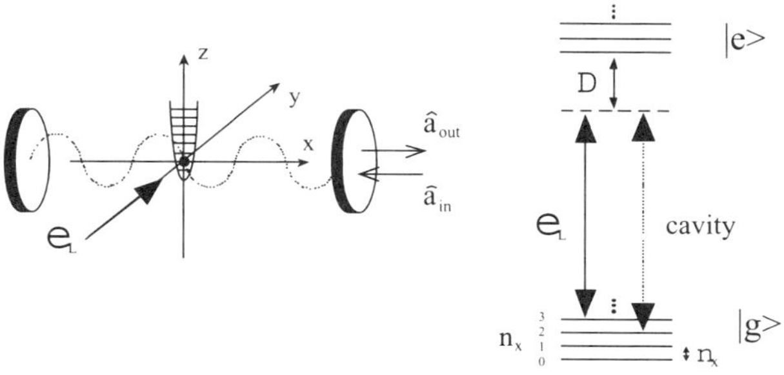

Figure 1. Schematic of proposed setup and excitation scheme for state transfer between the motion of a trapped atom and a quantized cavity mode of the electromagnetic field. All input and output to the cavity mode is assumed to be through just one mirror.

2 Motion-Light Coupling

Briefly, we consider a single two-level atom (or ion) tightly confined in a harmonic trap located inside a high-finesse optical cavity. The atomic transition of frequency ω_a is coupled to a single mode of the cavity field of frequency ω_{cav} and also to an external (classical) laser field of frequency ω_L and strength $\mathcal{E}_L$. The physical setup and excitation scheme are depicted in Figure 1. The cavity is aligned along the x-axis, while the field $\mathcal{E}_L$ is incident along the y-axis. Both the cavity field and $\mathcal{E}_L$ are far from resonance with the atomic transition, but their difference frequency is chosen so that they drive Raman transitions between neighboring motional Fock states (i.e., $\omega_{cav} - \omega_L = \nu_x$, with ν_x the x-axis trap frequency).

A number of assumptions are made in order to achieve the desired motion-light coupling:

1. Atomic spontaneous emission is neglected and the internal atomic dynamics adiabatically eliminated.

2. The size of the harmonic trap, located at a *node* of the cavity field, is taken to be small compared to the optical wavelength (Lamb-Dicke regime), enabling the approximations $\sin(k\hat{x}) \simeq \eta_x(\hat{b}_x + \hat{b}_x^\dagger)$ and $\mathcal{E}_L(\hat{y}, t) \simeq \mathcal{E}_L(t)e^{-i\phi_L}$, where η_x ($\ll 1$) is the Lamb-Dicke parameter and $\hat{x} = (\hbar/2m\nu_x)^{1/2}(\hat{b}_x + \hat{b}_x^\dagger)$.

3. The trap frequency ν_x and cavity field decay rate κ are assumed to satisfy $\nu_x \gg \kappa \gg |(g_0\eta_x/\Delta)\mathcal{E}_L(t)|$, where g_0 is the single-photon atom-cavity mode coupling strength, and $\Delta = \omega_{cav} - \omega_L$. The first inequality allows a rotating-wave approximation to be made with respect to the trap oscillation frequency, while the second inequality enables an adiabatic elimination of the cavity field mode.

Under these conditions, the motional mode dynamics in the x direction can be approximated by the simple quantum Langevin equation

$$\dot{\tilde{b}}_x = -\Gamma(t)\tilde{b}_x + \sqrt{2\Gamma(t)}\,\tilde{a}_{in}(t)\,, \tag{1}$$

where $\tilde{b}_x = e^{i\nu_x t}\hat{b}_x$ and $\Gamma(t) = [g_0\eta_x|\mathcal{E}_L(t)|/\Delta]^2/\kappa$. The operator $\tilde{a}_{in}(t) = e^{i\omega_{cav}t}\hat{a}_{in}(t)$ obeys the commutation relation $[\tilde{a}_{in}(t), \tilde{a}_{in}^\dagger(t')] = \delta(t-t')$ and describes the quantum noise *input to the cavity field* (in a frame rotating at the cavity frequency). From the linear nature of (1), it follows that the statistics of a (continuous) light field incident upon the cavity can be "written onto" the state of the atomic motion. For example, this effect can be used to prepare a (pure) squeezed state of the motion of a trapped atom by driving the cavity with broadband squeezed light from a degenerate parametric amplifier.[1] Also, entanglement between separate light fields (for example, generated by a nondegenerate parametric amplifier) can be transferred to entanglement between the motional states of separate atoms.[2]

3 Motional State Transfer Between Distant Sites

Eq.(1) describes a simple coupling of the motional mode to propagating light modes external to the cavity. From the input-output theory of optical cavities,[5] it can be shown that the cavity output field is given, under the assumptions made above, by

$$\tilde{a}_{\rm out}(t) = -\tilde{a}_{\rm in}(t) - \sqrt{2\Gamma(t)}\,\tilde{b}_x(t)\,. \tag{2}$$

Following work by Cirac *et al.*[6] on the transmission of a qubit between two nodes of a quantum network, one can show that if the output field from one of our atom-cavity configurations is incident on a second such atom-cavity configuration, with the coupling between systems being *unidirectional*, then with suitably tailored laser pulses $\mathcal{E}^1_{\rm L}(t)$ and $\mathcal{E}^2_{\rm L}(t)$ applied at the two sites one may realise the motional state transfer

$$|\phi\rangle^1_x\,|0\rangle^2_x \to |0\rangle^1_x\,|\phi\rangle^2_x\,, \tag{3}$$

where $|\phi\rangle_x$ is an *arbitrary* quantum state describing the motion along the x-axis.[1] Armed with this capability, we are able to distribute quantum states of a material oscillator, and generate entanglement, between distantly-separated locations.

4 Distribution of Motional State Entanglement

A variety of schemes exist for coherently manipulating and entangling orthogonal motional modes in a single-atom trap,[4,7] typically involving laser-induced stimulated-Raman transitions between internal atomic states. In fact, it is in principle possible to prepare an arbitrary state of the motion in two dimensions: $|\psi\rangle_{xz} = \sum_{n,m} c_{nm}|n\rangle_x \otimes |m\rangle_z$, where $|n\rangle_{x,z}$ are Fock states of the motion in the x and z directions.

The procedure for distributing entanglement is then straightforward: one prepares an entangled state of the vibrational modes of an atom trapped inside a cavity at one site, after which the x mode is coupled to the cavity mode and thence to the external light field. With appropriate time-dependent coupling we have seen that the properties of the x mode can be transferred to the corresponding mode of a distant cavity-confined atom. That is, we can perform the transformation

$$\left(\sum_{n,m} c_{nm}|n\rangle^{(1)}_x \otimes |m\rangle^{(1)}_z\right) \otimes |0\rangle^{(2)}_x \to |0\rangle^{(1)}_x \otimes \left(\sum_{n,m} c_{nm}|n\rangle^{(2)}_x \otimes |m\rangle^{(1)}_z\right), \tag{4}$$

where $|n\rangle^{(i)}_{x,z}$ is a vibrational Fock state of the atom at location i. Hence, the x and z modes of the atoms at locations 2 and 1, respectively, become entangled.

5 Practical Requirements

The validity of, and constraints set by, the assumptions made in deriving the motion-light state exchange model that is central to this work are discussed and examined in more detail elsewhere.[4] Briefly, required conditions are of (i) strong coupling optical cavity QED,[8] such that $g_0^2/(\kappa\gamma) \gg 1$, where γ is the atomic spontaneous decay rate, (ii) strong confinement of the atoms with minimal motional-state decoherence,[9] and (iii) high-efficiency cavity input-output light coupling, with low light propagation losses.

Given these conditions, the possibilities offered by the scheme for distributed entanglement are many and varied. In addition to potential applications in quantum communication and computing, entangled and delocalized states of the form shown above would offer fascinating opportunities for fundamental tests of quantum mechanics using massive particles.

Acknowledgments

ASP gratefully acknowledges support from the Marsden Fund of the Royal Society of New Zealand. HJK is supported by the National Science Foundation, by DARPA via the QUIC Institute which is administered by ARO, and by the Office of Naval Research.

References

1. A.S. Parkins and H.J. Kimble, *J. Opt. B: Quantum Semiclass. Opt.* **1**, 496 (1999).
2. A.S. Parkins and H.J. Kimble, *Phys. Rev.* A **61**, 052104 (2000).
3. A.S. Parkins and H.J. Kimble, in *Frontiers of Laser Physics and Quantum Optics, Proceedings of the International Conference on Laser Physics and Quantum Optics,* eds. Z. Xu *et al.* (Springer, Berlin, 2000). See also quant-ph/9909021.
4. A.S. Parkins and E. Larsabal, *Phys. Rev.* A **63**, 012304 (2001).
5. C.W. Gardiner and P. Zoller, *Quantum Noise* (Springer-Verlag, Berlin, 2000).
6. J.I. Cirac, P. Zoller, H.J. Kimble, and H. Mabuchi, *Phys. Rev. Lett.* **78**, 3221 (1997).
7. See, e.g., S.A. Gardiner, J.I. Cirac, and P. Zoller, *Phys. Rev.* A **55**, 1683 (1997); B. Kneer and C.K. Law, *ibid.* **57**, 2096 (1998); D.J. Wineland *et al.*, *Physica Scripta* **T76**, 147 (1998).
8. See, e.g., J. Ye, D.W. Vernooy, and H.J. Kimble, *Phys. Rev. Lett.* **83**, 4987 (1999); C.J. Hood *et al.*, *Science* **287**, 1447 (2000); P.W.H. Pinkse *et al.*, *Nature* **404**, 365 (2000).
9. See, e.g., Ch. Roos *et al.*, *Phys. Rev. Lett.* **83**, 4713 (1999); Q.A. Turchette *et al.*, *Phys. Rev.* A **61**, 063418 (2000).

TOMOGRAPHY AND ITS ROLE IN QUANTUM COMPUTATION

W. J. MUNRO[†], D. F. V. JAMES[‡], A. G. WHITE[*] AND P. G. KWIAT[**]
[†] *Hewlett-Packard Laboratories, Filton Road, Stoke Gifford, Bristol BS34 8QZ, UK*
[‡] *Theory Division, P-23, Los Alamos National Laboratory, Los Alamos, NM, USA*
[*] *Centre for Quantum Computer Technology, University of Queensland, Brisbane, QLD, Australia*
[**] *Department of Physics, University of Illinois, Urbana-Champain, IL, USA.*

Quantum computation depends on quantum entanglement, a correlation between subsystems that cannot occur classically. A variety of theoretical measures exist for quantifying the degree entanglement in such schemes, all of which are functions of the system density matrix. How can the entanglement be measured experimentally ? Using quantum tomography techniques developed for two photon entangled states, the density matrix can be reconstructed from the appropriate experimental data. In this case the state tomography gives the complete characterization of the physical system (for the relevant degree of freedom, such as spin or polarization). It gives information on both the degree of nonclassical correlation, that is entanglement, as well as the amount of decoherence in the system. In this proceedings we discuss the general state tomography procedure required to characterize a few qubit quantum computer, for any architecture.

1 Introduction

One of the distinguishing features of quantum mechanics, not found in classical physics, is the possibility of entanglement between subsystems. It lies at the core of many applications in the emerging field of quantum information science[1], such as quantum teleportation[2] and quantum error correction[3]. Quantum entanglement refers to correlations between the results of measurements made on component subsystems of a larger physical system which cannot be explained in terms of correlations between local classical properties inherent in those same subsystems. Alternatively, an entangled state cannot be prepared by local operations and local measurements on each subsystem. Thus one often says that an entangled composite system is nonseparable. Formally, the state of a composite system, pure or mixed, is *separable* if the state has an ensemble decomposition in terms of product states. A separable state has no quantum entanglement, and a nonseparable state is entangled.

The nonclassical nature of quantum entanglement has been recognized for many years[4,5] but only recently has considerable attention been focused on trying to understand and characterize its properties precisely. We now have a good understanding of entanglement for a pair of qubits[6], however, how does one determine the extent to which a real physical few qubit system is entangled? What measurements are actually required? There are a number of possible techniques but arguably the simplest (if not the most efficient) is to perform appropriate measurements to reconstruct the density matrix and then use the theoretical measures currently known. This reconstruction technique, known as Quantum state tomography, has a long history. Arguably, the first such experimental technique for determining the state of quantum system was devised in 1852 by George Stokes[7]. He found that four parameters allow one to uniquely determine the polarization state of a light beam. Such techniques can be applied to any ensemble of two-level quantum mechanical systems and allows one to determine the density matrix describing this ensemble. More recently, experimental techniques for the measurement of the more subtle quantum properties of light have been the subject of intensive investigation (see ref. 8 and references their in). Tomographic techniques, in which the density matrix of a quantum state have been applied to experiments such as the homodyne measurement of the Wigner function of a single mode of light [9] and of the density matrix of the polarization degrees of freedom of a pair of entangled photons[10,11].

For the characterization of a few qubit quantum computer, quantum state tomography provides invaluable information on the system. The degree of entanglement and the degree of mixture (entropy for instance) can be calculated. If one were to consider quantum process tomography (tomography associated with the evolution of the state) then effects such as decoherence can be investigated. Two caveats must be made: firstly, there must be a large enough number of copies of an identically prepared quantum system to allow to a reasonable approximation the reconstruction of the state (this may be time consuming in many architectures where the system must be re-initialized after each measurement); secondly, more measurements are preformed in the reconstruction that what is likely to be needed to get the degree of entanglement. However we believe these disadvantages are outweighed by the other information one can obtain from the reconstructed states.

We structure this paper as follows, we first introduce a suitable notation to describe arbitrary n qubit states and then describe a set of simple measurements necessary to reconstruct the state of the system. Several examples are used to illustrate the technique.

2 Multi Qubit states

Let us consider an arbitrary n qubit state (shown schematically in Figure (1)). This n

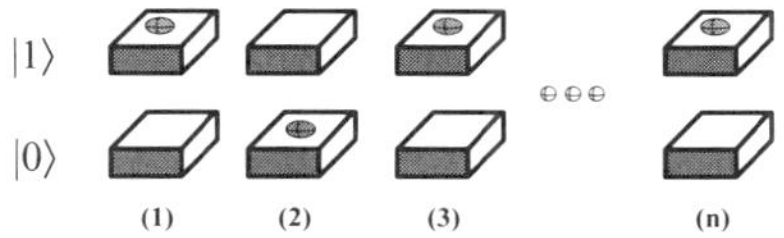

Figure 1. Schematic representation of an n qubit state.

qubit state can be mathematically described by a density matrix of the form,

$$\rho = \frac{1}{2^n} \sum_{i_1,\ldots,i_n=0}^{3} c_{i_1,\ldots,i_n} \lambda_{i_1}^{(1)} \otimes \lambda_{i_2}^{(2)} \otimes \ldots \otimes \lambda_{i_n}^{(n)}, \tag{1}$$

where the λ_i matrices[12] are given by,

$$\lambda_0 = \begin{pmatrix} 1 & 0 \\ 0 & 1 \end{pmatrix} \quad \lambda_1 = \begin{pmatrix} 0 & 1 \\ 1 & 0 \end{pmatrix} \quad \lambda_2 = \begin{pmatrix} 0 & -i \\ i & 0 \end{pmatrix} \quad \lambda_3 = \begin{pmatrix} 1 & 0 \\ 0 & -1 \end{pmatrix}, \tag{2}$$

and the j superscript in $\lambda_i^{(j)}$ labels the qubit. The $c_{i_1,\ldots,i_n}$ are the coefficients that specify the state. There are 4^n of these that need to be determined however the normalization criterion ($Tr(\rho) = 1$) ensures that $c_{0,\ldots,0} = 1$ leaving $4^n - 1$ parameters to be determined or specified . Noting that,

$$\langle \lambda_{i_1}^{(1)} \lambda_{i_2}^{(2)} \ldots \lambda_{i_n}^{(3)} \rangle = \frac{1}{2^n} \; c_{i_1,\ldots,i_n}, \tag{3}$$

we now observe the procedure to reconstruct the state. By measuring all the expectation values $\langle \lambda_{i_1} \lambda_{i_2} \ldots \lambda_{i_n} \rangle$, for $i_1, i_2, \ldots, i_n = 0, 1, 2, 3$ one determines the coefficients $c_{i_1,\ldots,i_n}$ and hence the state. How we actually measure these expectation values$\langle\lambda_{i_1}\lambda_{i_2}...\lambda_{i_n}\rangle$ depends heavily on the physical architecture. However one can always measure the probability of the system being in the ground state $|0\rangle$. ¿From such measurements we can then calculate expectation values like $\langle\lambda_{i_1}\lambda_{i_2}...\lambda_{i_n}\rangle$ (for $i_i = 0$ *or* 3). With appropriate single qubit rotations:

$$\hat{U}_j^k(\phi) = \exp\left[-ik\frac{\pi}{2}\left(|1\rangle_j\langle 0|e^{-i\phi} + |0\rangle_j\langle 1|e^{i\phi}\right)\right], \tag{4}$$

on the individual qubits (where j label the particular qubit, $k\pi$ is the length of the pulse and ϕ the polarization) followed by the ground state measurement, all the expectation values $\langle\lambda_{i_1}\lambda_{i_2}...\lambda_{i_n}\rangle$ (for $i_i = 0, 1, 2, 3$) can be determined and hence the state reconstructed. It is easy also to write the reconstructed state directly in terms of these single qubit rotations and ground measurements.

In any real situations, the information used to reconstruct the state will contain uncertainties due to small experimental errors. This errors make it possible that this reconstruction procedure for the state will not produce a physically acceptable state. While the resulting density matrix will be trace preserving and Hermitian, it may process small negative eigenvalues. Using a maximum likelihood technique[13] physically acceptable density matrices can be obtained. Let us now illustrate this procedure with t wo examples.

2.1 Single qubit and two qubit r econstruction

A single qubit density matrix can be written as $\rho = \frac{1}{2}\left(I_2 + \sum_{i=1}^3 c_i\lambda_i\right)$, where I_2 is the 2×2 identity matrix, and the coefficients c_i are given by the measured expectation values $c_i = 2\langle\lambda_i\rangle$. In matrix form this is written as,

$$\rho = \frac{1}{2}\begin{pmatrix} 1+\langle\lambda_3\rangle & \langle\lambda_1\rangle - i\langle\lambda_2\rangle \\ \langle\lambda_1\rangle + i\langle\lambda_2\rangle & 1-\langle\lambda_3\rangle \end{pmatrix}. \tag{5}$$

As an example, if one were to measure the correlations $\langle\lambda_1\rangle, \langle\lambda_2\rangle, \langle\lambda_3\rangle$ and find them all zero, the system is the maximally mixed state $\rho = I_2/2$ (diagonal elements only).

Let us consider the simplest states that may contain entanglement, namely t wo qubit states. Such states can be expressed in the form,

$$\rho = \frac{I_2\otimes I_2}{4} + \sum_{\substack{i_1,i_2=0\\ i_1=i_2\neq 0}}^{3} c_{i_1,i_2}\;\; \lambda_{i_1}^{(1)}\otimes\lambda_{i_2}^{(2)}. \tag{6}$$

So by measuring the moments $\langle\lambda_{i_1}^{(1)}\lambda_{i_2}^{(2)}\rangle$, the coefficients c_{i_1,i_2} are determined and hence the density matrix specified. As an example consider the result of the measurement of $\langle\lambda_{i_1}^{(1)}\lambda_{i_2}^{(2)}\rangle$ where the only nonzero zero measurements are given by $\langle\lambda_1^{(1)}\lambda_1^{(2)}\rangle = -\langle\lambda_2^{(1)}\lambda_2^{(2)}\rangle = \langle\lambda_3^{(1)}\lambda_3^{(2)}\rangle = \gamma/4$ and $\langle\lambda_0^{(1)}\lambda_0^{(2)}\rangle = 1/4$. The state is then the Werner state[14] given by,

$$\rho = \begin{pmatrix} \frac{1+\gamma}{4} & 0 & 0 & \frac{\gamma}{2} \\ 0 & \frac{1-\gamma}{4} & 0 & 0 \\ 0 & 0 & \frac{1-\gamma}{4} & 0 \\ \frac{\gamma}{2} & 0 & 0 & \frac{1+\gamma}{4} \end{pmatrix}.$$

With this reconstructed state properties like the degree of entanglement and entropy can

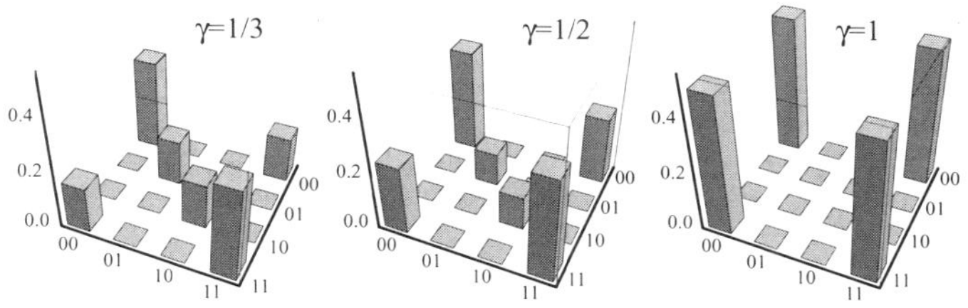

Figure 2. Graphical representation of the two qubit state reconstruction for the Werner state. For $\gamma = 1/3$ the state is separable, while for $\gamma = 1/2, 1$ the state is entangled with $\gamma = 1$ corresponding to the maximally entangled Bell state.

now be calculated. For instance it is straightforward to calculate the entanglement of formation (EOF)[15] and entropy S. At $\gamma = 1/2$ for the example given, we find

$$E_F(\hat{\rho}(\gamma = 1/2)) \approx 0.12 \qquad S(\hat{\rho}(\gamma = 1/2)) \approx 0.77$$

For systems with several more qubits the tomography procedure can be easily implemented however the number of measurements increases as 4^n. Currently their is no analytical formula for the degree of entanglement, however numerical techniques do exist and can be calculated from the density matrix. The degree of mixture is straightforward to calculate.

To summarize, in this proceedings we have shown a simple method by which the state of a few qubit quantum architecture can be reconstructed and hence the degree of entanglement determined. This characterization will be essential for early proof of principle quantum computation experiments.

Acknowledgements

We thank K. Nemoto for encouraging discussions.

References

1. *Introduction to Quantum Computation and Information*, edited by H.-K. Lo, S. Popescu, and T. Spiller (World Scientific, Singapore, 1998).
2. C. H. Bennett, *et. al*, Phys. Rev. Lett.**70**, 1895 (1993).
3. P. W. Shor, Phys. Rev. A **52**, R2493 (1995).
4. A. Einstein, B. Podolsky, and N. Rosen, Phys. Rev. **47**, 777 (1935).
5. J. S. Bell, Physics (N.Y.) **1**, 195 (1965).
6. W. K. Wootters, Phys. Rev. Lett. **80**, 2245 (1998).
7. G. C. Stokes, *Trans. Cambr. Phil. Soc.* **9**, 399 (1852).
8. U. Leonhardt, *Measuring the quantum state of light* (Cambridge University Press, 1997).
9. D. T. Smithey, M. Beck, M. G. Raymer and A. Faridani, Phys. Rev. Lett. **70**, 1244 (1993).
10. A. G. White, D. F. V. James, P. H. Eberhard, P. G. Kwiat, Phys. Rev. Lett. **83**, 3103 (1999).

11. A. G. White, D. F. V. James, W. J. Munro, P. G. Kwiat, "Exploring Hilbert space", in preparation (2001).
12. We have used λ matrix notation, which is the same as 2-dimensional Pauli matrices, but allows generalisation to *qunits*.
13. D. F. V. James, W. J. Munro, A. G. White, P. G. Kwiat, "Characterising multiple qubits",in preparation (2001).
14. R. F. Werner, Phys. Rev. A **40**, 4277 (1989).
15. C.H. Bennett, D.P. Vincenzo, J.A. Smolin and W.K. Wootters, Phys. Rev. A **54**, 3824 (1996).

QUANTUM CONTROL AND ENTANGLEMENT ENGINEERING WITH COLD TRAPPED ATOMS

P. S. JESSEN, D. L. HAYCOCK AND G. KLOSE

Optical Sciences Center, University of Arizona, Tucson, AZ 85721

I. H. DEUTSCH AND G. K. BRENNEN

Department of Physics and Astronomy, University of New Mexico, Albuquerque, NM 87131

Neutral atoms offer many advantages for quantum computation, including excellent isolation from the decohering environment and the existence of proven laboratory techniques for atom trapping and quantum control. Currently the most mature trapping technology is offered by optical lattices, arrays of microtraps formed by the AC Stark shift in laser standing waves, which provide an excellent platform to develop qubit implementations and universal quantum logic gates. We discuss a series of experiments that demonstrate some of the basic laboratory tools required for quantum computation in this system, including accurate quantum state preparation, coherent control of mesoscopic quantum states, and complete quantum state reconstruction. Ultimately, quantum computing may require new types of programmable trap arrays that allow atoms to be individually manipulated and read out. We suggest a possible trap architecture for a scalable neutral atom quantum computer.

Keywords: Optical lattice, Coherent control, Quantum state reconstruction

1 Quantum Computing with Neutral Atoms

To harness the power of coherent quantum dynamics is a formidable task and represents one of the great challenges for science in the 21st century. This is particularly true of the type of *many-body* quantum control required for quantum computing. Laboratory implementation with any real system must contend with a conflict inherent in the fundamental physical requirements:

- The individual information-carrying units (qubits) must be strongly coupled to each other and to an external field that drives the system through the computation.
- Conversely the qubits must be extremely well isolated from the surrounding environment, in order to preserve the entangled superposition states that arise during computation.

In general the mechanism by which qubits interact is also responsible for coupling to the environment, and increased coupling between qubits comes at the cost of increased decoherence.

Numerous realizations of quantum computing have been considered, each of which makes compromises with respect to the demands above. For a summary of current proposals see[1], and also the remainder of this issue. By far the most impressive achievements in control and manipulation of quantum coherence are found in the area of atomic and optical physics. The group of D. J. Wineland has used trapped ions to demonstrate universal quantum gates[2] and true entanglement of up to four qubits[3], and liquid state NMR has been used to simulate many aspects of quantum computing[4]. These systems have provided the so far only practical experience with the difficulties of scaling from single- to multi-qubit systems, and have spurred new fundamental insights such as the role played by entanglement[5]. This rapid progress derives to a large degree from the existence of proven laboratory technologies and a community who for years have worked with high-precision spectroscopy and quantum control in simple systems with long coherence times.

With these considerations in mind, we consider laser trapped neutral atoms a promising system with which to explore the practical implementation of quantum information processing. Among the most important advantages we note:

- Neutral atoms in the electronic ground state couple extremely weakly to the environment, and correspondingly long coherence times are routinely seen in e. g. laser spectroscopy.
- The atomic qubits can be easily and precisely manipulated by controlling the trap properties. Important tools such as quantum state preparation, coherent control and quantum state reconstruction have already been demonstrated in the laboratory.
- Neutral atoms couple weakly to each other, except when interatomic coupling is introduced through weak dipole or collisional interactions. Coupling *on demand* greatly simplifies the many-body control problem and helps suppress decoherence.

The major challenge of quantum computing in this system is to implement precise two-qubit quantum control protocols, or quantum logic gates. Current proposals by Brennen et al.[6] and by Jaksch et al.[7] accomplish this by encoding qubits in the ground hyperfine state of individual atoms, and superimposing the atomic qubits to induce coupling through dipole-dipole interactions or ground state collisions. For a discussion of entangling quantum logic gates based on dipole-dipole interactions see the article by Brennen et al. in this issue.

2 Laser Traps for Alkali Atoms

Far-off-resonance laser traps offer a unique platform for quantum control and quantum information processing. If the light field is intense and detuned far from atomic resonance then the rate of decoherence from spontaneous light scattering becomes negligible, while the large field amplitude maintains a substantial trapping potential. By far the best developed trapping technology is based on "optical lattices", periodic arrays of wavelength-sized traps formed by the AC Stark shift in a set of interfering laser beams. Specializing to the case of alkali atoms and a trap operating near the $nS_{1/2} \rightarrow nP_{3/2}$ transition (D2 resonance line), at detunings much larger than the hyperfine splitting but less than the fine structure splitting, the optical potential can be written:

$$\hat{U}(z) = U_J(z) + g_F \mu_B \hat{\mathbf{F}} \cdot \mathbf{B}_{fict}(z)$$

$$U_J(\mathbf{x}) = \frac{2U_1}{3} |\vec{\varepsilon}_L(\mathbf{x})|^2 , \quad \mathbf{B}_{fict}(z) = -i \frac{U_1}{3\mu_B} [\vec{\varepsilon}_L^*(\mathbf{x}) \times \vec{\varepsilon}_L(\mathbf{x})] \tag{1}$$

where $\hat{\mathbf{F}}$ is the angular momentum operator for the hyperfine ground state F, U_1 is the light shift for a transition with unit oscillator strength, and $\vec{\varepsilon}_L(\mathbf{x})$ is the local polarization of the trapping light field[8]. Casting the light shift potential in terms of a scalar component $U_J(\mathbf{x})$ (independent of the magnetic quantum number m_F) and a fictitious magnetic field $\mathbf{B}_{fict}(z)$ suggest that we encode qubits in a pair of hyperfine ground states $|F_\uparrow = I + \frac{1}{2}, m_F\rangle$ and $|F_\downarrow = I - \frac{1}{2}, -m_F\rangle$, whose g_F-factors have equal magnitude and opposite sign, so that they have nearly identical light shift and see the same trapping potential. This serves to decouple the spin- and space degrees of freedom, and makes the qubits less susceptible to dephasing from fluctuating light shifts and external magnetic fields.

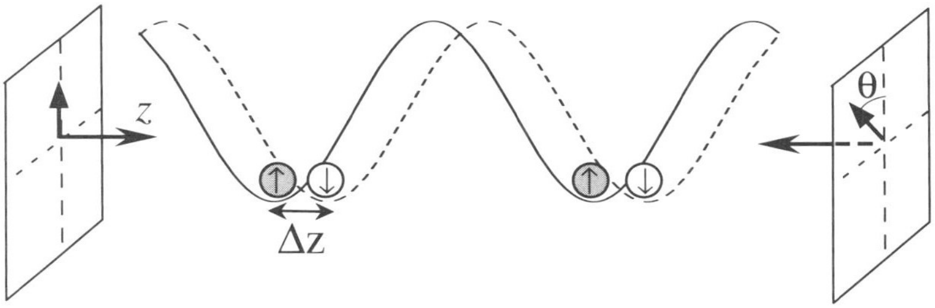

Figure 1. 1D Lin-θ-lin optical lattice. Counterpropagating laser beams create standing waves of opposite helicity, with an offset depending on the relative angle θ of the linear polarization. For blue detuning, the resulting optical potential traps spin-up(down) atoms at the nodes of the σ_+ (σ_-) standing waves, allowing dynamical control of the atomic wavepackets.

To illustrate how an optical lattice might be used to trap and manipulate atomic qubits, consider a simple configuration of counterpropagating laser beams whose linear polarizations form a relative angle θ (the 1D lin-θ-lin configuration). The resulting light field can be decomposed into standing waves of positive and negative helicity (Fig. 1), whose nodes are separated by θ / k, $\mathbf{E}_\pm(z) = E_0 \sin(kz \pm \theta)\mathbf{e}_\pm$. Rotation of the polarizations to parallel slides the standing waves on top of each other, while a rotation to $\theta = 90°$ separates the waves so the nodes of one overlap the antinodes of the other. According to Eq. 1 we can encode a (+) and a (–) species of qubits, which are trapped in the σ_+ and σ_- polarized standing waves respectively. To do so we can for example choose logical basis states $|1\rangle_\pm = |F_\uparrow, m_F = \pm 1\rangle \otimes |n = 0\rangle$ and $|0\rangle_\pm = |F_\downarrow, m_F = \mp 1\rangle \otimes |n = 0\rangle$, where $|n = 0\rangle$ is the vibrational ground state of the optical potential well. In principle any pair of atoms, one from each species, can then be superimposed through a suitable rotation of the lattice polarization beyond 360°. When two atoms are superimposed a quantum logic operation can be performed between them, through conditional excitation of atomic dipoles as discussed by Brennen et al.

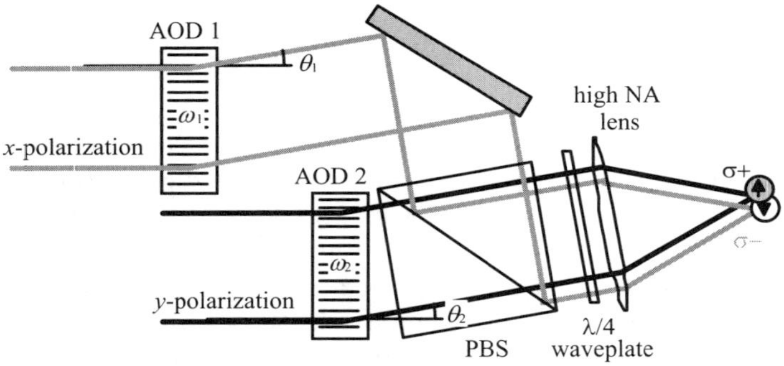

Figure 2. Schematic illustration of a programmable trap array for quantum processing. The positions of individual traps in the focal plane are independently controlled through the drive frequencies and deflection angles of the Acousto-Optic Deflectors (AOD's). Beams from the AOD's are combined on a Polarization Beam Splitter, and converted to opposite circular polarization. A number of independent traps of the same handedness can be created from a single input beam by feeding a number of different frequencies to the corresponding AOD. Large 2D arrays of traps can be generated with crossed AOD's in each beam path.

The general principle of trapping different species of atomic qubits in separately controlled σ_+ and σ_- polarized laser fields translates to trap geometries other than optical lattices. Consider for example the setup in Fig. 2, where two beams are independently controlled by Acousto-Optic Deflectors (AOD's), combined and focussed by a large N.A. lens, to form two independent Far-Off-Resonance Traps (FORT's)[9] in the focal plane. The two traps have opposite circular polarization, each hold an atomic qubit of the (+) or (−) species, and play roles analogous to the σ_+ and σ_- potential wells in an optical lattice. However, movement of the traps is fully programmable through the rf drive frequencies fed to the AOD's, and the atoms can be brought together and separated as required. The FORT's provide only weak localization in the axial direction, but stronger axial confinement can easily be provided by a superimposed 1D optical lattice. In a possible vision of a large quantum processor, each of the AOD's in Fig. 2 is replaced with a crossed pair, similar to the configuration used in laser scanners. Modern AOD's provide hundreds of optical resolution elements, and one might easily create, say, a 20×20 array of traps whose movement can be individually controlled through the frequency spectrum fed to the AOD's. Each trap might be loaded sequentially from a single-atom MOT source[10] and initialized with Raman sideband cooling. An important feature is the large separation between traps in the array, which allows direct optical addressing of individual qubits for computation and readout. This removes an important constraint of the optical lattice geometry, where atoms are spaced by half the wavelength of the trapping laser and therefore cannot be optically resolved.

3 Quantum Control Experiments in Optical Lattices.

Over the past several years we have conducted an experimental and theoretical program to develop Cesium atoms in far-off-resonance optical lattices as a model system for exploration of quantum control and quantum coherent dynamics. Work by our group and others have demonstrated the resources available with the optical lattice technology: an exceptional flexibility in the design of the optical potential, combined with quantum control at a level which is beginning to rival that found in ion traps and cavity QED systems. Among the most significant accomplishments are the initialization of atoms in the vibrational ground state of individual lattice potential wells by resolved-sideband Raman cooling[11], observation of mesoscopic quantum coherence in optical double-well potentials[12], and the reconstruction of the entire density matrix for the ground state angular momentum of cold atoms[13]. The demonstration of these techniques in the laboratory sets the stage for experiments to implement and evaluate the performance of single- and two-qubit quantum gates.

The preparation of a well defined, pure quantum state is a desirable starting point for any experiment in quantum control. In[11] we described a scheme for sideband cooling of optically trapped atoms which relies on Raman transitions between the vibrational manifolds associated with a pair of magnetic sublevels of the $6S_{1/2}(F=4)$ hyperfine ground state of Cs. Compared to ion traps, where one typically relies on transitions between different hyperfine ground states, our method offers the considerable advantage that Raman coupling between the magnetic sublevels can be designed into the lattice potential itself. In our original experiment we demonstrated that $\sim 10^6$ atoms could be individually trapped in independent potential wells of a two-dimensional (2D) optical lattice, and cooled to a mean vibrational excitation of $\bar{n}=0.01$ per degree of freedom. Following this work, other groups have now demonstrated sideband cooling in 3D lattices, and at very high densities where nearly half of the lattice potential wells were occupied by an atom[14].

Quantum coherence and quantum tunneling between localized but separated states of a particle in a double-well potential has long served as a paradigm for non-classical dynamics. We have developed a unique laboratory realization of this "toy" model, based on laser cooled Cs atoms in a double-well magneto-optical lattice. Expanding upon the discussion in section 2 above, one can show that the lowest adiabatic potential of a 1D lin-θ-lin magneto-optical lattice with an external field $B_x \neq 0$ forms a periodic array of double-wells. To accurately model the system we can calculate the bandstructure and Bloch states of the lattice, from which we obtain both the left/right localized states of the ground doublet in individual double-wells, as well as the frequency at which the system Rabi flops between them. The system states correspond to spinor wavepackets with highly entangled internal and motional degrees of freedom (Fig. 3d), allowing us to use the spin as a "meter" for the center-of-mass motion. For our parameters the adiabatic (Born-Oppenheimer) approximation breaks down, and the system dynamics cannot be described in terms of a particle moving on the adiabatic potential. As a result the onset of quantum tunneling and the distinction between classical and quantum dynamics becomes both rich and subtle[15].

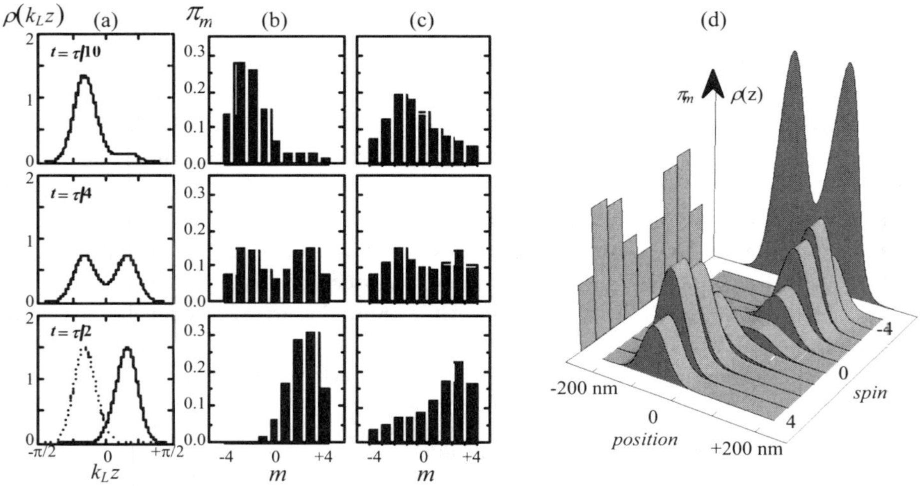

Figure 3. Spinor wavepackets during a Rabi oscillation between left and right localized states. At a quarter period the state is a Schrödinger cat like superposition. (a) Center-of-mass probability densities obtained from the calculated spinors by tracing over the spin degree of freedom. The dotted curve indicates the center-of-mass wavepacket at $t = 0$. (b) Magnetic populations obtained by tracing over the center-of-mass degree of freedom. (c) Measured populations. (d) Center-of-mass probability density for each magnetic sublevel in the "cat" state, showing the correlation between spin and spatial degrees of freedom. The marginal distributions in spin and space are projected onto the walls.

We have used a number of techniques developed in our lab over the past few years, to prepare Cs atoms in the magneto-optical double wells and to follow their quantum coherent evolution. Laser cooling, optical pumping and state selection in an accelerating lattice was used to produce a sample of atoms in a nearly pure ground state of the $m_F = 4$ potential. This state was evolved into a localized spinor state in a symmetric double-well potential, by ramping B_z to zero and increasing B_x to the desired value. Rabi oscillations between the two localized states of the double-well were subsequently detected by measuring the time-dependent magnetic populations via Stern-Gerlach analysis (Fig. 3a-c). The Stern-Gerlach measurement was accomplished by releasing the atoms from the lattice and quickly applying a bias field B_z to keep the quantization axis well defined, then letting the atoms fall to a probe beam in the presence of a magnetic field gradient and extracting the populations from the separate arrival distributions for the magnetic sublevels. We found excellent agreement between the measured Rabi frequencies and a bandstructure calculation with no free parameters, over a wide range of parameters. Direct examination of the magnetic populations during Rabi oscillation were consistent with a "Schrödinger cat" –like superposition at the appropriate time. In this first generation experiment the Rabi oscillations dephased relatively fast due to variations in the Rabi frequency across the atomic sample. We have made substantial progress towards better control of the magnetic fields, lattice polarization and intensity, and expect the next generation of the experiment to achieve much longer dephasing times. This opens the door to a variety of new experiments in quantum chaos[15] and quantum feedback control[16].

The ability to measure an unknown state of a quantum system forms a necessary complement to quantum state preparation and control. Reconstruction of a (generally mixed) state from repeated measurements of physical observables is, however, a nontrivial task which does not always permit a solution. Among the small number of physical systems for which this inverse problem has been solved are light fields, molecular vibrations, the motion of trapped ions, and free atomic waves[17]. More recently multi-particle quantum states have been measured for spin-1/2 systems in NMR, and for polarization entangled photon pairs[18]. We have now demonstrated a method whereby one can experimentally reconstruct the complete density matrix for a large angular momentum atomic ground state.

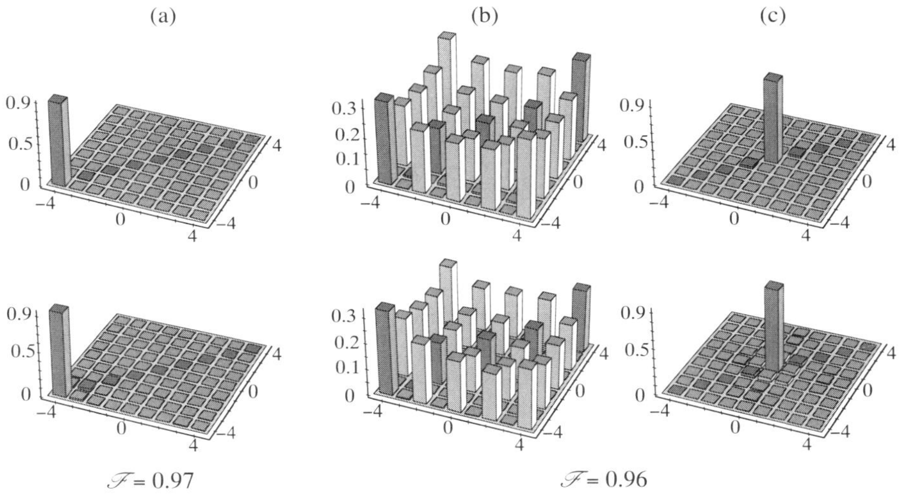

Figure 4. Examples of input (top) and reconstructed states (bottom). All plots show the magnitude of the matrix elements. (a) Input state $m_F = -4$, prepared by optical pumping with σ_- polarized light. (b) Input state prepared by optical pumping with linear polarization along $\hat{y}$. This produces a state close to $m_F = 0$ in a coordinate system with quantization axis along $\hat{y}$, but with large coherences in the basis used for the reconstruction. (c) Same input and reconstructed state as (b), but shown in the rotated basis where the state is close to $m_F = 0$.

Our reconstruction scheme relies on repeated Stern-Gerlach measurements with respect to many different quantization axes. These measurements can be performed in a straightforward manner, by choosing different directions for the bias magnetic field which determines the quantization axis for the Stern-Gerlach analyzer. The algorithm requires that one measures $2F+1$ magnetic populations for $4F+1$ different directions of the quantization axis, characterized by polar angles θ and azimuthal angles φ[19]. It is then possible to relate the measured populations to the unknown elements of the density matrix by means of a non-square matrix whose elements are determined by the angles θ and φ, and to invert the problem numerically. Straightforward inversion of this linear system with noisy experimental data occasionally results in a non-positive and therefore non-physical density matrix. A better approach is to express ρ in terms of $(2F+1)^2-1$ real valued variables in a way that ensures Hermiticity and positivity, and to do a best fit to the measured data. This method finds the best *physical* density matrix to fit the measurements, and often gives a better match between the input and reconstructed state. Our experiments use laser cooled Cesium atoms in the $6^2S_{1/2}(F=4)$ hyperfine ground state, optically pumped to produce suitable input test states. We can produce a further range of test states by Larmor precession, or by using the (presumably) mixed states that result directly from laser cooling. Stern-Gerlach measurements are performed as described above. Fig. 4 shows a few examples of input and reconstructed states. The test states are verified by a Stern Gerlach measurement with quantization axis parallel to that of the σ or π polarized optical pumping beam. As evident from the figure the input and reconstructed states are nearly indistinguishable by visual inspection. A quantitative measure for the match is the fidelity F $=[Tr\sqrt{\rho_i^{1/2}\rho_r\rho_i^{1/2}}]^2$, where ρ_i and ρ_r are the input and reconstructed density matrices. We typically achieve fidelities of F ~ 0.95 or better for a wide range of inputs.

In summary we have implemented several elements from the quantum optics and atomic physics "toolbox" in the neutral atom-optical lattice system, including quantum state preparation, control and measurement. In the near future we hope to extend our capabilities by developing schemes for continuous measurement, and to explore such possibilities as ensemble feedback control[16]. We are also working to implement Raman sideband cooling in double-well magneto-optical lattices, where it can be used to perform reservoir engineering[20]. One might then use our techniques for quantum state reconstruction as a basis for quantum process tomography[21], and study the fundamental process of decoherence in mesoscopic quantum systems. Such experiments will contribute significantly to our understanding and practical experience with coherent control in open quantum systems, and will provide a solid basis for future efforts to implement neutral atom-based quantum information processing.

Acknowledgements

We acknowledge helpful discussions with G. Smith, K. Cheong, J. Grondalski, P. Alsing and S. Ghose. This work was supported by NSF grant No. 9732612 and 9871360, by ARO grant No. DAAD19-00-1-0375, and JSOP grant No. DAAD19-00-1-0359.

References

1. Fortschr. Phys. **48** (2000) (Special Issue on Experimental Proposals for Quantum Computation).
2. Q. A. Turchette et al., Phys. Rev. Lett. **81**, 3631 (1998).
3. C. A. Sackett et al., Nature **404**, 256 (2000).
4. D. G. Corey, A. F. Fahmy and T. F. Havel, Proc. Natl. Acad. Sci. USA **94**, 1634 (1197); N. A. Gerschenfeld and I. L. Chuang, Science **275**, 350 (1997).
5. S. L. Braunstein et al., Phys. Rev. Lett. **83**, 1054 (1999).
6. G. K. Brennen et al., Phys. Rev. Lett. **82**, 1060 (1999).
7. D. Jaksch et al., Phys. Rev. Lett. **83**, 1975 (1999).
8. I. H. Deutsch and P. S. Jessen, Phys. Rev. A **57**, 1972 (1998).
9. J. D. Miller et al., Phys. Rev. A **47**, R4567 (1993).
10. D. Frese et al., Phys. Rev. Lett. **85**, 3777 (2000).
11. S. E. Hamann et al., Phys. Rev. Lett. **80**, 4149 (1998).
12. D. L. Haycock et al., Phys. Rev. Lett. **85**, 3365 (2000).
13. G. Klose, G. A. Smith and P. S. Jessen, submitted; e-print LANL archive: quant-ph/0101017.
14. A. J. Kerman et al., Phys. Rev. Lett. **84**, 439 (2000); D. J. Han et al., Phys. Rev. Lett. **85**, 724 (2000).
15. I. H. Deutsch et al., J. Opt. B: Quantum Semiclass. Opt. **2**, 633 (2000).
16. See A. C. Doherty et al., Phys. Rev. A **62**, 012105 (2000), and references therein.
17. D. T. Smithey et al., Phys. Rev. Lett. **70**, 1244 (1993); T. J. Dunn, I. A. Walmsley and S. Mukamel, Phys. Rev. Lett. **74**, 884 (1995); D. Liebfried et al., Phys. Rev. Lett. **77**, 4281 (1996); C. Kurtsiefer et al., Nature **386**, 150 (1997).
18. I. L. Chuang, N. Gerschenfeld, and M. Kubinec, Phys. Rev. Lett. **80**, 3408 (1998); A. G. White et al., Phys. Rev. Lett. **83**, 3103 (1999).
19. R. G. Newton and B. L. Young, Ann. Phys. **49**, 393 (1968).
20. C. J. Myatt et al., Nature **403**, 269 (2000).
21. I. L. Chuang and M. A. Nielsen, J. Mod. Opt. **44**, 2455 (1997).

QUANTUM INFORMATION PROCESSING WITH NEUTRAL ATOMS IN OPTICAL LATTICES

S. L. ROLSTON

Atomic Physics Division, National Institute of Standards and Technology, Gaithersburg, MD 20899-8424 USA

Neutral atoms trapped in periodic optical potentials (optical lattices) offer a possible route toward quantum computation. The atoms can be kept well isolated from the environment, yet can be easily manipulated with the tools of atomic physics. This paper will give an overview of the prospects for the implementation of quantum information processing in such systems.

Atomic physics systems offer an attractive approach to the challenges presented by quantum computation: many particle quantum systems that are at the same time well isolated from the environment and yet exquisitely controllable. Much of atomic physics research has dealt with two level systems, obvious candidates for qubits. The knowledge that has been accumulated has led to numerous tools for the manipulation of such systems, including radio frequency and optical techniques, techniques based on adiabatic passage, spin echoes, and quantum state tomography, to name just a few. The cesium atomic clock, now operating at an accuracy of 2×10^{-15} is an example of the exquisite control of a two-level system that is possible within atomic physics.[1] Neutral atoms in particular are immune to decoherence-inducing electric fields, which seem to be a current source of difficulty in the much more developed ion-trap implementations of quantum information processing. This paper will deal exclusively with neutral atom approaches to quantum computation, and within that context, I will only consider optical lattice traps. One should note that there is rapid progress in the area of magnetic microtraps for neutral atoms that may also have some bearing on neutral atom quantum computation implementations. I will present the different options that are available and identify some of the scientific and technological challenges that lie ahead. A recent paper by David DiVincenzo[2] on the physical implementation of quantum computing offers an ideal framework in which to evaluate the prospects for quantum computation with neutral atoms. He presents five requirements that are necessary for the construction of a scalable quantum computer. For each of these requirements, I will summarize the relevant issues and technologies available for neutral atoms trapped in optical lattices.

1 Scalable qubits

Scalability in optical lattices is not an issue. Optical lattices routinely trap 10^6 to 10^7 atoms. Operating at essentially T=0, there is no signal-to-noise issue like the one that prevents room temperature NMR from scaling. In neutral atom systems, there are essentially two possibilities for storing quantum information: the internal degrees of freedom (hyperfine and/or Zeeman levels) or the external degrees of freedom (bound vibrational states of the individual lattice potentials). The first two proposals to suggest neutral atoms in optical lattices[3,4] as a system for quantum computation both used the internal degrees of freedom. In the scheme from the Deutsch group[3], the two qubits were encoded in four $|F, m_f\rangle$

(hyperfine quantum numbers) such that the qubits would be insensitive to magnetic field variations, to first order. The scheme from the Zoller group[4] used two $|F, m_f\rangle$ levels. Any scheme that uses internal states for the qubits has available the well-developed technology of RF and Raman spectroscopy to produce the necessary single bit rotations.

Another possible optical lattice implementation we are evaluating at NIST uses the vibrational states of a lattice of double well potentials created by interfering laser with optical frequencies ω and 2ω. The qubit would be encoded in the ground and first excited vibrational states. In this case single bit rotations would be accomplished with Raman transitions between these vibrational levels. One advantage of using the external degrees of freedom is that one could choose to use an atom with J=0, so there would be no Stark or Zeeman shifts to couple to the environment.

2 Initialization to $|00000\cdots\rangle$

The real challenge with this requirement is with respect to the external degrees of freedom. Even for hyperfine qubits, there will inevitably be significant coupling between the internal and external degrees of freedom, such that the atoms must be kept in the ground state of the lattice. (Techniques for manipulating the internal states are well developed and do not present a significant challenge). To initialize the atoms into the ground state of the lattice, there are two possible approaches: laser cooling with Raman sideband cooling, or adiabatic loading from a BEC. Both approaches appear quite viable. The Jessen group has demonstrated[5] ≥ 95 % population of the ground state of a 2-D lattice using Raman sideband cooling, modifying a technique developed by the ion trap community. Recent work[6,7] has extended this approach to 3-D with populations of 80% for a 3-D lattice.

Our group at NIST has begun to explore loading lattices from a BEC. A Bose Einstein condensate consists of many atoms (typically $1 - 3 \times 10^6$) all occupying the same ground state wave function of a trap (the effective potential is determined by the trap potential and mean field interactions). Since it is a stationary state, the phase is uniform across the condensate, which would typically covers many thousands of 3-D optical lattice sites. If the periodic potential of the lattice is turned on slowly, the adiabatic theorem causes the condensate wave function to map on to the ground state band of the optical lattice, with the quasimomentum equal to zero. Preliminary measurements of our sodium BEC loaded into a 1-D optical lattice with a simple 20 μs ramp of the optical intensity indicate a ground state band population of greater than 99%. With longer ramp times and more sophisticated turn-on profiles, we expect to increase this number. In this 1-D configuration each site contains many thousands of atoms and is in itself not suitable for quantum computation. (Recent work[8] that observes number squeezing in a 1-D optical lattice suggests the intriguing possibility of a fixed number of atoms in each site, which might be suitable for quantum computing purposes.) A theoretical estimate[9] by the Zoller group predicts the existence of a quantum phase transition that in three dimensions will result in exactly one atom per lattice site, with each atom in the ground state. This is a form of the Mott-insulator transition, and we have an experimental program under way at NIST to look for this transition.

3 Coherence

Perhaps the most significant challenge for any quantum computation implementation will be to reduce the levels of decoherence to a place where error correction techniques can be useful. Current theoretical estimates put the threshold for fault tolerant quantum computation at 10^{-4} or so[10], but we can hope and expect that this threshold will be relaxed as more progress is made in the field. There are a number of sources of decoherence and dissipation that are possible for atoms trapped in optical lattices, but they are quite well understood. Among the possible sources are inelastic collisions, time varying magnetic fields, laser amplitude noise, laser position noise, and spontaneous emission. Spontaneous emission is a completely calculable decoherence source that will most likely be manageable. The depth of the optical potential that forms the lattice is proportional to I/δ (where I is the laser intensity, and δ is the laser detuning), while the spontaneous rate scales as I/δ^2 . Clearly the optical potential can be held constant while reducing the spontaneous emission rate by simultaneously increasing the intensity and detuning. Optical lattices and single focus dipole traps have been made with scattering rates less than one photon per second. Since gate operation speeds will inevitably be limited by oscillation frequencies within the lattice (typically 100-1000 kHz), decoherence due to spontaneous emission may already be below the fault tolerance threshold. There is a caveat, in that some of the proposed gates require specific laser detunings to create the appropriate polarization-dependent lattices, or involve some spontaneous emission during the two-qubit gate operation.

The most likely source of decoherence and dissipation will be laser noise. Heating rates were calculated[11] in an analysis of both laser amplitude noise and beam pointing noise in a single-focus dipole trap (laser frequency noise will be negligible). This analysis is applicable to optical lattices, where the case will be more favorable because of the higher oscillation frequencies in a lattice over that of a dipole trap. If we assume we want less than one heating event per 10^5 oscillation frequencies, and a 100 kHz oscillation frequency, this analysis yields fractional laser amplitude noise requirements of $S_a(200kHz) \leq 10^{-13}$ Hz^{-1} and laser beam pointing noise of $S_x(100kHz) \leq 10^{-17}\mu\text{m}^2/\text{Hz}$. Recent work[12] on a CO_2 laser dipole trap reported an amplitude noise well below this requirement, and trap lifetimes of 300 s. The pointing noise should also be achievable, especially since the optical lattice frequency is well above acoustic frequencies, the primary source of beam pointing instabilities. We can in addition bring to bear the formidable technologies of optical interferometry to minimize these problems. If the qubit is encoded in hyperfine levels, heating can degrade gate fidelity due to coupling between external and internal degrees of freedom during the gate operation. It is probably best to treat a transition to a vibrationally excited level as the loss of the qubit from the computational basis. Cooling of this qubit could return the qubit to the basis, where error correction routines could then correct for what would now look like a bit flip (although the cooling must be done without adversely affecting neighbors - a clear challenge). If the qubit is encoded in the vibrational states, amplitude noise will produce a direct decoherence of the qubit (the noise on the lattice will cause the phase of the upper state to evolve with a random component), but preliminary estimates suggest this will be a tractable problem. It may also have a significant effect on the fidelity of two-qubit gates that are based on tunneling. One significant possibility for "dissipation" is that an atom is lost entirely from the lattice due to an inelastic collision. If we can develop a way of reloading this site , then once again this just looks like a bit flip error that can be corrected with error correction coding.

4 Universal quantum logic gates

There is extensive literature available on a variety of possible gates that have been proposed: dipole-dipole interactions[3]; elastic ground state collisions[4]; enhanced dipole-dipole interactions in Rydberg atoms to produce fast gates[13]; magnetic dipole-dipole interactions[14]; and motional states in double wells with elastic ground state collisions[15]. The field is still quite young and the best neutral atom quantum logic gate has probably not yet been proposed. The experimental systems required for implementing these concepts are in general quite flexible and somewhat generic, which will allow the evaluation of different schemes without much difficulty, by simply changing laser frequencies and directions, for example.

5 Measurement

One of the potential strengths of optical lattices is their inherent parallelism. Perhaps new algorithms will be developed that can exploit this. If not and individual addressing is required, this will be a major challenge. The inherent difficulty is that the spacing in a typical optical lattice is sub-wavelength, and optical resolution is limited to about a wavelength. One obvious solution is to use a long-wavelength laser to form the lattice (e.g. 10.6 μm CO_2 laser) and a short wavelength laser for the readout. The downside to this is that most schemes for logic gates require tight confinement for good gate speed, which will be compromised to some extent with a long-wavelength lattice. Another possibility is to use multiple wavelengths to create "superlattices" that have both a fine period (for tight confinement) and a coarse period for large separation and individual addressing. Another possibility is to borrow techniques from MRI and apply a gradient that brings only one site into (Raman) resonance. Although this would require a very large magnetic field gradient to get wavelength resolution, using the AC Stark shift of a focused laser should produce the require spatial resolution. In the end the measurement issues will be intimately connected with the qubits, the gates used, and error coding techniques that might use blocks of qubits.

6 Connection to optical qubits

In DiVincenzo's list of requirements for a quantum computer, he lists two extra (optional) requirements related to the ability to transmit quantum information, i.e. a connection to an optical qubit. This will be important if ideas about quantum networks are to be brought to fruition. Connections between atomic and optical qubits have been proposed[16,17] using cavity QED. Optical lattices are perhaps the best system available to mate with the required high finesse optical cavities, which will allow in principle the conversion of quantum information stored in atoms to be transferred to photons, where it can be transmitted.

At the moment, it seems that it would take a quantum computer just to divine the final form of a successful physical implementation of a truly useful quantum computer. Atomic physics-based systems offer the many years of atomic physics research on well-controlled and isolated atoms as flexible building blocks for quantum information processing systems. While there are many challenges ahead, there are just as many opportunities.

Acknowledgments

The NIST-Gaithersburg experimental quantum information group consists of Kristian Helmerson, Brian King, Paul Lett, Steven Peil, William Phillips, Trey Porto, and Steven Rolston. The loading of a BEC into a 1-D optical lattice work also included participation by Johannes Denschlag and Jesse Simsarian. Theoretical analysis of quantum gates using vibrational states is being carried out by Eric Charron, Paul Julienne, Fred Mies, Eite Tiesinga, and Carl Williams. I would like to acknowledge funding from NIST, ARDA-NSA, and the ONR.

References

1. http://www.bldroc.gov/timefreq/cesium/fountain.htm
2. D. DiVincenzo,*Fortschr. Phys.* **48**, 771 (2000).
3. G. K. Brennen, *et al.*, *Phys. Rev. Lett.* **82**, 1060 (1999).
4. D. Jaksch, *et al.*, *Phys. Rev. Lett.* **82**, 1975 (1999).
5. S. E. Hamann, *et al.*, *Phys. Rev. Lett.* **80**, 4149 (1998).
6. Andrew J. Kerman, *et al.*,*Phys. Rev. Lett.* **84**, 439 (2000).
7. Dian-Juan Han, *et al.*,*Phys. Rev. Lett.* **85**, 724 (2000).
8. C.Orzel, A.K.Tuchman, M.L.Fenselau, M.Yasuda and M.A.Kasevich,*to be published.*
9. D. Jaksch, *et al.*, *Phys. Rev. Lett.* **81**, 3108 (1998).
10. J. Preskill, *Phys. Today* **52**, 24 (1999).
11. T. A. Savard, *et al.*, *Phys. Rev. A***56**,R1095 (1997).
12. K. M. O'Hara, *et al.*, *Phys. Rev. Lett.* **82**, 4204 (1999).
13. D. Jaksch, *et al.*,*Phys. Rev. Lett.* **85**, 2208 (2000).
14. L. You and M. S. Chapman, PRA **6205**, 2302 (2000).
15. Carl Williams, *private communication.*
16. T. Pellizzari, *Phys. Rev. Lett.* **79**, 5242 (1997).
17. J. I. Cirac, *et al.*, *Phys. Rev. Lett.* **78**, 3221 (1997).

QUANTUM INFORMATION PROCESSING IN OPTICAL LATTICES

G.K. BRENNEN AND I.H. DEUTSCH
Department of Physics and Astronomy, University of New Mexico, Albuquerque, New Mexico 87131, USA

P.S. JESSEN
Optical Sciences Center, University of Arizona, Tuscon, Arizona 85721, USA

We study the deterministic entanglement of a pair of neutral atoms using pairwise dipole-dipole couplings in a state dependent trap. By bringing atoms together pairwise and inducing state selective interactions, multiparticle entanglement of atoms in an optical lattice can be realized. The main source of decoherence is spontaneous emission which can be made negligible if the two atom interactions are performed fast compared to the scattering rate. A figure of merit is derived for the fidelity of entanglement including molecular interactions.

1 The Tao of Quantum Computing

Implementing quantum information processing in a physical system is a problem of coherent control. That is, one desires a means to prepare an initial quantum state (encoding the input), guide it through some well defined path (the computation) to the desired final quantum state (the output), and finally read out the final state with good fidelity. In the current paradigm of quantum computation, this is achieved with great efficiency if the Hilbert-space has a natural tensor product structure so that the transformation from input to output can be accomplished through simple operations on the individual subsystems (usually qubits), and interactions between them. Quantum computation is thus a problem of coherent control on a *many-body* system. An important theoretical result of the last decade proves that an *arbitrary* transformation on this many-body system can be achieved with solely single-qubit rotations plus a single entangling unitary acting only in the two-body sector. Thus, the problem of implementing a quantum algorithm is reduced to the problem of implementing controlled one-body and two-body interactions (the universal logic-gate set).

Though the challenge of implementing quantum information processing has been reduced to basic building blocks, any system is faced with a conflict inherent in the fundamental physical requirements:

- The qubits must be strongly coupled to each other and to an external field that drives their evolution through a precise sequence of steps.
- Conversely, the qubits must be extremely well isolated from their physical surroundings, because uncontrolled interactions with the environment destroy the entangled superposition states necessary for quantum parallelism through the process of decoherence.

Generally coherent dynamics and incoherent dynamics are driven by the same physical coupling mechanism and therefore any attempt to increase the coupling between qubits always brings with it an increased coupling to the environment. This inherent conflict, which we call the "Tao of Quantum Computing", highlights the daunting realities of laboratory experiments. The key challenge is then to find or engineer a quantum many-body system so that coherent evolution and decoherence can take place on very different time scales.

2 Neutral Atoms in Optical Lattices

By far, the best examples of coherent control that have been demonstrated in the laboratory to date are in Atomic-Molecular-Optical system (AMO). The primary example is the atomic-clock whose oscillator is based on the ultra-long-lived coherence of the ground-state hyperfine manifold. Examples of coherent control in AMO physics being applied to quantum information processing include ion traps, cavity QED, and correlated two-photon interferometry. Progress in the areas derives to a large degree from the existence of proven laboratory technologies and a community of physicists who for years have worked with ultra-high precision spectroscopy and quantum control in simple systems with long coherence times.

While the supreme coherent behavior of AMO systems is generally accepted by the scientific community, it is often stated that these systems are not *scalable* beyond a few qubits. We note here that there is nothing *intrinsic* in AMO implementations that prevents scalability. What is required is a basic unit that can be repeated with high-fidelity, such that the addition of another component does not degrade the behavior of the whole.

We have identified trapped ultra-cold neutral atoms as a nearly ideal system for storing and manipulating quantum information. This system has several advantages:

- Neutral atoms in traps are extremely well isolated from the decohering environment, as evidenced by the robustness of macroscopic quantum systems such as BEC.
- Ground state neutral atoms couple weakly to each other, except when interactions are deliberately introduced through external fields or by direct overlap of the atomic wave packets. The ability to couple qubits only when required leads to dramatic simplification of many-qubit control and suppresses the spread of errors.
- Neutral atom traps can be designed to allow simultaneous, parallel preparation and coherent control of the qubits in ensembles containing millions of atoms and is thus *scalable*. This resource is particularly important as we look towards error correction and fault-tolerant protocols, which require operation on a massively parallel set of qubits, and towards new paradigms beyond the serial quantum circuit model.

For quantum information studies, the best developed and most flexible atom trapping technology is based on "optical lattices", extended periodic arrays of microscopic potential wells formed by the AC-Stark shift in laser standing waves. The fundamental source of decoherence in the optical lattice is spontaneous photon scattering, which can be suppressed through the use of intense, far-detuned trapping fields, to a point where the coherence time can be many seconds.[1] One of the central strengths of this system is that laboratory demonstrations have been carried out for such crucial capabilities as the preparation of atoms in the vibrational ground state of an optical lattice by resolved-sideband Raman cooling[2] (state initialization), coherent control of mesoscopic spinor wavepackets with highly entangled spin and center-of-mass degrees of freedom,[3] and experimental reconstruction of the quantum density matrix for the angular momentum states in the hyperfine ground manifold[4] (state read out).

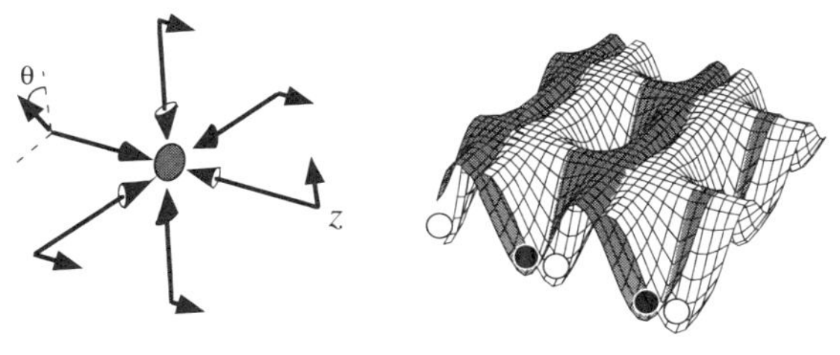

Figure 1. Three dimensional optical lattice formed by three beam pairs. Potential trapping surfaces are dynamically varied via polarization rotation, allowing coherent control.

We thus see that the majority of the requirements for implementing quantum circuits are in place. The crucial missing components are high-fidelity logic gates and the ability to address individual qubits. We take the former as the problem of fundamental importance. Individual qubit addressing is a technical problem that may be alleviated by alternative trap design, or perhaps through some gradient field mapping. Moreover, alternatives to the standard quantum circuit model or through special encodings [5] indicate that universal operation may not require individual addressing at all, and we anticipate new approaches will be developed. The problem of logic gate, on the other hand, is a critical element. Single-qubit rotations can be carried out via standard laser spectroscopy techniques as in ion traps. The key hurdle is thus the two-qubit entangling operation. The remainder of this presentation focuses on the prospect for achieving this in the optical lattices.

3 Entangling Dipole-Dipole Interactions

We have considered the possibility of utilizing optically excited electric dipoles to mediate a unitary two-body interaction between the trapped atoms. To see that this is possible, consider the following scaling argument. For two atoms separated at a distance $r \ll \lambda$ this interaction scales as $V_{dd} \sim \langle d \rangle^2 / r^3$, where $\langle d \rangle$ is the induced dipole moment per atom. Of course the real dipoles excited by a resonant field can spontaneously emit, and so lead to increased decoherence and destruction of the entanglement we seek to generate (the Tao of Quantum Computing).

However, this process is bounded by twice the single-atom scattering rate, $\Gamma' \sim k^3 \langle d \rangle^2 / \hbar$ where k is the wavenumber of the driving field. We can define a figure of merit as the ratio of the coherent to incoherent energy scales, $\kappa = V_{dd} / \hbar \Gamma' \sim (kr)^{-3}$. If the atomic wavepackets are very tightly confined relative to the optical wavelength, κ is large, and the dipole-dipole interaction is coherent.

We thus consider the following protocol. Through an appropriate geometry, atoms can be trapped in a three dimensional optical lattice and can be dynamically moved together pairwise to a well defined internuclear separation (see Fig.1).[6] An external laser field (referred to as the "catalysis") then excites optical dipoles for a prescribed time until the entangling operation is performed. Afterwards the atoms are separated and the field is removed so that the atoms no longer interact.

The simple scaling model discussed above treats the atoms as point particles. In reality they are localized in finite wave packets set by the trapping details. In addition, when the atom are separated by distances small compared to the optical wavelength, the excited state energy level structure is strongly perturbed; we are no longer exciting independent dipoles, but instead exciting a molecule. Thus, whereas the catalysis may be consider to be far from atomic resonance, for some internuclear radius (the "Condon point") it will be on-resonance with the pseudo-molecule leading to poor fidelity. This is show schematically in Fig.2. The design of the logic gate must balance then the need to bring the atoms as close together as possible to achieve a large dipole-dipole interaction, but at sufficient separation so that there is negligible probability to be at the Condon radius.

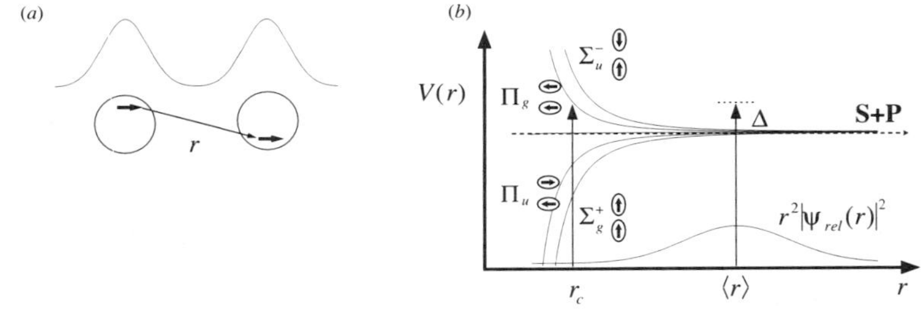

Figure 2. Dipole-Dipole interaction between atomic wavepackets. (a) Spherical atomic wavepackets separated along z. (b) Molecular potentials for spinless atoms as a function of relative coordinate r. Orientation of the dipoles corresponding to the various molecular states is indicated.

Our initial work assumed independently excited dipoles.[6] While valid for large separations, this is not the regime of highest fidelity. Proper treatment requires including of the hyperfine interaction in molecular potential since we encode the logical basis in these combined nuclear-electronic levels. Such a model has considerable (though tractable) complexities given the plethora of potentials. As a first step we consider here including only the *fine-structure* in the molecular potentials (setting the nuclear spin to zero). Though not a model of the protocol we plan to implement in the laboratory, it is the simplest situation which afford a logical basis, and will capture many of the salient features of the more complete model.

Consider then two alkali atoms driven by the catalysis on the D1 resonance $nS_{1/2} \rightarrow nP_{1/2}$. The logical basis is stored in the electron spin in the ground state,

$$\left\{ \begin{array}{l} \mid S_{\frac{1}{2}}, -\frac{1}{2}\rangle \otimes \mid S_{\frac{1}{2}}, -\frac{1}{2}\rangle, \mid S_{\frac{1}{2}}, -\frac{1}{2}\rangle \otimes \mid S_{\frac{1}{2}}, \frac{1}{2}\rangle, \\ \mid S_{\frac{1}{2}}, \frac{1}{2}\rangle \otimes \mid S_{\frac{1}{2}}, -\frac{1}{2}\rangle, \mid S_{\frac{1}{2}}, \frac{1}{2}\rangle \otimes \mid S_{\frac{1}{2}}, \frac{1}{2}\rangle \end{array} \right\} \equiv \{|\downarrow\downarrow\rangle, |\downarrow\uparrow\rangle, |\uparrow\downarrow\rangle, |\uparrow\uparrow\rangle\}. \quad (1)$$

This ground state basis can be treated as degenerate for the atomic separations of interest as the ground state potentials scale as C_6/r^6, while the dipole-dipole potentials scale as C_3/r^3. For this logical encoding, atoms brought together in a state dependent manner would be delocalized with interactions occuring only in the subspace $\{|\uparrow\downarrow\rangle, |\downarrow\uparrow\rangle\}$. This is akin to the proposal by Jaksch *et al.*[7] for entangling atoms in optical lattices using cold collisions. Using such delocalized atomic states suffers from the possibility of phase errors due to inhomogeneities in the magnetic field and in the following, for sake of argument, we treat the atoms as localized in state independent traps with a variable separation.

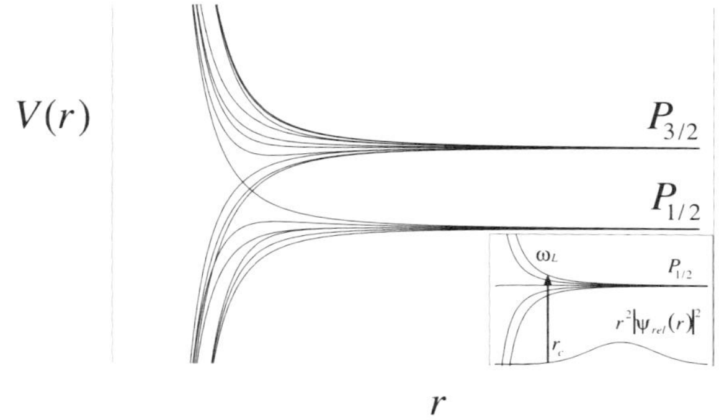

Figure 3. The Movre-Pichler molecular potentials as a function of internuclear separation r. The inset shows the relative coordinate probability density with support at separations too large to fit on the scale of the full potentials plot. The laser which induces the interaction is indicated tuned blue of the $P_{1/2}$ atomic resonace but with $\hbar\Delta \ll V_{fs}$, resonant with a molecular potential at the Condon radius r_c which is *far* in the tails of the probability distribution.

Our model includes fine structure splitting of the excited state molecular potentials using the so called Movre-Pichler potentials for couplings between states in the $S_{1/2}$ manifold to states in the $P_{1/2}$ and $P_{3/2}$ manifolds.[8] There are three regions of consideration. For the dipole-dipole interaction strength much weaker than the fine structure splitting, $V_{dd} \ll V_{fs}$, the molecular basis set is denoted by the projection of the total electronic angular momentum on the internuclear axis and the free atomic states are split by V_{dd}. For the intermediate regime $V_{dd} \sim V_{fs}$, there is mixing between states which asymptotically approach the $P_{1/2}$ and $P_{3/2}$ manifolds and the good quantum number becomes the projection of the total orbital angular momentum on the internuclear axis. Finally, at very small internuclear separations, electron exchange processes dominate. In our regime the relative coordinate probability density at these short distances is negligible and can be ignored.

Treating the dipole-dipole interaction as a perturbation to the trapping potential, the excited state molecular potentials can be adiabatically eliminated. The reduced Hamiltonian in the logical basis is

$$H_{ij} = \frac{\hbar\Omega^2}{4}\left\langle \sum_{|e\rangle} \frac{c_{ie}^*(\mathbf{r})c_{ie}(\mathbf{r})(\Delta_e(r) - i\Gamma_e(r)/2}{\Delta_e(r)^2 + \Gamma_e(r)^2/4} \right\rangle_{rel}. \tag{2}$$

where Ω is the Rabi frequency of the catalysis laser, c($\mathbf{r}$) are the position dependent coupling coefficients, $\Gamma(r)$, $\Delta(r)$ are the position dependent linewidths and detunings. The average is taken over the relative coordinate probability distribution of the atomic pair.

4 Quantum Logic Gates

For dipoles induced by a π-polarized catalysis beam, the Hamiltonian, Eq.(2), will introduce light shifts diagonal in the logical basis, Eq.(1), as well as exchange couplings $|\downarrow\uparrow\rangle \leftrightarrow |\uparrow\downarrow\rangle$. A Bell-State can be formed by initially preparing the product state $|\uparrow\downarrow\rangle$ when the atoms are separated, bringing the atoms together and turning on the dipole-dipole interactions for a time $\tau = \frac{\pi}{4|Re[H_{\uparrow\downarrow,\downarrow\uparrow}]|/\hbar}$, in which case the entangling unitary U (up to single bit gates) in the logical basis is

$$U(\phi) = \begin{pmatrix} 1 & & \\ & \frac{e^{i(\phi+\frac{\pi}{4})}}{\sqrt{2}} \begin{pmatrix} 1 & -i \\ -i & 1 \end{pmatrix} & \\ & & 1 \end{pmatrix}. \quad (3)$$

This unitary is similar to the universal two bit $\sqrt{SWAP}$ gate, except for the phase ϕ which depends on the strength and sign of the dipole-dipole interaction. In fact U cannot generally be mapped to the $\sqrt{SWAP}$ gate by single bit gates applied before and after (except for the cases $\phi = 0, \pm\pi/2, \pi$)[a]. However, by changing the catalysis beam intensity and the sign of detuning, a second gate can be applied with phase $-\phi$ in which case the following sequence produces the universal $CPHASE$ gate,

$$CPHASE = -ie^{i\sigma_z^1\pi/4} \otimes e^{-i\sigma_z^2\pi/4}U(\phi)e^{i\sigma_z^1\pi/2}U(-\phi). \quad (4)$$

4.1 Errors

The dominant source of errors for this quantum logic protocol is spontaneous emission induced by the catalysis beam during the entangling gate. Light induced collisional losses such as photoassociation to an attractive molecular state leading to radiative escape are exponentially suppressed when the catalysis detuning is chosen far in the tails of the relative coordinate probability distribution. Ground state collisions, either inelastic or elastic spin exchange processes occur on a time scale long compared to the gate time and are negligible for atomic separations beyond the ground state scattering length. The fidelity for producing a maximally entangled state from a product state is given by $F = \min_{|i\rangle} |\langle i \mid U^\dagger V_{real} \mid i\rangle|^2$ where V_{real} is the true non-unitary operator corresponding to Eq.(3), and the minimum is taken over all possible input states. We find

$$F = \exp\Big(-\frac{Im[H_{\uparrow\uparrow,\uparrow\uparrow}]\pi}{2\,|Re[H_{\uparrow\downarrow,\downarrow\uparrow}]|}\Big)\cosh^2\Big(\frac{Im[H_{\uparrow\uparrow,\uparrow\uparrow}]\pi}{4\,|Re[H_{\uparrow\downarrow,\downarrow\uparrow}]|}\Big). \quad (5)$$

A plot of fidelity as a function of detuning and internuclear separation is shown in Fig.4 using the fine structure splitting of ^{133}Cs. Near zero detuning, the atoms are resonant with free atomic states and the fidelity is poor. For large enough detunings, $|\Delta| > 5000\Gamma$ the fidelity reaches a maximum at around $\Delta z = 6z_0$ where the support the wavefunction is far enough away from Condon points but the dipole-dipole potential which falls off as $1/r^3$ remains substantial. For a localization of $\eta_z = 0.05$, the fidelity is maximum at $\delta = 2 \times 10^4\Gamma, \Delta z = 5.6z_0$ where $F = 0.934$. Here the outermost Condon point is at $r_c = 3.7 \times 10^{-2}\lambda$, well outside the support of the wavefunction. If we consider the trapping laser to perform the gate, then at this localization and detuning, the intensity $I = 2.87 \times 10^6 I_0 = 4 \times 10^3 W/cm^2 = 1W$ focused to a $158\mu m$ spot size, gives a gate time of $\tau = 67.2/\Gamma = 1.29 \times 10^{-5}s$. As discussed in Sec. 3, this model assumed atoms trapped in state

[a]This can be easily be seen by the fact that $U = D(\phi/2)\sqrt{SWAP}D(\phi/2)$ where $D(\phi/2)$ is a diagonal gate than generally cannot be rotated to the identity by single bit rotations.

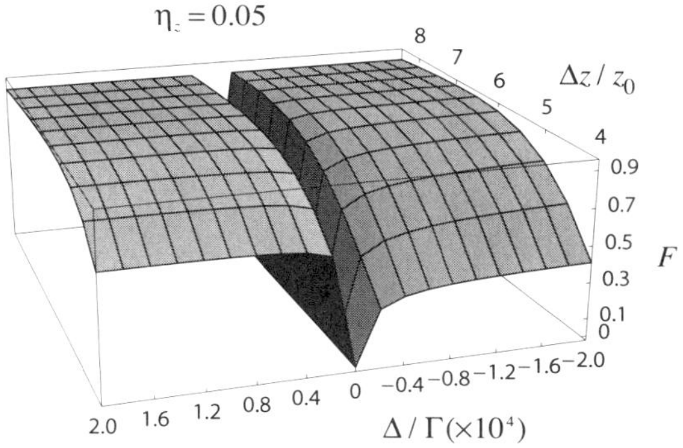

Figure 4. Entanglement fidelity as a function of detuning and well separation for spherical traps with 1D localization $\eta_z = k_L z_0$.

independent traps, however, the fidelity is nearly as good (to within 10^{-3}) for atoms in state dependent traps. Much improved fidelity results with tighter localization where the atoms look more like point dipoles, but this comes at the expense of greater laser power as $I \propto \eta^{-4}$.

5 Future Directions

We have presented a scheme for creating on demand entanglement between neutrals using induced dipole-dipole interactions. The calculated fidelity for atoms with zero nuclear spin is rather modest and sufficient to reliably perform only a few gates. The next step is to include hyperfine structure in molecular potentials. We expect this fidelity to be better as the logical states are split by the ground state hyperfine splitting which should allow large differential shifts to accumulate a $CPHASE$ gate over a time much faster than the rather weak exchange coupling for zero nuclear spin.

Much of this research falls under the rich and open realm of molecular coherent control and in particular demonstrates the ability to use optical lattices to probe dimer dynamics. When the molecular interactions are well characterized it might be possible to design alternative protocols for quantum logic. In contrast to off-resonant phase shifts discussed above, one can imagine working in the opposite limit, where one rapidly induces a coherent Rabi 2π-rotation between the target t wo-bit state and a resolved auxiliary molecular state, thereby accumulating a π conditional phase shift. An option is to use a long range excited bound molecular state such as one of the bound states the 0_g^- state in Na_2 which are well resolved by at least 50 MHz.[9] An alternative scheme based on the same principle is to make use of bound states in the *ground* electronic manifold. For example, through application of a magnetic field, a bound-state of t wo atoms associated with one hyperfine state can be made resonant with a scattering state associated with the other in a event known as a Feschbach resonance. By performing a 2π-pulse between the free atom near the dissociation limit and the ground-state resonance, one can in principle implement a high-fidelity $CPHASE$ gate. Reflexively, quantum information processing can be used to infer the physical properties of molecular dimers.

Acknowledgements

We acknowledge helpful discussions with Carl Williams, Paul Julienne, John Grondalski, and Shohini Ghose. Part of this work was carried out at the 2000 workshop of the Benasque Center for Physics. IHD and GKB acknowledge support from the National Science Foundation (Grant No. PHY-9732456) and the Office of Naval Research (Grant No. N00014-00-1-0575).

References

1. P.S. Jessen and I.H. Deutsch, Adv. Atom. Mol. Opt. Phys. **37** 95 (1996).
2. S.E. Hamman *et al.*, Phys. Rev. Lett. **80** 4149 (1998).
3. D.L. Haycock *et al.*, Phys. Rev. Lett. **85** 3365 (2000).
4. G. Klose, G.A. Smith, and P.S. Jessen, submitted to Phys. Rev. Lett., quant-ph/0101017.
5. S. Benjamin, Phys. Rev. A **61** 020301R (2000).
6. G.K. Brennen *et al.*, Phys. Rev. Lett. **82** 1060 (1999); I.H. Deutsch, G.K. Brennen, and P.S. Jessen Fortschr. Phys. **48** 926 (2000).
7. D. Jaksch *et al.*, Phys. Rev. Lett. **82** 1975 (1999).
8. M. Movre and G. Pichler, J. Phys. B **10** 2631 (1977).
9. C.J. Williams, E. Tiesinga, and P.S. Julienne, Phys. Rev. A **53** R1939 (1996).

ENCODED UNIVERSALITY IN PHYSICAL IMPLEMENTATIONS OF A QUANTUM COMPUTER

D. BACON,[1,2] J. KEMPE,[1,3] D. P. DIVINCENZO,[4] D. A. LIDAR,[1*] AND K. B. WHALEY[1†]

Departments of Chemistry[1]*, Physics*[2] *and Mathematics*[3]*, University of California, Berkeley 94270*

[4]*IBM Research Division, T.J. Watson Research Center, Yorktown Heights, New York 10598*

We revisit the question of universality in quantum computing and propose a new paradigm. Instead of forcing a physical system to enact a predetermined set of universal gates (e.g., single-qubit operations and CNOT), we focus on the intrinsic ability of a system to act as a universal quantum computer using only its naturally available interactions. A key element of this approach is the realization that the fungible nature of quantum information allows for universal manipulations using quantum information encoded in a subspace of the full system Hilbert space, as an alternative to using physical qubits directly. Starting with the interactions intrinsic to the physical system, we show how to determine the possible universality resulting from these interactions over an encoded subspace. We outline a general Lie-algebraic framework that can be used to find the encoding for universality ,and give examples relevant to solid-state quantum computing.

1 Introduction

Our generous universe comes equipped with the ability to compute. In physics, determination of the *allowable* manipulations of a physical system is of central importance. Computer science, on the other hand, has arisen in order to *quantify* what resources are needed in order to perform a certain algorithmic function. For computer science to be applicable to the real world, this quantification should be limited by what physics has determined to be allowable manipulations. Thus we arrive at the realization that because information is physical, computer science should be built on primitives which respect the laws of physics.

An important property of classical information which carries over to the quantum world is the fungible nature of information. A resource is fungible if interchanging it with another resource does not destroy the value of the resource. Whether we represent a classical bit by the presence or absence of a chad on a punch-card[1] or in the average orientation of a million electron spins, the intrinsic value of the information (the value of the bit) is untouched: it does not depend upon the medium in which it is represented. The fungible nature of information has been key to the exponential growth of the computer revolution: it does not matter that the information is being confined to ever smaller components on silicon chips. So, too, for quantum information: the plethora of experimentally proposed systems from which a quantum computer could be built is made possible by the fungibility of quantum information.

Another example of the fungible nature of quantum information is the idea that one can *encode* quantum information. The theory of quantum error correcting codes, for instance, makes use of this to allow computation in the presence of noise.[2] In this paper we argue that we can harness the fungible nature of quantum information in another important aspect, namely in universality constructions. In particular, we propose that instead of taking a physical system and adding external Hamiltonians and interactions to it so that a

*CURRENT ADDRESS: DEPT. OF CHEMISTRY, UNIV. OF TORONTO, ONTARIO, CANADA
†CORRESPONDING AUTHOR

certain universal set of gates may be performed, one examines instead the intrinsic system interactions for their potential universality. The key to this approach is the recognition that the ability to encode quantum information allows for certain interactions to be universal over an encoded Hilbert subspace, even though they are not universal over the entire Hilbert space. Our task is then to search for the encodings providing such universality, given the intrinsic system interactions. Such encoded universality is common in quantum error correction but appears to have not been recognized in the discussions concerning the physical construction of a quantum computer.

2 Universality

The action of gates $\mathbf{G}$ on a circuit corresponds to unitary evolution of the system, with the gates given by the evolution operators derived from the system Hamiltonian $\mathbf{H}$: $\mathbf{G}(t) = \mathtt{T}\left\{\exp[-i\int^t \mathbf{H}(\tau)d\tau]\right\}$. Here $\mathtt{T}$ denotes the time ordering operation. A finite set $\mathcal{G}$ of gates is defined to be *universal* if: i) each $\mathbf{U}_i \in \mathcal{G}$ acts on a constant number of qubits only, and ii) any quantum circuit (performing any desired unitary operation) can be simulated to arbitrary accuracy efficiently by a circuit with gates in $\mathcal{G}$.

Somewhat surprisingly, most discussions of universality in quantum computing are cast in the language of gates, rather than Hamiltonians. As physicists we find, however, that in order to understand encoded universality, it is often more useful to think about a gate in terms of the generating Hamiltonian instead of the resulting gate. Given the notion of a universal set of gates, various physical implementations of such a quantum computer have been proposed. Naturally, in order to qualify as a valid quantum computer, each physical implementation must be shown to possess a universal set of gates arising from controllable interactions of the given physical system. To date, with some exceptions,[3,4,5] all physical proposals for a quantum computer have implemented a set of universal gates via a route in which: (i) the basic qubit is identified in the physical system, (ii) single-qubit unitary gates are shown to be possible on these qubits, and (iii) some two-qubit gate acting between qubits is shown to be possible. The latter is usually a CNOT or a controlled phase. For a few systems this standard model does provide a natural path towards building a quantum computer, but for other implementations severe device engineering must be performed in order to force the system to achieve this universal set of gates.

For instance, in many solid state implementations a fast intrinsic interaction is the exchange interaction (Heisenberg coupling)

$$\mathbf{E}_{ij} = \vec{\sigma}^i \cdot \vec{\sigma}^j, \tag{1}$$

(here $\vec{\sigma}^i$ is the vector of Pauli spin matrices $(\sigma_x, \sigma_y, \sigma_z)$ on the i-th qubit). We call $\mathbf{E}_{ij}$ the exchange interaction because addition of a global phase and rescaling gives a Hamiltonian which exchanges the qubits ($|x\rangle|y\rangle \to |y\rangle|x\rangle$).[6] The gates which can be obtained by tuning the exchange interaction strength J are the unitary gates $\mathbf{U}_{ij}(t) = \exp(itJ\mathbf{E}_{ij})$. Given n qubits coupled via exchange interactions it can be shown[4] that these gates alone cannot be used to implement an arbitrary quantum circuit on n qubits. Thus designers of solid state quantum computers have had to supplement the exchange interaction by single-qubit rotations in order to make their set of interactions universal over all n qubits. However, adding this single-qubit capability generally leads to considerable device complexity and to disadvantages such as slow single-qubit gate times and possible additional sources of decoherence resulting from addition of auxiliary, non-intrinsic interactions to the system.

It was recently realized[3,4,5] that the exchange interaction by itself can be used to simulate a quantum circuit if a particular encoded space of three or more qubits is used to represent the quantum information. Thus while the exchange interaction on n qubits cannot be used to implement an arbitrary quantum circuit on n qubits, it can be used to simulate a quantum circuit on cn qubits, where c is a constant representing the encoding efficiency of such a code. The fact that the exchange interaction *alone* can be used to simulate a quantum circuit by using encoded quantum information allows one to avoid implementing the generally difficult single-qubit gates.

This example illustrates a new approach to universality very different from the conventional paradigm of using single qubit-gates supplemented by a two-qubit gate. The latter has consituted the primary guideline in recent years for deciding whether a physical system might be used as a quantum computer, without regard to the fact that this is just one of many possible paths to universality. Nevertheless, other approaches have also been proposed. Thus, it was shown[7,8] that *almost any* two-qubit unitary gate along with the exchange gate acting between n qubits is universal over all n qubits. We propose here a radically different route to achieve universality, namely, that starting from the natural interactions given by the physics of the proposed qubit system (e.g., $\mathbf{E}_{ij}$), the underlying potential of that interaction for encoded universality be investigated. For a given physical system, the natural interactions of choice should be determined by various factors including their speed, the ease with which they can be implemented, and their robustness towards decoherence processes. The path we take, then, is to proceed from the intrinsic interactions to encoded universality, rather than to artificially modify or build the interactions to fit a standard model.

3 General Formalism

The general formalism will be illustrated here with the example of the exchange interaction. We omit most of the proofs and details, which can be found in Ref. 4.

Assume that we are given a set of interactions on some d-dimensional Hilbert space – Hermitian Hamiltonians $h_i(t)\mathbf{H}_i$ where $h_i(t)$ are controllable coupling constants – that we want to use for our universal quantum computer. Which unitary gates can be approximated by turning these interactions on and off (perhaps in parallel)? The answer to this[7,8,9] is that every unitary gate in the Lie group corresponding to the Lie algebra *generated* by the Hamiltonians $\mathbf{H}_i$ can be approximated to arbitrary accuracy by a sequence of gates obtained through the temporal evolution of $h_i(t)\mathbf{H}_i$. This is true because of the following properties:

$$e^{i(\alpha\mathbf{A}+\beta\mathbf{B})} = \lim_{n\to\infty}(e^{i\alpha\mathbf{A}/n}e^{i\beta\mathbf{B}/n})^n \tag{2}$$

$$e^{[\mathbf{A},\mathbf{B}]} = \lim_{n\to\infty}(e^{-i\mathbf{A}/\sqrt{n}}e^{i\mathbf{B}/\sqrt{n}}e^{i\mathbf{A}/\sqrt{n}}e^{-i\mathbf{B}/\sqrt{n}})^n. \tag{3}$$

To approximate to a given accuracy, say $\exp(i(\alpha\mathbf{A}+\beta\mathbf{B}))$, by a sequence of gates with generators $\mathbf{A}$ and $\mathbf{B}$ one just truncates Eq. (2) for large enough n. Every element in the Lie algebra, that is the algebra formed via linear combination $(\alpha\mathbf{A}+\beta\mathbf{B})$ and via the Lie commutator $(i[\mathbf{A},\mathbf{B}])$, can be approximated to given accuracy using these results. Thus all gates in the Lie group corresponding to this Lie algebra generated by the $\mathbf{H}_i$ can be approximated to arbitrary accuracy by using the controllable coupling constants $h_i(t)$. The inverse of this statement is also true.

Let $\mathcal{A}$ be the Lie algebra generated by the $\mathbf{H}_i$. The Lie group generated by the Lie algebra $\mathcal{A}$ is a subgroup of the d-dimensional unitary group $U(d)$ and thus is a compact Lie group. An important property of compact Lie groups is that they, along with the Lie algebra which generates the Lie group, are completely reducible. A representation of a Lie algebra is irreducible if the action of the representation on its vector space does not possess an invariant subspace. An algebra which is completely reducible can be written as a direct sum of irreducible representations (irreps):

$$\mathcal{A} \cong \bigoplus_{J} \bigoplus_{\lambda=1}^{n_J} \mathcal{L}_J, \tag{4}$$

where $\mathcal{L}_J$ is the Jth irrep which appears with degeneracy n_J and is d_J-dimensional. For the operators $\mathbf{H}_i$ this implies that there is a basis for the Hilbert space on which $\mathbf{H}_i$ acts as

$$\mathbf{H}_i = \bigoplus_{J} \mathbf{I}_{n_J} \otimes (\mathbf{L}_J)_i \tag{5}$$

where $(\mathbf{L}_J)_i$ are elements of the Jth irrep $\mathcal{L}_J$ and $\mathbf{I}_d$ is the d-dimensional identity operator. Clearly, understanding the irreps of the Lie algebra generated by the Hamiltonians $\mathbf{H}_i$ will allow us to understand the suitability of this Lie algebra for simulating a quantum circuit. In particular, Eq. (5) implies that with a *suitable encoding* the $\mathbf{H}_i$ can produce the action of a specific irrep $\mathcal{L}_J$. This action may or may not be useful for quantum computation, an issue we address below.

It is often helpful in understanding the universality of an interaction to find the *commutant* $\mathcal{A}'\,\{\mathbf{X} : [\mathbf{X}, \mathbf{A}] = 0, \forall \mathbf{A} \in \mathcal{A}\}$ of the Lie algebra $\mathcal{A}$. It should be clear from Eq. (5) that every element $\mathbf{K} \in \mathcal{A}'$ is reducible to:

$$\mathbf{K} = \bigoplus_{J} \mathcal{M}_{n_J} \otimes \mathbf{I}_{d_J} \tag{6}$$

where $\mathcal{M}_{n_J}$ is an n_J-dimensional (complex) square matrix, and d_J is the dimension of the Jth irrep. This structure, dual to $\mathcal{A}$, offers a useful way for seeing how the encoding arises: $\mathcal{A}'$ splits into a sum of degenerate irreps. The degeneracy of a particular representation gives the dimension of the space over which the encoding is made.

We should point out a non-trivial aspect of the question of "universality given a set of interactions", which makes our task more difficult. This is the tensor product nature of quantum computation. We recall that a set of interactions is universal if it can be used to efficiently simulate a quantum circuit. An important property of the quantum circuit model is that at its most basic level it possess a tensor product of single qubit systems. Hidden within this seemingly trivial fact is the notion of locality: we believe that physics is local and hence we require our model of computation based on physics to preserve this locality. For the purposes of building a universal quantum computer this implies that we must, at some point, introduce a suitable tensor product structure in order to *efficiently* simulate a quantum circuit. The above discussion of representation theory for Lie algebras has been solely in terms of some abstract d-dimensional Hilbert space. We now admit two tensor product structures: first, on our physical system, and second, on the encoded qubits from which encoded universality will be constructed. The first tensor product structure is simply that implied by our physical system and is forced on us by the locality of physics. For example, if we are using the spins of single electrons on a quantum dot, the tensor product is just the natural one of these spin-qubits. The nature of this tensor product structure is also manifest in the set of interactions which will be present in the real world, and is represented in the Hamiltonian as operators acting on separate degrees of freedom.

The second tensor product is required by the fact that we are going to simulate a quantum circuit. Given a set of interactions, we know that these can be used to produce the action of a given set of irreps $\mathcal{L}_J$. However, without a mapping from the subspace on which the irrep acts to the tensor product structure of a quantum circuit, this is not enough to quantify the suitability of the interactions for universality. The exact cutoff where this tensor product structure comes into play is somewhat arbitrary. We will take the position here that we are searching for *small encodings*. That is, we seek to minimize the number of physical qubits used to encode a logical qubit. This logical qubit will be formed by a block of nearest neighbor qubits with the defining property that single encoded-qubit operations are possible within each such block. Given a set of such encoded qubits we then form a tensor product between pairs of blocks. We refer to this process of forming a tensor product code as "conjoining". These considerations are illustrated below.

4 Example of Isotropic Exchange

As an example of our general discussion consider the Heisenberg model for spin-spin interactions. The most general form of this is given by a sum of fully anisotropic pair-wise spin-spin couplings

$$\mathbf{H}_{ij} = J_{ij}^X \sigma_x^i \sigma_x^j + J_{ij}^Y \sigma_y^i \sigma_y^j + J_{ij}^Z \sigma_z^i \sigma_z^j. \tag{7}$$

We consider in detail here the isotropic limit in which the exchange couplings between a given ij pair are all equal, *i.e.*, $J_{ij}^X = J_{ij}^Y = J_{ij}^Z \equiv J_{ij}$, so that we are dealing then with the Heisenberg Hamiltonian

$$\mathbf{H}_{\mathrm{Hei}} = \sum_{i \neq j} J_{ij} \mathbf{E}_{ij} \tag{8}$$

where we assume that the coupling parameters J_{ij} can be turned on/off at will. To show how this intrinsic coupling can lead to encoded universality we analyze its action on three qubits.

The Lie algebra $\mathcal{L}_E$ generated by $\mathbf{E}_{ij}$ on three qubits allows us to implement the Hamiltonians in the set $\{\mathbf{E}_{12}, \mathbf{E}_{23}, \mathbf{E}_{13}, \mathbf{T} \equiv i([\mathbf{E}_{12}, \mathbf{E}_{23}]\}$. A better basis for the Lie algebra is given by the set of operators $\mathbf{H}_0 = \mathbf{E}_{12} + \mathbf{E}_{23} + \mathbf{E}_{13}$, $\mathbf{H}_1 = \frac{1}{4\sqrt{3}} (\mathbf{E}_{13} - \mathbf{E}_{23})$, $\mathbf{H}_3 = \frac{1}{12} (-2\mathbf{E}_{12} + \mathbf{E}_{23} + \mathbf{E}_{13})$, $\mathbf{H}_2 = i[\mathbf{H}_3, \mathbf{H}_1]$. We then find that

$$[\mathbf{H}_0, \mathbf{H}_\alpha] = 0 \tag{9}$$

for all α and

$$[\mathbf{H}_\alpha, \mathbf{H}_\beta] = i\epsilon_{\alpha\beta\gamma} \mathbf{H}_\gamma \tag{10}$$

with $\alpha, \beta, \gamma \in \{1, 2, 3\}$. $\mathbf{H}_0$ is an abelian invariant subalgebra of this Lie algebra and thus factors out as a global phase. The set $\{\mathbf{H}_1, \mathbf{H}_2, \mathbf{H}_3\}$, on the other hand, act as the generators of $su(2)$. Thus the exchange interaction between three qubits can be used to implement a single encoded-qubit $su(2)$. More precisely, we find that the $\mathbf{H}_\alpha$ for $\alpha \in \{1,2,3\}$ generates the algebra

$$\mathcal{L}_E^{(3)} = \left(\bigoplus_{i=1}^{4} \mathcal{S}_1\right) \oplus \left(\bigoplus_{i=1}^{2} \mathcal{S}_2\right) \tag{11}$$

where $\mathcal{S}_d$ is the d-dimensional irrep of $su(2)$. Note the degeneracy of the corresponding irreps. Corresponding to this decomposition the $\mathbf{H}_\alpha$ act as

$$\mathbf{H}_\alpha = \mathbf{0}_4 \oplus \left(\frac{1}{2}\sigma_\alpha \otimes \mathbf{I}_2\right) \tag{12}$$

for $\alpha \in \{1,2,3\}$ where $\mathbf{0}_4$ is 4-dimensional zero operator (the 1 dimensional irreps all act as 0) and $\mathbf{I}_2$ is the two-dimensional identity operator. The action of the exchange is thus identical to that of an $su(2)$ operator on a single qubit when working over the encoded space defined by the above decomposition. If we encode our logical qubits as

$$\begin{aligned} |0_L\rangle &= \frac{1}{\sqrt{2}}(|010\rangle - |100\rangle) \\ |1_L\rangle &= \sqrt{\frac{2}{3}}|001\rangle - \sqrt{\frac{1}{6}}|010\rangle - \sqrt{\frac{1}{6}}|100\rangle \end{aligned} \tag{13}$$

then we find that the action of $\mathbf{H}_\alpha$ is $\frac{1}{2}\sigma_\alpha$. Due to the degeneracy of the irrep, other encodings are also possible. Note that these states are nothing but $J = 1/2$ total angular momentum states of 3 spin-1/2 particles with a given projection along a certain axis, and can thus be found using elementary addition of angular momentum.[4]

Having shown that the action of the exchange interaction on three qubits can produce the effect of a single 2-dimensional representation of $su(2)$, it is natural to induce a tensor product structure between blocks of three qubits in order to simulate a quantum circuit. We thus *choose* an encoding scheme in which a single qubit is identified with 3 physical qubis. This is not the only choice of tensor product structure. In fact, any tensor product between sets of $k \geq 3$ qubits can be used to construct a universal gate set. We define the encoding efficiency of our code as the number of encoded bits of our simulated quantum computer ($e = \log_2(d)$ where d is the dimension of the encoded information) divided by the total number of qubits on which the encoding exists: $E = e/n = \log_2(d)/n$. Thus, for a scheme using the exchange interaction between three qubits, where we have a $d = 2$ dimensional encoding on $k = 3$ qubits, we find that $E = 1/3$. A peculiar property of the exchange interaction is that it has an asymptotic encoding efficiency of unity. In particular, if we take n qubits and use the exchange interaction to compute on these qubits then the dimension of the encoded space on which we can compute scales in such a way that the encoding efficiency is at least $E(n) = 1 - \frac{3}{2}\frac{\log_2 n}{n}$. [4] As $\lim_{n\to\infty} E(n) = 1$ we loose almost nothing in terms of encoding efficiency when we use the exchange interaction. We note, however, that one must still introduce a tensor product structure at some level in order to achieve universal quantum computation.

Once one has introduced a tensor structure for the encoding, it is necessary to show that the natural couplings of the system can produce a non-trivial action between the tensor components of this encoded tensor product structure. In the above case, where we have shown that we can obtain full control over an encoded qubit, what we now need to show is that a non-trivial action between the encoded qubits can be enacted. This is nothing more than a map from the encoded universality to the fully universal set of physical gates consisting of single qubit gates supplemented by a non-trivial coupling between the qubits, i.e., to the standard model. This point of the universality proof is typically the most daunting, but really amounts to nothing more than understanding the Lie algebra generated by the interaction over two tensored encoded qubits. For the exchange interaction it can be shown that nearest neighbor exchange interactions can be used to produce such an action.[4] In particular, we find that the effect of the exchange interaction on 6 qubits becomes

$$\mathcal{L}_E^{(6)} = \left(\bigoplus_{i=1}^{7} \mathcal{S}_1\right) \oplus \left(\bigoplus_{i=1}^{5} \mathcal{S}_5\right) \oplus \left(\bigoplus_{i=1}^{3} \mathcal{S}_9\right) \oplus \mathcal{S}_5 \tag{14}$$

When we conjoin the two three-qubit codes, we find that these codes lie entirely within the last two irreps of this decomposition. This in turn implies that non-trivial interactions between the encoded irreps are possible.[4] In fact the original single encoded qubit $su(2)$ is also contained within this decomposition. The important point, however, is that on the tensor product between encoded qubits we can indeed couple these encoded qubits in such a way as to achieve a map to the unencoded, fully universal set of gates. Thus we have established that our encoding can efficiently (to within a factor of three in spatial resources) simulate the unencoded set of universal gates.

A similar analysis can be applied to the anisotropic Heisenberg model (XY model). There it turns out that the smallest encoding one can achieve is a logical qutrit (3-level system) into three physical qubits.[10] For both the isotropic and anisotropic Heisenberg models these universality results are general, i.e., they can be shown to hold for arbitrary numbers of qubits, n.[4,10] However, the approach of conjoining small blocks that encode qubits via a tensor product can be generally used to construct the required mapping of universality for arbitrary interactions, even when a general proof appears inaccessible.

5 Outlook

We have shown that the commonly assumed paradigm of universal quantum computation requiring a set of both single-qubit and two-qubit gates can be replaced by a more general notion of encoded universality. This approach allows universality of quantum computation to be achieved via encodings specific to the natural or convenient physical interactions in a system. In the example given here we have discussed an instance where encoded universality can be achieved using only a *single* underlying physical interaction, namely the exchange interaction. This result is of particular significance for solid state implementations of quantum computation. The approach of coupling blocks containing small numbers of physical qubits (each forming one encoded qubit) via tensor products appears to be applicable to any interaction.

The results for encoded universality that we have described here are primarily existential, and do not address the very practical concerns of the overhead in temporal and spatial resources which may result from encoding. The practical question is then: what is the trade-off in resources of time (gates) and space (qubits) resulting from the universality encoding for a given set of physical interactions? Finding a universality encoding generally immediately reveals the spatial overhead, *i.e.*, the number of physical qubits required to perform the encoding, and may also provide information about the asymptotic overhead.

Thus, for the exchange interaction, the maximum spatial overhead is a factor of three, and becomes arbitrarily small asymptotically.[5] The overhead in temporal resources, *i.e.*, how many physical gates are required to implement a given set of encoded universal gates, is somewhat more challenging to deal with. General issues of optimization of physical gates for circuits of encoded qubits have been addressed.[11] It has recently been shown explicitly that the encoded universal operations deriving from the exchange interaction can be realised within $\sim$ 10 clock cycles.[5] However there does not exist to our knowledge any general methodology to find the optimal gate sequence for a specific two-qubit (or two-qutrit) gate.

The finding of encoded universality for the Heisenberg Hamiltonians suggest that this approach will be very useful for consideration of solid state quantum computation. Establishment of results concerning the existence of encodings for general classes of Hamiltonians is therefore now of interest. The new paradigm of encoded universality clearly opens up significant new opportunities for exploring the architectural realization of quantum computation.

Acknowledgements:- This work was supported in full by the National Security Agency (NSA) and Advanced Research and Development Activity (ARDA) under Army Research Office (ARO) contracts DAAG55-98-1-0371 (KBW) and DAAG55-98-C-0041 (DPD).

References

1. US Supreme Court case no. 00-949, George W. Bush vs Albert Gore, Jr.
2. A.M. Steane. Quantum Error Correction. In H.K. Lo, S. Popescu, and T.P. Spiller, editor, *Introduction to Quantum Computation and Information*, page 184. World Scientific, Singapore, 1999.
3. D. Bacon, J. Kempe, D.A. Lidar, and K.B. Whaley. Universal Fault-Tolerant Computation on Decoherence Free Subspaces. *Phys. Rev. Lett.*, 85:1758, 2000.
4. J. Kempe, D. Bacon, D. Lidar, and K.B. Whaley. Theory of Decoherence-Free Fault-Tolerant Universal Quantum Computation. *Phys. Rev. A*, 63:042307 (2001).
5. D. P. DiVincenzo, D. Bacon, J. Kempe, G. Burkard, and K. B. Whaley. Universal quantum computation with the exchange interaction. *Nature*, 408:339, 2000.
6. D.A. Lidar, D. Bacon, J. Kempe, and K.B. Whaley. Protecting quantum information encoded in decoherence-free states against exchange errors. *Phys. Rev. A*, 61:052307, 2000.
7. D. Deutsch, A. Barenco, and A. Ekert. Universality in quantum computation. *Proc. Roy. Soc. London Ser. A*, 449:669, 1995.
8. S. Lloyd. Almost Any Quantum Logic Gate is Universal. *Phys. Rev. Lett.*, 75:346, 1995.
9. D. P. DiVincenzo. Two-bit gates are universal for quantum computation. *Phys. Rev. A*, 51(2):1015, 1995.
10. J. Kempe et al., to be published.
11. G. Burkard, D. Loss, D. P. DiVincenzo, and J. A. Smolin. Physical optimization of quantum error correction circuits. *Phys. Rev. B*, 60:11404, 1999.

QUANTUM ALGORITHMS AND THE MINIMIZATION PROBLEM

LLOYD C.L. HOLLENBERG
School of Physics, University of Melbourne, Victoria 3010, AUSTRALIA.

The global minimization problem occurs at the heart of many computer simulations of physical (and non-physical) systems. It is therefore of great interest to consider how one might perform minimization tasks on a quantum computer. A quantum algorithm framework is described which uses assumed information about the spectrum to reduce the search space to the neighbourhood of the minimum.

1 Introduction

In recent years we have witnessed an explosion of interest in quantum computing on both experimental and theoretical fronts, and everything between. Here we focus our attention on the use of quantum computation in the minimization problem which is important in many areas of study. Quantum algorithms for minimization have been developed since Grover discovered the fast searching algorithm[1]. Most algorithms are thus based on gradual amplitude amplification which is quadratically faster than exhaustive search[2]. A heuristic quantum algorithm was developed by Hogg and Portnov[3] based on problem-tuned phase amplification. There also exists ideas employing adiabatic evolution[5]and quantum annealing[4].

If an algorithm were found which could find the global minimum efficiently (i.e. is polynomial bounded) then all **NP**-complete problems would be solvable. However, since it has been proved that Grover's algorithm[1] is optimal on unstructured search problems[6,7,8], it would seem that this is unlikely to be possible. On the other hand, it is not clear that an exploitation of structure may not lead to faster algorithms for certain problems. In this paper a quantum algorithm framework for minimization is presented which attempts to use "global" structure of the problem in a hybrid sense.

2 A hybrid algorithm

The problem is introduced as the minimization of an energy function $E(x)$ over variables x, with $N = \dim\{x\}$: the task is to find the global minimum $E_{\min}(x_{\min})$. Assume a classical calculation has roughly mapped the density of states with the result $\rho_C(E)$. From $\rho_C(E)$ we can compute the cumulative states function, $n_C(E)$:

$$n_C(E) \equiv \sum_{E'}^{E} \rho_C(E')\, dE'. \tag{1}$$

We store the variables x in register R_x, and the energy values in a second register, R_E. To simplify the argument we assume the x variables are integers: $x \in [0, N-1]$ where $N \equiv 2^{Q_x}$ defines Q_x as the number of qubits in R_x. The state of the computer is initialized as:

$$|\Psi_0\rangle = \frac{1}{\sqrt{N}} \sum_{x=0}^{N-1} |x\rangle\, |0\rangle. \tag{2}$$

The evolution of the computer in the first instance designed to compute and store in R_E not $E(x)$, but rather the bin-function, $E_{\text{bin}}(x)$:

$$E_{\text{bin}}(x) \equiv \begin{cases} E_l, & E(x) \le E_l \\ E_h, & E(x) > E_l, \end{cases} \tag{3}$$

where the "low" and "high" values E_l and E_h may be adjusted externally. The bin position E_l is chosen using $n_C(E)$ so that $n_l \equiv n_C(E_l) \simeq N/4$ (and therefore $n_h \simeq 3N/4$). In the binning process the total quantum state of the two registers can now be re-partitioned into two orthonormal states:

$$|\Psi\rangle = \frac{1}{\sqrt{N}} \sum_{x=0}^{N-1} |x\rangle \otimes |E_{\text{bin}}(x)\rangle = \sum_{b=l,h} c_b\, |\phi_b\rangle \otimes |E_b\rangle, \tag{4}$$

where, for $b = l, h$ (n_b is the number of members of the set $\{x_j^b : E_{\text{bin}}(x_j^b) = E_b\}$),

$$|\phi_b\rangle = \frac{1}{\sqrt{n_b}} \sum_{E_{\text{bin}}(x_j^b)=E_b} |x_j^b\rangle, \quad c_b = \sqrt{\frac{n_b}{N}}. \tag{5}$$

Creating the histogram state with approximately $N/4$ states in the lower bin of interest allows us to take advantage of the special one-step search property of Grover's operator,

$$U_G = -\,(1 - 2\,|\Psi\rangle\langle\Psi|)\,(1 - 2|E_l\rangle\langle E_l|)\,. \tag{6}$$

Evolving the state according to one application of U_G gives:

$$|\Psi'\rangle = U_G|\Psi\rangle = \sum_{b=l,h} c'_b\, |\phi_b\rangle \otimes |E_b\rangle, \tag{7}$$

with

$$c'_b = \begin{cases} c_l\,(3 - 4c_l^2), & b = l \\ c_h(1 - 4c_l^2), & b = h. \end{cases} \tag{8}$$

Perfect partitioning of $n(E)$ into two bins with $c_l = 1/2$ results in exact projection of the lower quarter of states, $|\phi_l\rangle \otimes |E_l\rangle$. This is the goal we wish to achieve – systematic projection of the lowest quarter of the spectrum – but errors in $n_C(E_l)$ will thwart any attempt. As is obvious from Eq.8, the most crucial bin positioning is the target bin, E_l. To analyse the effect of errors in $n_C(E_l)$ on the bin positions, and hence the deviation of c_l from the ideal, the bin-error, ϵ, is introduced as:

$$n_l \equiv \frac{N}{4}(1 + \epsilon) \to c_l = \frac{1}{2}\sqrt{1+\epsilon}. \tag{9}$$

The bin-error is related to the error in $n_C(E)$. For $\epsilon << 1$, the probabilities of measuring states $|E_l\rangle$ in the energy register are:

$$p_l \equiv c_l'^2 = 1 - \frac{3\epsilon^2}{4} + O(\epsilon^3), \qquad p_h \equiv c_h'^2 = \frac{3\epsilon^2}{4} + O(\epsilon^3). \tag{10}$$

These equations show explicitly how we can tolerate small errors in the position of E_l. The state $|E_l\rangle$ would be found with probability $p_l = 1 - 3\epsilon^2/4$. A measured value of $E \neq E_l$

would tell us immediately that we have not succeeded in removing the higher states (or that the lowest bin was empty). After a successful measurement of $E = E_l$, the x states in the R_x register (which was undisturbed) has been reduced to the subset with $E(x) \leq E_l$. The R_x register may now be made to evolve once more with the energy register, but with a new set of bin parameters, which will again lead to the removal of all but the lower quarter of energy states. The evolution-measurement process corresponds to one step of the state reduction hybrid algorithm. With exact positioning of the $E_l^{(i)}$ (no repeats) the number of steps, k, required to reduce from $E_{\max}$ to $E_l^{(k)} \simeq \Delta E$ is $k = \log_4(N/n(\Delta E))$.

Errors associated with the bin positions that would mean repeating the sequence from the start when one measures $E \neq E_l^{(i)}$ at some iteration. Obviously, for the algorithm to proceed with virtually no repeats requires $|\epsilon| << \sqrt{2/3k}$. However, one can show that the more useful condition $\prod_i^k p_l^{(i)} \geq \frac{1}{M}$ $(M > 1)$ can be achieved with $|\epsilon| \simeq \sqrt{\frac{\log M}{3k}}$ with $O(\frac{1}{k})$ corrections.

3 A non-hybrid algorithm

In any analysis of the overall efficiency of the hybrid algorithm the computational complexity of the classical input must somehow be included in the reckoning (hidden quantum costs are yet another issue). The strategy we will follow is to incorporate a search for the correct $E_l^{(i)}$ position, a process which we assume requires $R^{(i)}$ "bin-search" steps. As we move the value of the lower bin position, let us denote by $E_l^{(i)}(j)$ the jth approximation in the bin-search process. For each j we need to know the positioning error so that the next selected value $E_l^{(i)}(j+1)$ will be closer to the true $E_l^{(i)}$. One has access to the positioning error, $\epsilon_l^{(i)}(j)$ (or equivalently the number of states in the lower bin, $n_l^{(i)}(j)$), indirectly through the probability $p_l^{(i)}(j)$ of measuring $|E_l^{(i)}(j)\rangle$. For each step j we take a sample consisting of $M^{(i)}(j)$ measurements of the energy register, and thus determine an approximation to $p_l^{(i)}(j)$. That this procedure is possible to carry out efficiently is due to the fortunate fact that the evolution of $p_l^{(i)}(j)$ always has the same form (from the expression for c_l').

At the ith iteration it suffices to start the bin-search at $E_l^{(i-1)}$ and ensure that $M^{(i)}(j)$ is large enough to detect the evolution of $p_l^{(i)}(j)$ as $E_l^{(i)}(j)$ is changed. Near the exact $E_l^{(i)}$ the number of sampling points can be increased to achieve a certain accuracy, $\epsilon^{(i)}$. The systematic binning process has ensured that we are no longer dealing with probabilities of the order of $1/N$. Each measurement in the current iteration involves the measurement of the lower bin of each preceding iteration – a fast "cascade" of projections down from the $|x\rangle$ initialization – the previous bin positions $E_l^{(i)}$ have all been determined to high accuracy and so the cascading process will only rarely lead to a repeat.

The quantum minimization algorithm is as follows:

1. ith iteration:
   - (a) $j = 0$: set $E_l^{(i)}(0) = E_l^{(i-1)}$
   - (b) jth move: shift to $E_l^{(i)}(j) = E_l^{(i)}(j-1) - \delta E(j)$
   - (c) Make $M^{(i)}(j)$ measurements, each requiring i steps from the previous iterations:
     - (i) $|x\rangle$ initialization
     - (ii) Cascade: for $p = 1 \ldots i-1$

- $E_{\rm bin}^{(p)}(x)$ evaluation
- $U_G^{(p)}$ evolution
- $|E_l^{(p)}\rangle$ measurement ($p_l^{(p)} \simeq 1$)
- If $E \neq E_l^{(p)} \rightarrow$ go to i.

(iii) Measurement: $|E_b^{(p)}\rangle \rightarrow$ add to probability data sample.

Determine $p_l^{(i)}(j)$ from data sample and hence the next move $\delta E(j+1)$ accordingly.

(e) Repeat from (b) until $p_l^{(i)}(j) \simeq 1$ within tolerance

2. Set $i \rightarrow i+1$ and go to 1, while $i \leq k$.
3. Stop when $E_l^{(k)} \simeq \Delta E$.
4. Use Grover search, or continue moving $E_l^{(i>k)}$ down, to find the global minimum.

The algorithm is initiated with $E_l^{(1)}(0)$ arbitrary, but preferably in the region of the maximum energy. Factoring in all the previous reduction steps, the total number of measurement steps, $S_{\rm tot}$, required to reduce the superposition to the neighbourhood of the global minimum is $S_{\rm tot} = \sum_{i=1}^{k} i\, S(i)$, where the total measurement steps count for each iteration is given by $S(i) = \sum_{j=1}^{R^{(i)}} M^{(i)}(j)$. An analysis gives $S_{\rm tot} \simeq \frac{\beta}{2} k^3$, where $\beta >> 1$ and depends on the problem in question.

4 Conclusion

The complexity of the non-hybrid algorithm implies that exploitation of density of states information can be absorbed into a polynomial correction, which means that the Grover limit must assert itself somewhere in the algorithm. Most likely this occurs in the dynamic implementation of the Grover operator, $U_G^{(i)}$. However, the basic algorithm framework may still prove to be useful for specific problems for which structure might possibly be helpful in constructing $U_G^{(i)}$. Work on problems along these lines is in progress.

The author thanks M. Nielson, R. Clark, G. Milburn, B. McKellar and C. Wellard for providing critical feedback and support. This work was supported by the Special Research Centre for Quantum Computer Technology.

References

1. L.K. Grover, *Phys. Rev. Lett.* **79**, 325 (1997).
2. C. Dürr and P. Høyer, "A Quantum Algorithm for finding the Minimum", quant-ph/9607014.
3. T. Hogg and D. Portnov, "Quantum Optimization", quant-ph/0006090.
4. J. Brooke, D. Bitko, T.F. Rosenbaum and G. Aeppli, Science **284** (1999) 779.
5. E. Farhi, J. Goldstone and S. Gutman, "Quantum Computation by Adiabatic Evolution", quant-ph/0001106.
6. C.H. Bennett, E. Bernstein, G. Brassard, and U. Vazirani, quant-ph/9701001.
7. M. Boyer, G. Brassard, P. Høyer and A. Tapp, Fortschr. Phys. **46** (1998), 493.
8. C. Zalka, "Grover's Qusntum Searching Algorithm is Optimal", quant-ph/9711070.

SIMPLE QUANTUM ALGORITHMS WITH AN ELECTRON IN A PENNING TRAP

G. CIARAMICOLI, I. MARZOLI, P. TOMBESI, D. VITALI

Dipartimento di Matematica e Fisica and INFM, Università di Camerino, 62032 Camerino, Italy
E-mail:vitali@campus.unicam.it

We show how a single trapped electron in a Penning trap offers the opportunity to realize universal quantum logic gates within the present experimental possibilities. Three-qubit networks con be realized by using the quantized cyclotron and axial motion, and the electron spin.

Recently an electron in a Penning trap has drawn the attention as a potential candidate for quantum logic operations [1]. This claim is based on the very low decoherence and on the experimental accuracy achieved so far. Just to mention a few facts, this system has been designed to perform the most precise measurement of the electron g factor [2] and other fundamental constants. Radiative damping is negligible, also because of cavity effects which prevent emission of synchrotron radiation [3,4]. Moreover, transitions induced by blackbody radiation becomes extremely unlikely below 1 K [3]. All this, very recently, has lead to the nondestructive observation of Fock states for the cyclotron motion [3].

In our scheme we encode the three qubits in the cyclotron, axial and spin degrees of freedom of the electron. This choice is determined by the fact that these are the only motions that can be prepared in a truly quantum regime (the cyclotron motion has been recently prepared in its ground state while it seems possible in near future to cool down also the axial motion in its ground state).

The electron state preparation and manipulation relies on the anharmonicity of the cyclotron and axial motion due to small relativistic and electrostatic corrections. Once we can resolve among different cyclotron and axial transitions, using appropriate interactions among the different motions, conditional logic gates, like the CNOT, become feasible. Furthermore, it is possible to realize one-qubit rotations. With these elements in hand, one can already build up a small network. Indeed, three qubits are enough to implement the Deutsch-Jozsa algorithm, the Grover algorithm, the quantum Fourier transform and error correction schemes to test the performance of this prototype of quantum processor.

A single electron in a Penning trap can be described in fact in terms of four independent degrees of freedom, three bosonic (cyclotron, magnetron, and axial, with annihilation operator a_c, a_m, a_z, respectively) and the spin, that is,

$$H_0 = -\hbar\omega_m \left(a_m^\dagger a_m + \frac{1}{2}\right) + \hbar\omega_c' \left(a_c^\dagger a_c + \frac{1}{2}\right) + \hbar\omega_z \left(a_z^\dagger a_z + \frac{1}{2}\right) + \frac{\hbar}{2}\omega_s\sigma_z. \tag{1}$$

The magnetron motion is unstable; however, this instability is fully under control, because the magnetron motion is weakly coupled to the environment. Hence, the electron does not roll down the potential hill, but remains well confined inside the trap for several months. The three motions build up a well defined hierarchy in the frequency range $\omega_m \ll \omega_z \ll \omega_c'$. Typical values [4] of the eigen frequencies are $\omega_m/(2\pi) \simeq 12$ kHz, $\omega_z/(2\pi) \simeq 64$ MHz and $\omega_c'/(2\pi) \simeq 164$ GHz.

The energy levels of the electron are therefore

$$E_0(n,k,l,s) = -\hbar\omega_m \left(l+\frac{1}{2}\right) + \hbar\omega_c' \left(n+\frac{1}{2}\right) + \hbar\omega_z \left(k+\frac{1}{2}\right) + \frac{\hbar\omega_s}{2} s\,, \tag{2}$$

where the four quantum numbers $l = 0,1,2,\ldots$, $n = 0,1,2,\ldots$, $k = 0,1,2,\ldots$, and $s = \pm 1$ define, respectively, the magnetron, cyclotron, axial, and spin states.

In typical experimental configurations [4], the axial harmonic oscillator is coupled to an external detection circuit, which heats up this motion. However, switching off the interation with the external circuit and applying sideband cooling, it should be possible to prepare the axial motion in the ground state. The same conclusion applies to the cyclotron motion, which has been recently prepared in its lowest Fock state. The situation is, instead, quite different for the magnetron motion whose temperature, controlled by sideband cooling technique, is generally of 4.2 K, with $\langle l \rangle \approx 10^3$.

We recall that if the trap is decoupled from the external circuit, the lifetime of the first axial and cyclotron energy levels is about of 10^6s and 10s respectively. We will focus only on the true quantum motions of the electron, namely the cyclotron and axial oscillators and the spin, with the aim of using them as registers for quantum information. The electron spin, being a two-level system, lends itself quite naturally to encode the logical states $|0\rangle$ and $|1\rangle$. Instead, it is not so obvious to implement a qubit in a multilevel system like a harmonic oscillator with equally spaced energy levels. This is, unfortunately, the case of the cyclotron and axial motions. A possible way to circumvent this problem is to introduce some kind of anharmonicity in order to address a specific transition between two energy levels.

Actually, before perturbing this simple system from outside, we can consider the intrinsic sources of anharmonicity. Such an analysis has been already done [4]. The main corrections to the Hamiltonian Eq. (1) are produced by the relativistic effects and by the departures of the actual electrostatic field from the ideal quadrupole potential. These corrections make the three oscillators, describing the electron motion inside the trap, anharmonic, i.e., their energy levels are no longer equally spaced.

For the cyclotron oscillator the transition frequency between neighboring levels of quantum numbers $n+1$ and n, in typical experimental configurations, becomes

$$\omega_c(n,s) \equiv \frac{E(n+1,k,l,s) - E(n,k,l,s)}{\hbar} \approx \omega_c' - \delta(n+1) - \frac{\delta}{2} s\,, \tag{3}$$

where $\delta \equiv \hbar\omega_c^2/(mc^2)$. Equation (3) shows that the cyclotron transition frequency depends on the quantum numbers n and s. Hence, the relativistic effects turn on an interaction between the cyclotron and the spin motion. Consequently, under the same assumptions, we find that the spin flip frequency

$$\omega_s(n) \equiv \frac{E(n,k,l,+1) - E(n,k,l,-1)}{\hbar} \approx \omega_s - \delta\left(n+\frac{1}{2}\right) \tag{4}$$

depends on the cyclotron excitation n.

One has $\delta/2\pi \approx 200$ Hz, which is a very small shift compared to the characteristic cyclotron and spin frequencies, which are of the order of 160 GHz. However, cyclotron and spin individual transitions are resolvable because the corresponding natural linewidths, due to radiative decay [4], are much narrower.

The transition frequency between neighboring axial levels is

$$\omega_z(k) \equiv \frac{E(n,k+1,l,s) - E(n,k,l,s)}{\hbar} \approx \omega_k + \delta_1(2k+1), \tag{5}$$

where δ_1 depends upon the deviation from the ideal quadrupole potential, and it can be increased enhancing the potential of the compensation electrodes (which in this case act actually as decompensation electrodes). With the experimental parameters, one can have a frequency difference between neighboring transitions of axial energy levels of about 200 Hz. This difference is larger than the natural linewidth of the axial levels so that one can control single axial transitions applying electromagnetic pulses with narrow frequency width.

For the quantum control of the various degrees of freedom, external electromagnetic pulses can be applied. The cyclotron motion can be manipulated using an external microwave field. For example, by using a multifrequency pulse tuned to $\omega_c' - \delta/2$ and $\omega_c' - 3\delta/2$, with the appropriate intensity and duration, it is possible to control the various cyclotron states, leaving the spin unaffected.

The manipulation of the spin is instead obtained by applying, inside the trap, a small magnetic field **b**, which lies in the xy plane and oscillates at a frequency close to the spin resonance ω_s [4]. Here again using double-frequency pulses, with frequencies $\omega_s - \delta/2$ and $\omega_s - 3\delta/2$ and with appropriate durations, one can perform any spin rotation. Moreover, it is important to note that one can also execute transformations on the spin states controlled by the cyclotron state. This suggests the possibility of performing conditional logic operations, which involve the cyclotron and the spin states, in a very straightforward way. To act on axial energy levels transitions we can apply an oscillating potential in the trap containing the term $zV_2cos(\omega t - \beta)$, and in this way one simply realizes full control of the axial degrees of freedom.

However, the main prerequisite for quantum information processing is to realize quantum gates. A feasible way to achieve conditional quantum dynamics between all the different electronic degrees of freedom is to apply an inhomogeneus static magnetic field. Infact, with an appropriate dependence on the spatial coordinates, a static magnetic field can induce shifts on the transition frequencies depending on the state of the electron motions. If we consider the additional static magnetic field

$$\mathbf{\Delta B} = B_2 \left[\left(z^2 - \frac{x^2+y^2}{2} \right) \hat{k} - z(x\hat{\imath} + y\hat{\jmath}) \right], \tag{6}$$

we obtain the following transition frequencies,

$$\omega_c(n,s,k) \approx \omega_c' - \delta(n+1) - \frac{\delta}{2}s + \delta_2[k + \frac{1}{2}], \tag{7}$$

$$\omega_z(k,s,n) \approx \omega_k + \delta_1(k+1) + \delta_2(n + \frac{1}{2} + \frac{s}{2}), \tag{8}$$

$$\omega_s(n) \approx \omega_s - \delta\left(n + \frac{1}{2}\right) + \delta_2(k+1), \tag{9}$$

where $\delta_2 \propto B_2$. As we can see the transition frequencies depend on the quantum numbers describing the state of the electron motion. We note that, in addition to the degeneracies due to the relativistic effects, there are new degeneracies in the transition frequencies. Indeed the transitions $|k,0,\uparrow\rangle \leftrightarrow |k+1,0,\uparrow\rangle$ and $|k,1,\downarrow\rangle \leftrightarrow |k+1,1,\downarrow\rangle$ have the same frequencies $\omega_k + \delta_1(k+1) + \delta_1$. The same happens for the t wo transitions $|k,1,\uparrow\rangle \leftrightarrow |k+1,1,\uparrow\rangle$ and

$|k,2,\downarrow\rangle \leftrightarrow |k+1,2,\downarrow\rangle$, with the frequency $\omega_k + \delta_1(k+1) + 2\delta_1$ and so on. This transition degeneracy slightly complicates the scheme, because it implies that a given pulse can have additional, undesired effects. The consequence of this is that in some cases, additional pulses are needed to eliminate these undesired transitions.

Using the above interactions it is possible to implement all the possible one and two-qubit universal operations, in the case where the logical qubits are realized by the spin and by the two lowest states ($|n = 0,1\rangle$) of the axial and the cyclotron degrees of freedom. In the case of the axial and spin qubit one has simply to use an appropriate multifrequency pulse to realize the most general rotation. Transition degeneracy makes the situation less straightforward for the cyclotron qubit because in this case simple resonant multifrequency pulses also flip the spin states, so that they have to be supplemented with additional and appropriate spin flips.

Then in the network it is possible to set up all the various C-NOT gates and all of them can be achieved with few multifrequency pulses. It is possible to see that in this case transition degeneracy does not create too much problems and the average number of pulses per C-NOT gate is 3.3. An interesting aspect of the scheme is that three-qubit gates, such as the Toffoli gate, which is useful for quantum error correction codes, and the three-qubit conditional phase shift gate, used in the Grover algorithm [5], can be very easily implemented with two and one multifrequency pulse, respectively.

Using these facts it is possible to see that simple quantum algorithms that can be implemented with only three qubits could be easily implemented employing a limited number of pulses. For example, the Grover search within 8 objects can be implemented with 44 pulses, the 3-qubit quantum error correction code with 26 pulses, and the 3-qubit discrete Fourier transform with 20 pulses.

Another important condition for quantum information processing is quantum state detection. In typical single electron Penning trap experiments, what is commonly measured are axial frequency shifts. In the presence of inhomogeneous magnetic fields as that of Eq. (6), these shifts depend on the spin and cyclotron state, and therefore they can be used to measure directly the state of these two qubits. There is instead no way to directly measure the axial qubit. However, using multifrequency pulses as those employed for the gate operations, it is possible to map the axial states into spin and and cyclotron states, and therefore measure them too. To be more precise, one can first map the spin state involving also the second and third cyclotron excited states, for which quantum control has been already demonstrated [3]. In this way the spin qubit is "ready" and the axial state is transferred coherently to the spin qubit. In this way one has a perfect mapping of the the relevant eight-dimensional Hilbert state, in which each state can be measured. This mapping can be efficiently performed using 7 pulses only.

References

1. S. Mancini, A. M. Martins, and P. Tombesi, Phys. Rev. A **61**, 012303 (2000).
2. R. S. Van Dyck,Jr., P. Schwinberg, and H. Dehmelt, Phys. Rev. Lett. **59**, 26 (1987).
3. S. Peil and G. Gabrielse, Phys. Rev. Lett. **83**, 1287 (1999).
4. L. S. Brown and G. Gabrielse, Rev. Mod. Phys. **58**, 233 (1986).
5. L. K. Grover, Phys. Rev. Lett. **79**, 325 (1997).

SMALL SCALE QUANTUM ALGORITHMS

B. C. TRAVAGLIONE, G. J. MILBURN
Centre for Quantum Computer Technology,
University of Queensland,
St Lucia, Queensland, Australia
E-mail: btrav@physics.uq.edu.au

It has been shown theoretically that a system, obeying the laws of quantum mechanics, can be manipulated in such a way as to perform certain calculations exponentially faster than can be done on a classical computer. However, it remains to be seen if it is experimentally feasible to implement such systems. The first step towards this is implementing small scale quantum algorithms. We discuss modifications to Kitaev's algorithm, such that it can be implemented on relatively small quantum systems. We give the examples of an ion trap implementation and an optical implementation.

1 Introduction

The race is on to develop the technologies to build a quantum computer. Much of the fuel for this race stems from Shor's algorithm[1], which allows a quantum computer to factor integers exponentially faster than can currently be done on a classical computer. The factoring of large integers allows the cracking of the RSA encryption system[2]. However, to crack encryption codes, we will require a quantum computer with many hundreds of qubits. A stepping-stone to this ultimate goal will be the experimental implementation of small scale quantum algorithms. One algorithm which lends itself to small scale implementation is the generalized Shor's algorithm, known as Kitaev's algorithm[3,4,5].

I will begin by defining what I mean by a small scale quantum algorithm, and give some reasons as to why we would want to study them. I will then introduce and briefly discuss Kitaev's algorithm. This will lead me on to discuss some modifications to the algorithm which will allow it's application on 'small' quantum systems. Finally I will describe two proposed implementation schemes, one based on an ion trap quantum computer, the other on an optical system.

2 Small Quantum Algorithms

As we all know, the main goal of quantum computation is to perform calculations on a quantum computer which would take us a great deal longer to accomplish on a classical computer. The most renowned of the current quantum algorithms is Shor's factoring algorithm. A great deal of the interest in Shor's algorithm stems from the fact that it can be used to crack the RSA encryption system. However, a quantum computer which could crack current encryption codes would have to contain many hundreds of qubits.

As we have seen this week, building such a device will indeed be a daunting task. A stepping-stone to this ultimate goal is to experimentally implement quantum algorithms on systems with limited dimensionality. What do I mean by limited dimensionality? Consider an (as yet non-exist) quantum computer consisting of around six hundred and fifty qubits. This quantum computer would be operating in a Hilbert space of dimension approximately

10^{200}. To put that number into perspective, remember that there are no more than 10^{100} particles in the entire universe. We shall define here a small scale algorithm to be one that works on a system with a Hilbert space dimensionality of only a few thousand.

3 Kitaev's Algorithm

The algorithm that we have been looking at is Kitaev's algorithm, also know as the phase estimation algorithm. The goal of Kitaev's algorithm is to obtain an eigenvalue, and an eigenstate of a unitary operator, $\hat{U}$. This task is accomplished using two qubit registers; the index register, which is used to store an eigenvalue of $\hat{U}$, and the target register, on which $\hat{U}$ acts. I have shown here a quantum circuit diagram of the algorithm. By initializing

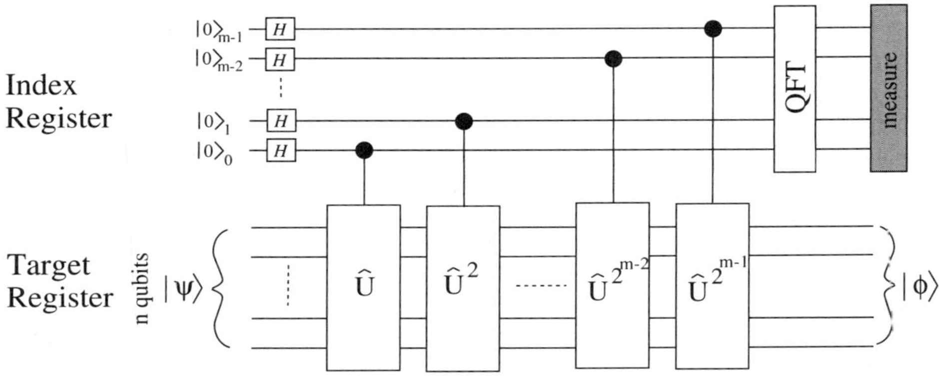

Figure 1. Quantum circuit diagram of Kitaev's algorithm.

the target system in an approximate eigenstate, and then applying the depicted sequence of gates, the index register will be measured in an eigenstate of $\hat{U}$ and the target system will be left in the corresponding eigenstate. For a more complete description of Kitaev's algorithm, please see either of the following references [3,6].

4 Modifying Kitaev's Algorithm

Kitaev and others have shown that the algorithm performs effectively when the index register has dimensionality greater than or equal to the square of the target system dimensionality. However, if we are wishing to perform the algorithm on small quantum systems, we do not have the luxury of using such a large index system. We were able to show[7] that Kitaev's algorithm *can* work if the index register dimensionality is far smaller than that of the target system. However, the effectiveness is dependent upon the initial approximate eigenstate of the target system, and the eigenvalue spectrum of the operator, $\hat{U}$. Having shown that the algorithm can be performed on a small quantum system, we investigated two different implementation schemes.

5 Ion trap Implementation

The first implementation scheme that we looked at was the ion trap quantum computer[8]. Here, we use the ground and first excited electronic levels of each ion as the index register qubits. The target system is the quantized motion of the ions.

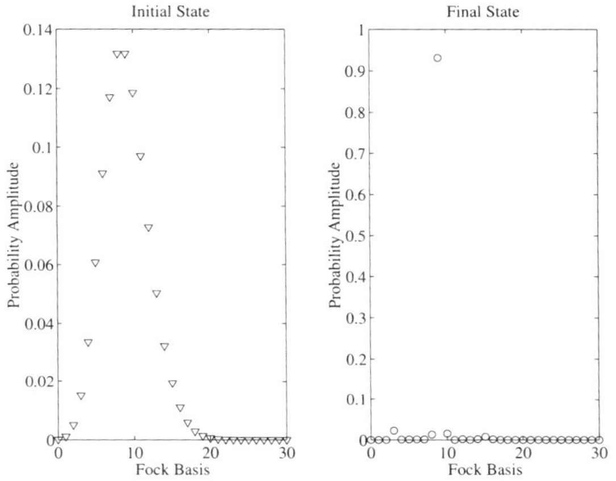

Figure 2. Fock state distributions for the target system initially in a coherent state with $\alpha = 3$, and the state of the system after applying the phase estimation algorithm and measuring the four index qubits.

Choosing $\hat{U}$ to be the number operator, we have shown numerically that the algorithm can generate phonon number states[7]. Fig. 2a shows the ions initially in a motional coherent state. After applying a series of laser pulses, and then measuring the electronic levels of each ion, the motion is reduced to a phonon number state (Fig. 2b). You will recollect that I mentioned the algorithm's performance is dependent on $\hat{U}$. Whilst eigenstates of the number operator can be effectively generated, the algorithm breaks down if we try to generate eigenstates of the displacement operator.

6 Optical Implementation

In the optical scheme, we move away from qubit systems, and use quantum simple harmonic oscillator modes for both the target system and the index system. The standard qubit computation basis is replaced by the position basis of the index mode. Choosing $\hat{U}$ to be $e^{i\hat{p}_B}$, we find that the algorithm can be approximated using parametric amplifiers (squeezers) and a beam splitter, as depicted in Fig. 3. We also find in this situation that the algorithm is equivalent to performing a quantum non-demolition measurement on the momentum quadrature of the target mode[9]. Although the effectiveness of the algorithm is limited by the amount of squeezing, we have shown numerically that interesting effects are attainable with experimentally realizable squeezing.

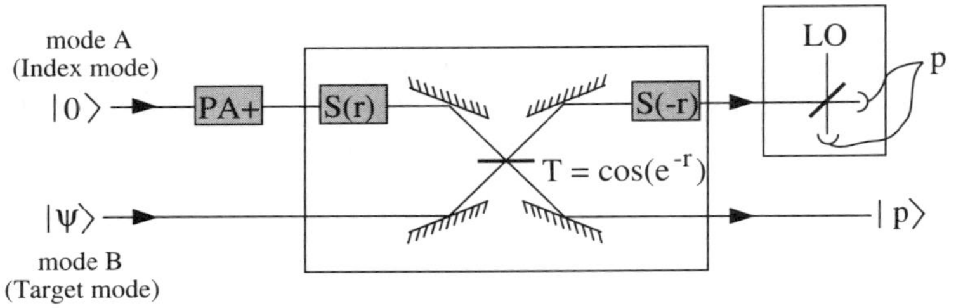

Figure 3. Optical circuit diagram of the continuous version of Kitaev's algorithm, with $\hat{U} = e^{i\hat{p}_B}$.

7 Conclusion

In conclusion, we have shown that a small scale version of Kitaev's algorithm can be implemented in both an ion trap quantum computer and an optical quantum computer. In the ion trap scheme, the algorithm can be used to generate motional number states, whilst in the optical scheme, the algorithm can be used to perform quantum non-demolition measurements.

Acknowledgments

During the preparation of this work BCT had many fruitful discussions with T.C. Ralph, and useful email exchanges with S. Lloyd.

References

1. P. W. Shor, Proc. 35th Annual Symposium on Foundations of Computer Science p. 124 (1994).
2. R. L. Rivest, A. Shamir, and L. Adleman, Comm. ACM **21**, 120 (1978).
3. A. Y. Kitaev, quant-ph/9511026.
4. D. S. Abrams and S. Lloyd, Physical Review Letters **83**(24), 5162 (1999).
5. S. Lloyd, *Hybrid quantum computing* (2000), to be published.
6. R. Cleve, A. Ekert, C. Macchiavello, and M. Mosca, Proc. Roy. Soc. London A **454**(1969), 339 (1998).
7. B. C. Travaglione and G. J. Milburn, Physical Review A **63**, 032301 (2001).
8. C. A. Sackett, D. Kielpinski, B. E. King, C. Langer, V. Meyer, C. J. Myatt, M. Rowe, Q. A. Turchette, W. M. Itano, D. J. Wineland, and C. Monroe, Nature **404**(6775), 256 (2000).
9. B. C. Travaglione, G. J. Milburn, and T. C. Ralph, *Optical implementation of Kitaev's algorithm* (2001), in preparation.

OBSERVATION OF PHOSPHORUS ATOMS ON SILICON SURFACES

S. R. SCHOFIELD, J. L. O'BRIEN, M. Y. SIMMONS, N. J. CURSON, R. G. CLARK

Centre for Quantum Computer Te chnology, School of Physics, University of New South Wales, Sydney Australia

I. ANDRIENKO AND S. PRAWER

Centre for Quantum Computer Te chnology, University of Melbourne, Parkville, Victoria, 3052

The Kane quantum computer architecture requires the formation of a regularly spaced array of P atoms embedded in silicon as qubits. A useful tool in the fabrication of such an array is the ability to detect single P atoms at and below the surface of the silicon substrate. Scanning tunnelling microscopy (STM) and current imaging tunnelling spectroscopy (CITS) are capable not only of detecting the locations of P atoms, but also of determining the electronic c haracteristics of the P atoms in the matrix. Here we present t wo results on the detection of P atoms silicon surfaces using these methods. In one set of experiments, phosphorus doped silicon (001) wafers were annealed at high temperature and the reconstructed surface was imaged to reveal P dimers which had formed on the surface. In a parallel set of experiments, phosphorus doped silicon (111) wafers were cleaved *in-situ* and the cleaved surface was imaged at atomic resolution revealing single P atoms.

1 Introduction

Since the proposal of a silicon based quantum computer was first put forward by Kane,[1] there has been considerable effort applied to determine how such a device may be built. One such method is to fabricate an atomically precise phosphorus array on the (001) surface of silicon using a scanning tunnelling microscope (STM) tip,[2,3] with subsequent silicon overgrowth to embed the phosphorus dopant array. Another is to use ion implantation to place the P atoms in silicon. The ability to identify individual phosphorus atoms on, and incorporated into, the silicon surface, is of crucial importance to both these fabrication strategies. In this letter we report STM results of the (001) and (111) silicon surfaces, and current imaging tunnelling spectroscopy (CITS) results of the (001) surface. In particular, we use these techniques to detect individual phosphorus atoms incorporated into silicon.

1.1 UNSW node experiments on Si(001)

xperiments on the Si(001)2×1 surface were performed at the University of New South Wales using an OMICRON variable temperature STM and electrochemically etched tungsten tips. The base pressure of the ultra-high vacuum (UHV) chamber was $< 4 \times 10^{-11}$ mbar. Phosphorus doped Si(001) wafers with resistivities of both 0.1 and 10 Ωcm were used. Sample preparation[4] was performed in vacuum without prior *ex-situ* treatment, by outgassing overnight at 850 K, followed by flashing to 1450 K. All STM measurements were performed at room temperature.

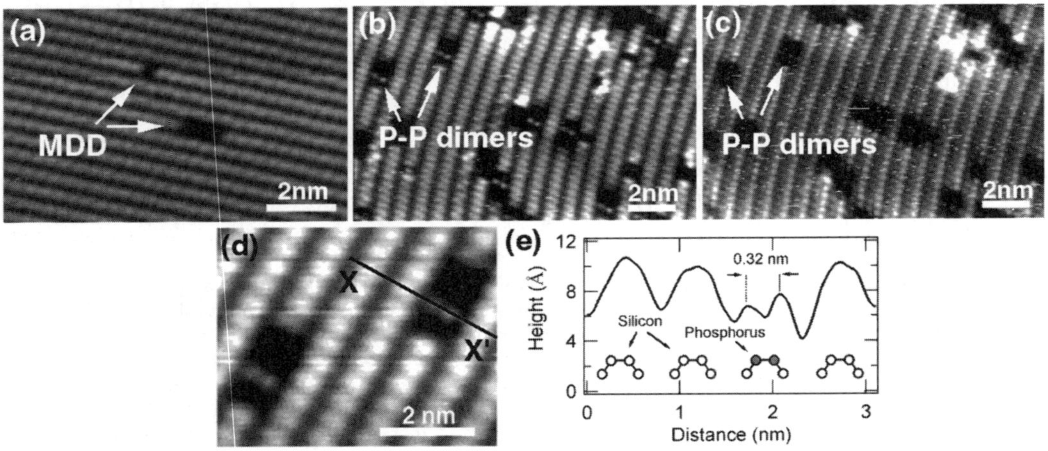

Figure 1. STM images of Si(001)2×1 surfaces. The image shown in (a) is of a newly prepared sample in which two missing dimer defects (MDD) are visible. Images (b) and (c) show a surface from the same wafer after extended annealing. We propose the pairs of circular protrusions in (b) and (c) are P-P dimers in defect sites. These two images show the same area of the surface, imaged at two different sample biases, (b) −1.3 V and (c) +1.3 V. Image (d) is an enlarged region of (b) and the line scan (e) is taken across the line $X - X'$ indicated in (d).

2 Results and Discussion

2.1 Annealed Si(001)

Figure 1(a) shows a filled state (negative sample bias) STM image of a clean Si(001)2×1 surface.

The "bean-shaped" protrusions visible in this image are Si–Si dimers, consisting of two Si atoms bonded together by a strong σ-bond and a weaker π-bond. When imaging Si–Si dimers, the tunnelling current in the filled state STM images arises from the π-bond, giving the dimers their characteristic bean-shape.[5] The area shown also contains two missing dimer defects (MDD), as indicated. Figures 1(b) and 1(c) show simultaneously acquired filled- and empty-state STM images of a Si(001)2×1 surface that has undergone a series of heat treatments to temperatures in the 600 – 1200 K range, over a period of weeks. Before this heat treatment the surface appeared much like the one shown in Fig. 1(a) and was cleaved from the same wafer. The filled state image, Fig. 1(b), contains not only the bean-shaped protrusions of Si–Si dimers, and the dark regions of MDD's, but also a third species that appear as pairs of circular protrusions located in MDD structures. The empty state image, Fig. 1(c), shows the same surface region as Fig. 1(b). In this image, the circular protrusions present in Fig. 1(b) appear quite dark and can only just be resolved, while the Si–Si dimers have their characteristic nodal appearance due to tunnelling into the π^*-antibonding state. In this letter, we speculate that the pairs of circular protrusions present in Figs. 1(b) and 1(c) are surface segregated P dopants that are located at MDD sites on the surface in P–P dimer configurations. Figure 1(d) shows a higher resolution filled state image where these protrusions are present. The line profile in Fig. 1(e) shows that the separation between two

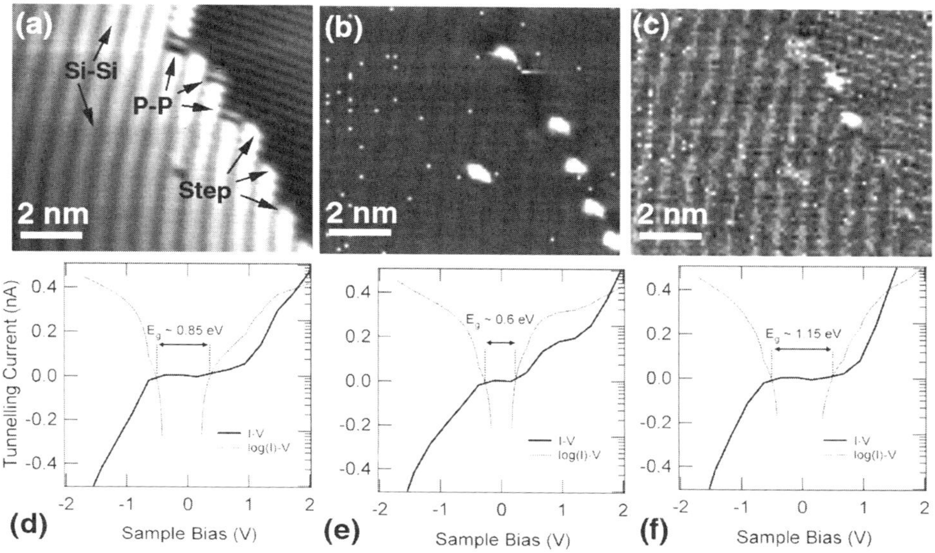

Figure 2. Current imaging tunnelling spectroscopy data of P-P dimers at a step edge on Si(001)2×1. The surface topography is shown in the STM image (a). Current images taken at +0.68 V and -1.17 V sample bias are shown in (b) and (c), respectively. Tunnelling spectroscopy curves are shown for (d) a surface Si–Si dimer, (e) a step edge Si–Si dimer and (f) a step edge P–P dimer.

protrusions in Fig. 1(d) is 3.2 Å, in agreement with the observation of P-P dimers by Wang *et al.*[6] In this work, Wang *et al.*,[6] studied a clean Si(001)2×1 surface that was saturation dosed with phosphine (PH_3) gas while heating the surface to 825 K. They observed pairs of circular protrusions with a separation between protrusions of 3.2 Å, and attributed this to tunnelling out of the P lone pair electrons of P-P dimers.[6]

In order to determine more information about these possible P–P dimers we have employed the CITS technique[7] to perform atomically resolved tunnelling spectroscopy of a Si(001)2×1 surface exhibiting these features.

In addition to observing P–P dimers in MDD's, we have also observed P–P dimers at step edges, as shown in Fig. 2. The STM image in Fig. 2(a) shows two P–P dimers at a single layer step edge on a Si(001)2×1 surface, and a third P–P dimer near the step edge. Figures 2(b) and 2(c) show current images at sample biases of +0.68 V and -1.17 V, respectively, acquired simultaneously with the STM image of Fig. 2(a). In the positive sample bias current image, Fig. 2(b), the Si–Si dimers at the step edge appear bright, while the P–P dimers at the step edge appear dark. Conversely in the negative sample bias current image, Fig. 2(c), the P–P dimers at the step edge appear bright while the Si–Si dimers at the step edge appear no different than the surrounding surface silicon dimers. Figures 2(d) – 2(f) show current-voltage (*I-V*) curves for three different surface species; Fig. 2(d) a surface Si–Si dimer, Fig. 2(e) a step edge Si–Si dimer, and Fig. 2(f) a step edge P–P dimer. The surface band gaps shown on these plots have been determined by the points where the magnitude of the tunnelling current drops below 0.01 nA. All these curves are quite distinct from one another. The *I-V* curves taken for the surface Si–Si dimer, Fig. 2(d), and the step

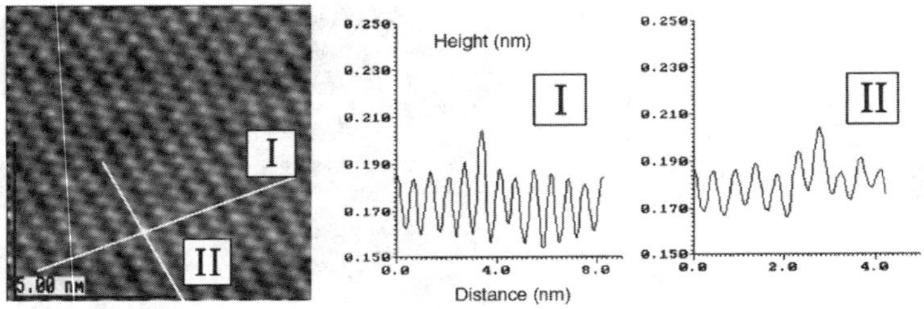

Figure 3. High resolution STM image of the Si(111)2×1 surface obtained by cleavage in UHV. Sample bias +2.0 V, tunnelling current 0.13 nA. Scan profiles through a bright feature (P atom) are also shown, see text for details.

edge Si–Si dimer, Fig. 2(e), are in excellent agreement with those measured in a study of the spectroscopy of step edges on Si(001)2×1 by by Komura *et al.*[8] However, the *I-V* curve for the P–P dimer is unlike any of the curves observed Komura *et al.*[8] The band gap of 0.85 eV for the Si–Si dimers is also in perfect agreement with the results of Wang *et al.*[6] Wang *et al.*[6] also studied the *I-V* characteristics of a completely P terminated Si(001) surface and found the surface band gap to be ~2.0 eV. We find the surface band gap for a P–P dimer at a step edge is 1.15 eV. It is expected that the surface band gap for isolated P–P dimers at step edges will differ from the band gap for P–P dimers of a monolayer P terminated surface, however, we note that in both cases the trend is for an increased band gap for P–P dimers over the band gap for Si–Si dimers. These results lend support to our assignment of the pairs of circular protrusions being surface P–P dimers, since their electronic signature is unlike the electronic signature of either surface of step edge Si–Si dimers as observed in this, and previous studies.

2.2 Cleaved Si(111)

Studies of single P atoms on a Si(111) surface were carried out at the University of Melbourne using a JEOL JAFM-4500XT scanning probe microscope in a UHV chamber with base pressure of $< 2 \times 10^{-10}$ mbar. Silicon samples 3 × 7 × 0.3 mm were cut from (111) n-type (P-doped) wafer having a resistivity of 1.0 Ωcm, and cleaved, *in-situ*, in the $[2\bar{1}\bar{1}]$ direction. Again, electrochemically etched tungsten tips were used. Samples were cleaved *in-situ* so that we could image the cleaved surface using the STM tip to investigate the presence of single P dopants in the silicon matrix.

Figure 3 represents a typical STM image of the cleaved Si(111) surface exhibiting the metastable Si(111)2×1 reconstruction, and where a single P atom is identified.

Scan profiles (I) across atomic rows and (II) along the row show that the P atom appears as a protrusion at positive sample bias voltage, as observed by Trappmann *et al.*[9] Since the surface is prepared by cleavage in UHV without further treatment, STM images show the inherent dopant distribution inside the host crystal. This means that atomic resolution STM of the cleaved Si(111)2x1 surface can provide information about the distribution of singly substitutional P atoms in the Si matrix.

3 Summary

In summary, using STM we have observed single phosphorus dopant atoms on both the (001) and (111) surfaces of silicon. The electronic signature of a P-P dimer at a Si(001)2×1 step edge has been observed using current imaging tunnelling spectroscopy. The techniques that we have identified here to locate and identify P atoms incorporated into Si surfaces will be of crucial importance for the reliable fabrication of the Kane quantum computer.

Acknowledgments

This work was funded by the Australian Research Council, and the US National Security Agency through the Army Research Office.

References

1. B. E. Kane, Nature **393**, 133 (1998).
2. J. L. O'Brien *et al.*, IQC01 conference proceedings (Th2) Sydney Australia 2001.
3. J. L. O'Brien *et al.*, to appear in Phys. Rev. B Rapid Comm. (see cond-mat/0104569)
4. B. S. Swartzentruber, Y. W. Mo, M. B. Webb, and M. G. Lagally, J. Vac. Sci. Technol. A **7**, 2901 (1989).
5. K. Hata, S. Yasuda, and H. Shigekawa, Phys. Rev. B **60**, 8164 (1999).
6. Y. Wang, X. Chen, and R. J. Hamers, Phys. Rev. B **50**, 4534 (1994).
7. R. J. Hamers, R. M. Tromp, and J. E. Demuth, Phys. Rev. Lett. **56**, 1972 (1986).
8. T. Komura, T. Yao, and M. Yoshimura, Phys. Rev. B **56**, 3579 (1997).
9. T. Trappmann, C. Sürgers, and H. von Löhneysen, Appl. Phys. A **68**, 167 (1999).

DEMONSTRATION OF SINGLE ION IMPACT ON THE Si(001) SURFACE USING A SELF-ASSEMBLED ORGANIC MONOLAYER OF 1,3,5,7-CYCLOOCTATETRAENE

I.ANDRIENKO, A.CIMMINO, D.N. JAMIESON and S. PRAWER

Centre for Quantum Computer Technology, School of Physics, University of Melbourne, Victoria 3010, Australia

We have investigated the use of self-assembled monolayers of 1,3,5,7-cyclooctatetraene (COT) for the registration of the passage of single ions implanted into silicon substrates for construction of arrays of single atoms with possible application to quantum computing in the solid state. This paper describes a method for the synthesis of COT monolayers bonded to a silicon substrate to give a film with a low defect density. It is also shown how topological changes to the film result from ion bombardment and that the observed structures are consistent with the impact of single ions. The images, obtained with scanning tunneling microscopy (STM) show that the film offers some advantages over other schemes.

1 Introduction

The Kane quantum computer will use phosphorus atoms embedded in silicon as the qubits.[1] One of the fabrication technologies for creating arrays of P atoms in a Si matrix is ion implantation.[2] Since the implantation of ions is essentially random, some method of registering the passage of a single ion is required. This can be accomplished by implantation of the ions through a thin surface layer consisting of a resist. Changes to the topological and/or electrical properties of the resist can be used to mark the site of the buried ion using high resolution scanning tunneling microscopy (STM) or atomic force microscopy (AFM).

Possible materials for the resist are C_{60} or polymethyl metacrylate (PMMA).[2] An alternative possibility is the use of self-assembled organic monolayers on Si. Some cyclic unsaturated organic molecules such as cyclopentene **(I)**, 1,5-cyclooctadiene **(COD, II)** and 1,3,5,7-cycooctatetraene **(COT, III)** form well-ordered, low defect density monolayers on Si. [3-6]

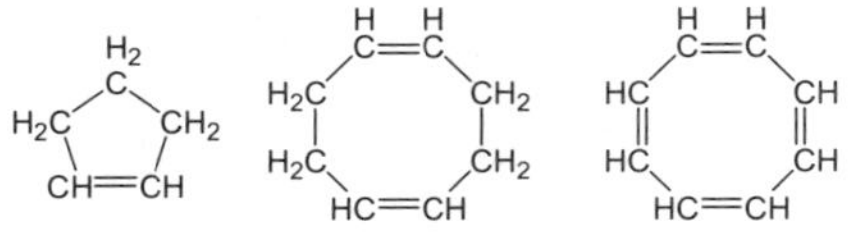

(I) (II) (III)

Figure 1. Some cyclic unsaturated organic molecules.

The monolayers are produced by so-called cycloaddition reactions between Si=Si dimers and C=C double bonds:

Figure 2. Cycloaddition reactions between Si=Si dimers and C=C double bonds.

Reactions (Fig. 2) take place at room temperature. Also, they are self-terminating, so that only a single monolayer is formed after a saturation exposure of the Si surface to vapours of the reagents.[5] Some of the most promising reagents for creating *ordered* monolayers on Si are COD and COT. The mechanism of their adsorption on the Si(001)2x1 surface has been studied thoroughly by Hovis and Hamers,[5,6] and in the case of COT it involves double cycloaddition of one molecule to two Si=Si dimers leaving two remaining C=C bonds exposed to the vacuum:

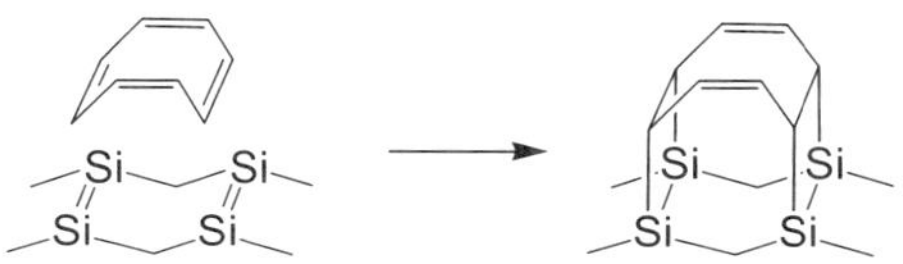

Figure 3. Double cycloaddition of one molecule to two Si=Si dimmers.

Process (Fig. 3) gives some advantages over other schemes like C_{60} and PMMA. In contrast to C_{60}, COT does not leave patches of bare silicon at the surface which cannot be distinguished from the damage produced by irradiation. COT molecules are small enough to be packed closely, and they are flexible to fit the Si(001)2x1 structure, so that each Si=Si dimer is saturated after the reaction (Fig. 3). Compared to PMMA, the COT monolayer is extremely thin, and even the smallest topological changes upon irradiation should be visible in high-resolution STM images. Thus, there is no need in chemical etching or current imaging to reveal irradiation damage to the film as it has been done previously with only limited success in PMMA films.[2]

2 Experimental section

The experiments were carried out in an ultrahigh vacuum (UHV) system consisting of different isolatable chambers. The preparation chamber contains heating stage, argon ion gun and liquid evaporation source with a variable leak valve. Pressure in this chamber can vary depending on the treatment procedure. The observation chamber contains the STM/AFM stage, and it is always kept at a base pressure of $<2x10^{-10}$ Torr.

The Si(001) samples were cut from polished wafers (P-doped, 1 Ω cm resistivity) and cleaned ultrasonically for 10 min. After insertion into the preparation chamber the samples were outgassed for extended period at 600 °C, briefly at 900 °C, and finally, annealed at 1050 °C for approximately 2 min and gradually cooled down over a period of approximately 4 min.[7] This procedure produced clean, well-reconstructed Si(001)2x1 surface if pressure in the chamber did not exceed $1.5x10^{-9}$ Torr during the procedure.

The COT was purchased from Sigma-Aldrich with guaranteed purity of 98 % and was used without additional purification. It was introduced into the preparation chamber through a leak valve. To obtain a monolayer, the Si(001)2x1 surface was exposed to 2 Langmuirs ($9x10^{-8}$ Torr for 22 s) of COT vapours at room temperature.

COT coated samples were irradiated with 2 keV Ar^+ ions. For irradiation, the chamber was filled with high purity argon at a pressure of $3x10^{-4}$ Torr, and an ion current of 0.33 nA was passed through the sample for a period of 4 s. With the effective surface area of 6 mm^2 such an exposure corresponds to the irradiation dose of approximately $1.4x10^{11}$ ions cm^{-2}.

Direct observation of adsorbed monolayers and irradiation damage was achieved using the JEOL JAFM-4500XT scanning probe microscope working in STM mode. Imaging was the most effective at sample bias voltage of about 2 V and tunneling current of 0.05-0.10 nA.

3 Results and discussion

The STM image in Figure 4(i) shows the Si(001) surface after saturation exposure to COT. The image was acquired in UHV. A single atomic height step is clearly visible. The adsorbed monolayer is quite disordered, but it has very low density of defects such as vacancies or holes. For irradiation, the sample must be transferred from UHV into an Ar atmosphere with a relatively high pressure ($\sim 10^{-4}$ Torr). To be sure that surface defects are not produced by this procedure, we obtained STM images of a COT coated surface exposed to high pressure of Ar, see Figure 4(ii). The monolayer is disordered, but at some places molecular rows are clearly visible. Numerous nm-size adsorbates appear at the surface, but there are no features like holes or pits. Figure 4(iii) shows the same surface after irradiation with Ar^+ ions (irradiation dose $1.4x10^{11}$ ions cm^{-2}). The image clearly shows defects produced by ion bombardment: nm-size holes marked by arrows. These holes appear *only after irradiation*, and their number is consistent with what we can statistically expect for a scan area 100 x 100 $nm^2 = 10^{-10}$ cm^2 under experimental conditions: 1.4 x 10^{11} ions cm^{-2} x 10^{-10} cm^2 = 14 ions. Figure 4(iv) shows scan profiles **(a-a)** along the row of adsorbed molecules at the initial surface (i) and **(b-b)** through the hole at the irradiated surface (iii). Corrugations along the row are about 2 Å in height which is in agreement with Ref. [6]. The depth of the hole is about 5-6 Å which means that ion bombardment results in removal not only of COT molecules but of a few underlying Si atoms as well.

COT coated samples were annealed in UHV at 1000 °C. As expected, COT reacted further with silicon. STM images (not shown here) revealed the presence of characteristic rectangular features which presumably are SiC islands.[8] Thus, the adsorbed organic layer cannot be removed simply by heating.

4 Conclusion

Irradiation of a self-assembled monolayer of COT on the Si(001)2x1 surface results in the formation of nm-size holes or pits observed by STM. This can be used for detection of the ion passage in Si after ion bombardment.

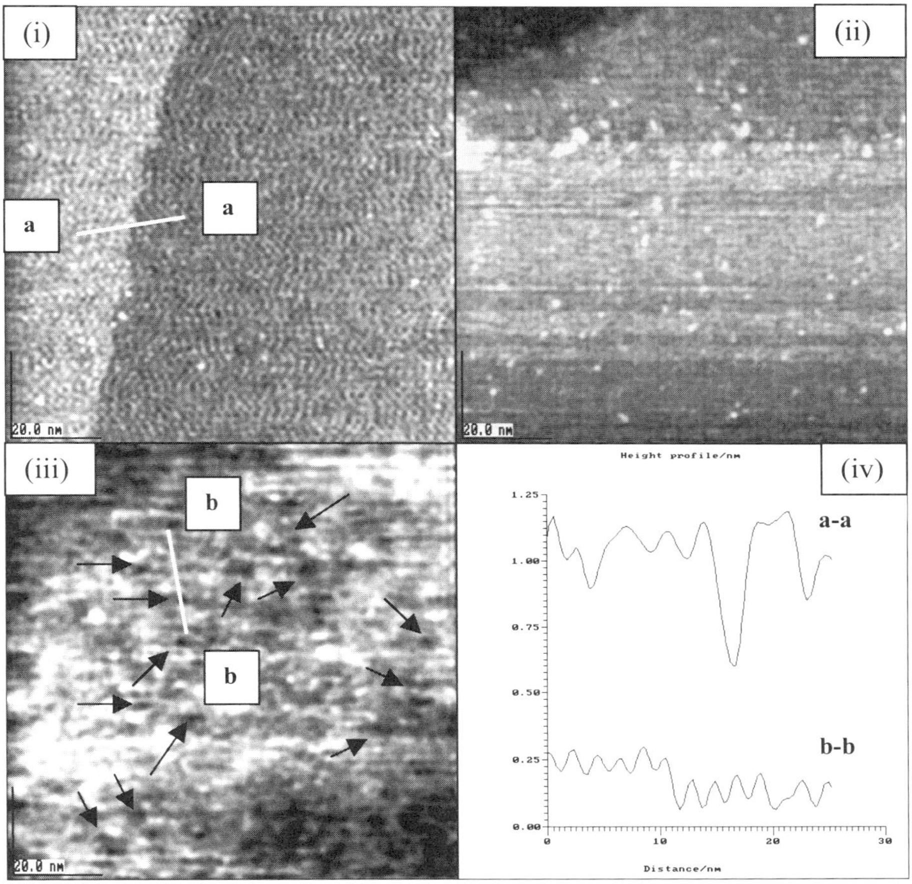

Figure 4. STM images: (i) the Si(001) surface after saturation exposure to COT; (ii) COT coated surface exposed to Ar atmosphere (p~10^{-4} Torr); (iii) the same surface after irradiation with 2 keV Ar^{+} ions; (iv) scan profiles: (a-a) along the row of adsorbed molecules at the initial surface (i) and (b-b) through the hole at the irradiated surface (iii).

References

1. B.E.Kane, Nature **393**, 133 (1998).
2. D.N.Jamieson, S.Prawer, I.Andrienko, D.Brett and V.Millar, Proc. 12th Int. Conf. on Ion Beam Modification of Materials, Porto-Alegre, Brazil, Sept. 4 – 8, 2000, p.O16.
3. J.S.Hovis, S.Lee, H.Liu and R.J.Hamers, J.Vac.Sci.Technol. B **15**, 1153 (1997).
4. J.S.Hovis, H.Liu and R.J.Hamers, Appl.Phys. A **66**, S553 (1998).
5. J.S.Hovis and R.J.Hamers, J.Phys.Chem. B **101**, 9581 (1997).
6. J.S.Hovis and R.J.Hamers, J.Phys.Chem. B **102**, 687 (1998).
7. R.J.Hamers, R.M.Tromp and J.E.Demuth, Phys.Rev. B **34**, 5343 (1986).
8. X.-D.Wang, T.Hashizume, H.Shinohara, Y.Saito, Y.Nishina and T.Sakurai, Phys.Rev. B **47**, 15923 (1993).

ADDRESSING SSQC FABRICATION ISSUES USING UHV-STM IMAGING

M.E. HAWLEY[1], G.W. BROWN[1], STEVEN SCHOFIELD[2],

Los Alamos National Laboratory, Los Alamos, NM, USA 87545, hawley@lanl.gov

A number of issues relating to fabrication of a silicon-based solid state computer based on the Kane model need to be addressed experimentally. One is the preparation of an extremely low defect density silicon substrate surface on which to put the array of spins while a second one is the stability of the array under deposition of an overlayer of silicon. Both of these issues can be addressed using ultra-high vacuum (UHV) scanning tunneling microscopy (STM). In fact, understanding the preparation of a low defect density Si(100) has enjoyed considerable attention from UHV-STM researchers over the last two decades.

We have chosen to revisit the issue of defects on the Si(100) in our ultra-high vacuum system in conjunction with launching a study to understand the stability of phosphorus (P) arrays during substrate heating. We have heated Si(100) samples to a range of temperatures and annealing times to study the density and structure of various defects. We have also begun attempting buried dopant imaging in P-doped Si(100). The results of recent work on the structure of a variety defects are presented, along with preliminary results on surface charge affects in the vicinity of surface features.

1 Introduction

Scanning tunneling microscopy (STM) has for some time been recognized as an extremely powerful technique for characterizing atomic positions and electrical properties of surface features observed on semiconductor and metal surfaces. In particular, STM has been used extensively to study the Si(100) 2 x 1 reconstructed surface, the most important surface used in semiconductor device fabrication. This surface consists of Si-Si dimers arranged in parallel rows that alternate between being parallel to (smooth terrace) or perpendicular to (rough terrace) the terrace edge. A great deal of work has been done over the last 20 years to study defect structures on this surface but systematic study has been done to correlate preparation conditions with residual surface defects.

In this paper, we revisit the issue of surface defects generated in the Si(100) 2 x 1 reconstructed surface during preparation in the context of using this surface as a template for the fabrication of spin arrays for a silicon-based solid state quantum computer (SSQC). Various buried defects, vacancies or impurities, are expected to have different degrees of detrimental effect on the operation of a SSQC. It is important to understand what effect these defects will have and to what extend these defects can be eliminated. The most common types of vacancy defects are discussed below.

Two important issues are addressed here. The first is the preparation of a low defect density Si(100) surface. Surface preparation is more of an art form than a science with slightly variations in method needed for different ultra-high vacuum (UHV) systems. The general steps for preparation of this surface are outlined below, but specific details published in the literature vary from group to group with sometimes contradictory results and conclusions. In this work, we present preliminary results from a systematic study to correlate preparation parameters and procedure with defect density and structure, with particular attention to those defects whose spin or charge state may be preserved during subsequent fabrication steps such as silicon overgrowth.

The second related issue we address is the existence, density, and character of buried charged features such as dopants that exist below the surface. STM is particularly well suitable for identifying charged features through the characteristic, associated band bending signature induced by the excess or reduced density-of-states (DOS) observed between tunneling tip and sample. It is important to be able to identify charged features in STM images in order to locate the phosphorus atoms under the silicon overcoat before electrode deposition and to avoid potentially detrimental crystal defects. The intent is to link charge with spin defects that can interfere with the qubit operation. This capability will be exploited to locate and understand the electronic properties of buried dopant or defects. Imaging of buried P-dopants has only been reported for cleaved Si(111) [1]. No buried dopant imaging on the Si(100) surface has been reported in the literature to date.This paper represents the first effort to associate some of these features with buried P-dopants on the Si(100) surface.

2 Defect Considerations for a Solid State Quantum Computer

Charged defects may have an associated spin but will only create problems for the solid state quantum computer architecture if they remain charged after the silicon MBE step. Vacancies will likely be filled in during MBE, but if the C-defect and associated complexes involve subsurface impurities, they may not disappear during the MBE step. It is, therefore, important to use the STM images to find a place on the surface without these complexes initially.

3 Experimental

Proper preparation of a clean, low defect density Si (001) 2 x 1 reconstructed surface requires an initial long degas with a high temperature flash (~ 1070 to 1150 °C for 0.5 to 2 minutes), followed by a rapid decrease in temperature to ~ 875 °C. The sample temperature is then decreased to ambient. In this study, we varied both flash time and temperature to determine the best conditions to minimize defects and the relationship between preparation parameters and types of defects generated. The chamber pressure was below 5×10^{-10} during the heat treatment.

All sample surfaces were characterized using a Omicron variable temperature UHV-STM. Constant current images were taken with a tungsten tip at 100 to 200 pA tunnel current at a sample biases of 1 to 2 volts.

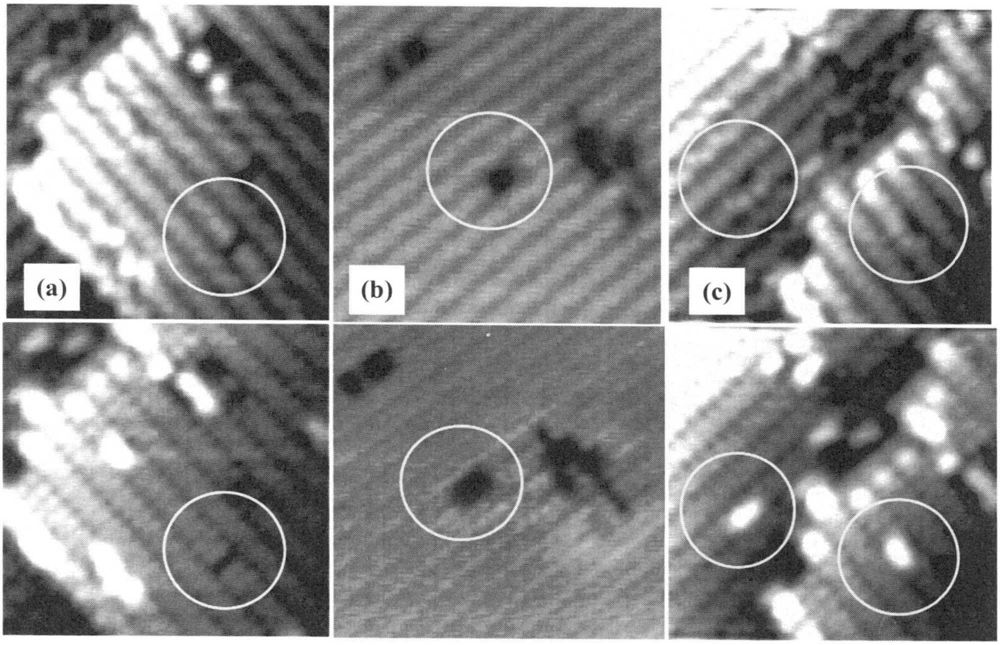

Figure 1. (a) "A" - Single vacancy defect appears as single dimer vacancy in both filled (top row) and empty (bottom row) states; (b) "B" - Divacancy_appears as adjacent dimer vacancies in filled and empty states; and (c) "C" - Complex appears as atomic vacancy in filled states and protrusions with depression in empty states.

4 Results

The three most common types of vacancy defect structures reported in the literature [2] and observed here are shown in Figure 1. They include A-type consisting of one missing silicon dimer in a row, B-type consisting of two missing adjacent dimers in a row, and C-type consisting of two missing silicon atoms from the same side of two adjacent dimers in a row.

Results of our own study of preparation parameter versus defects using a single sample-tip combination (Figure 2) suggests that longer flash times decreases defect densities. Our results also seem to suggest repeated flashes do not increase the number of defects.

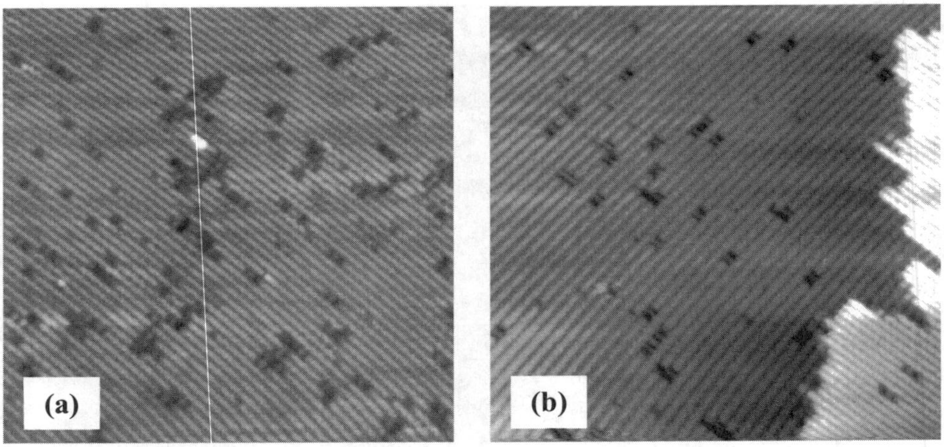

Figure 2. Two different ~ 1070 °C flash durations: a) sample was flashed for 1 minute and b) was flashed for 2 minutes. Note significant decrease in defect density with longer flash.

We have been able to see charge signatures on the Si(001)-2x1 surface for certain defect complexes. The density appears to be low but observation efficiency may depend on tip induced effects as well. None of the basic A, B, or C-type defects appear to be charged. However, we tentatively identified these features as resulting from complexes of either two C-defects or an A-defect together with two C defects. The charged defect complexes circled show strong band bending in their vicinity, consistent with previously published results [3]. The density of these charged feature appears to be low but may depend on tip induced effects as well. The doping level for this sample is only about 10^{-16} per cm^3 so only one in ten 100 μm x 100 μm square image is expected to show a charged feature associated with a P dopant. We do not know if any of these features is associated with a buried dopant. A typical pair of filled and empty state images showing charged defects is shown in Figure 3. The charge features are positive, showing up predominately in the empty state image.

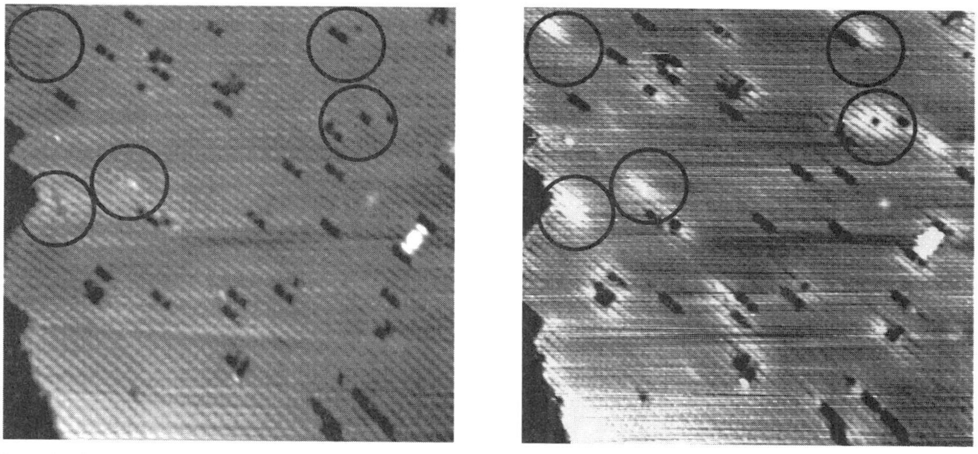

Figure 3. Filled and empty state images (-1.5V/+1.5V sample voltage at ~150 pA, left and right images respectively) of a P-doped Si (100) 2 x 1 reconstructed surface showing charge induced band bending (circles) from buried dopants or defects resulting in a positive charge at the surface observed as a brighter area in the centre of the STM image.

5 Summary and conclusions

An important capability with STM is imaging charged features at or below the silicon surface - both the qubits and detrimental defects will be seen. The results of our preliminary study on preparation parameters versus defect density seem to indicate that longer flashes may reduce the number of defects. Over the narrow range of temperatures used in this study, we observed no clear trend in flash temperature versus defect density.

We have been able to see charge signatures on the Si(001)-2x1 surface for certain defect complexes but further work is needed to understand the origin of features. Future work will include continued examination of charge signatures and careful examination of surface preparation steps to optimise the process for low defect densities. A more systematic study is planned. Near future studies will focus on samples with higher dopant density.

References

1. T. Trappmann, C Surgers, and H. v. Luhneysen, Appl. Phys. A **68**, 167 (1999).
2. R.J. Hamers and U.K. Kohler. J. Vac. Sci. Tech. **A7**, 2854 (1989).
3. V.A. Ukraintsev, Z. Dohnalek, and J.T. Yates, JR., Surf. Sci. **388**, 132 (1997).

STM INVESTIGATION OF SILICON SURFACES FOR FABRICATION OF THE KANE QUANTUM COMPUTER

S. R. SCHOFIELD,[1,2] J. L. O'BRIEN,[1,2] M. Y. SIMMONS,[1,2] R. G. CLARK,[1,2] A. S. DZURAK,[1,3] B. E. KANE,[4] M. E. HAWLEY[5], G. W. BROWN,[5] AND N. J. CURSON[1,2]

[1] *Centre for Quantum Computer Technology, School of Physics, University of New South Wales, Sydney 2052 Australia*

[2] *School of Physics, University of New South Wales, Sydney 2052, Australia*

[3] *School of Electrical Engineering & Telecommunication, University of New South Wales, Sydney 2052, Australia*

[4] *Laboratory for Physical Sciences, University of Maryland, College Park, MD 20740, USA*

[5] *Los Alamos National Laboratory, Los Alamos, NM 87545, USA*

The fabrication of the Kane quantum computer requires the ability to create an ordered array of phosphorus atoms in an otherwise pure silicon crystal. An intricate fabrication strategy has been employed to achieve these goals, requiring atomic precision imaging and manipulation of silicon based surfaces. Here we present detailed scanning tunnelling microscopy (STM) results at various stages of this fabrication process. In particular, we demonstrate our ability to prepare low defect density Si(001)2×1 surfaces and passivate these surfaces with either monolayer or submonolayer hydrogen coverages. The Si(001)2×1 surface is imaged at elevated temperatures and the results suggest that phosphorus arrays fabricated on the surface would survive their encapsulation in silicon provided the substrate is not heated above 750 K. These results highlight the stability of the silicon (001) surface for single atom engineering.

1 Introduction

The Kane quantum computer architecture[1] requires the creation of an ordered array of P atoms in an otherwise pure Si crystal. We have employed an intricate fabrication strategy, consisting of four major stages, to achieve this goal. These stages are, (i) to hydrogen passivate a clean, low defect density silicon (001)2×1 surface, (ii) to perform atomic scale hydrogen lithography of the surface prepared in stage (i), (iii) to expose the patterned silicon/hydrogen surface to phosphine (PH_3) gas and (iv) to encapsulate the phosphorus array in epitaxial silicon using molecular beam epitaxy (MBE). The clean Si(001)2×1 surface is highly reactive, while the hydrogen terminated Si(001)2×1:H surface is chemically inert. Thus, the hydrogen layer can be used as a lithographic resist layer, allowing PH_3 molecules to bond to the surface only in the regions where the hydrogen has been removed with the STM tip. In this letter, we present details of the first stage of our fabrication strategy and discuss the stability of the silicon surface at high temperatures, of relevence to stage four. In a companion letter,[2] we present details of stages (ii) and (iii) of the fabrication process.

2 Surface preparation

Experiments were performed using an OMICRON variable temperature scanning tunnelling microscope (STM) with electrochemically etched tungsten tips. The STM is mounted inside a stainless steel ultra-high vacuum (UHV) chamber with a base pressure $< 4 \times 10^{-11}$ mbar. Silicon samples, $2 \times 10 \times 0.5$ mm^3 in size, were cut from a (001)-orientated, n-type (P-doped) wafer having a resistivity of 0.1 Ωcm. Sample preparation was performed in vacuum

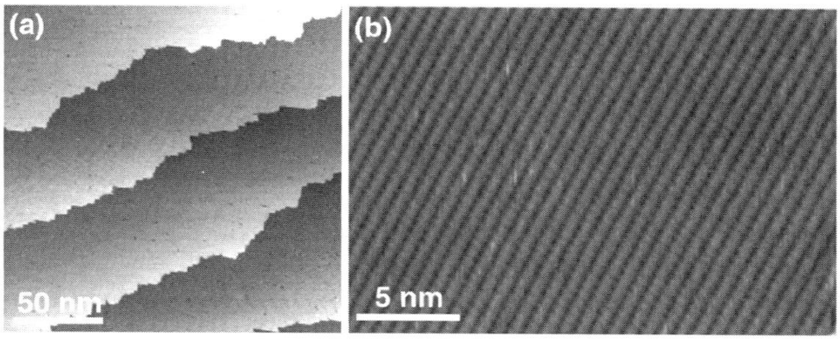

Figure 1. STM images of clean, low defect density Si(001)2×1 surfaces. Several atomically flat terraces are visible in (a). The higher resolution image, (b), shows a 24 × 17 nm^2 region of a single terrace that is completely defect free. In this image the individual Si–Si dimers can be seen. Tunnelling conditions were (a) –2 V, 0.2 nA and (b) –1 V, 0.4 nA.

without prior *ex-situ* treatment. The samples were first outgassed overnight by heating to 850 K using a resistive heating element. Following this, the samples were flashed to 1450 K by passing a direct current through the sample. After flashing, the samples were cooled from 1250 K to room temperature over a period of minutes. The sample temperature was monitored using an infrared pyrometer. Figure 1 shows images of a well-ordered Si(001)2×1 surface prepared in this manner.

The defect densities of the surface shown in these images are equal to the lowest reported in the literature.[3] Crucial to the fabrication strategy is the ability to form an effective hydrogen resist layer. Thus, a series of hydrogen adsorption calibration experiments have been performed. Atomic hydrogen was produced by dissociating molecular hydrogen (H_2) gas on a hot (~ 1800 K) tungsten filament, placed 15 cm from the sample. Figures 2(a) and 2(b) show filled and empty state STM images, respectively, of a partially hydrogen terminated Si(001)2×1 surface. Figure 2(c) is a schematic showing the atomic arrangement of the surface in the insets to these t wo images. The identified features are typical of STM images of partially hydrided Si(001)2×1 surfaces.[4] Figure 2(d) shows a filled state STM image of a completely hydrogen terminated Si(001)2×1 surface. This surface was formed by saturation dosing a clean Si(001)2×1 surface at a substrate temperature of 625 K. This surface consists predominantly of monohydride dimers, however dihydride and trihydride dimers can also be seen. A surface consisting of only monohydride dimers can be formed by saturation dosing a clean Si(001)2×1 surface with hydrogen, at a substrate temperature of 675 K. At this temperature, hydrogen desorbs from the di- and tri-hydride phases, leaving a surface of just the monohydride phase. Hydrogen desorption from the monohydride phase occurs at a temperature of 825 K.[5] These results demonstrate our ability to form a well-ordered hydrogen terminated Si(001)2×1 surface, thus completing stage (i) of our fabrication strategy.

3 High temperature studies

The fourth stage of the fabrication strategy, i.e., the molecular beam epitaxy (MBE) overgrowth of silicon, may require heating the samples above room temperature to ensure good

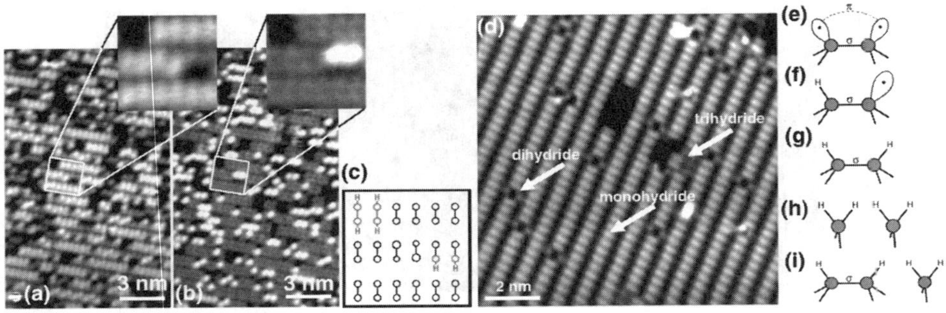

Figure 2. Hydrogen terminated Si(001)2×1 surfaces. Both filled- and empty-state images of a partially hydrogen terminated Si(001) surface are shown in (a) and (b), where the sample biases were −1.8 V and +1.8 V, respectively. Schematic (c) shows the atomic structure of the insets to (a) and (b). The image in (d) is of a completely hydrogen terminated Si(001) surface. This surface is predominantly monohydride, but di- and tri-hydride species are also present. Tunnelling conditions for (d) were −2.5 V and 0.4 nA. Further schematics are of (e) Si–Si, (f) hemi-hydride, (g) mono-hydride, (h) di-hydride, and (i) tri-hydride dimers.

quality crystal growth. Whether or not the adsorbed phosphine array, created in stages (ii) and (iii) of the fabrication strategy (see Ref. 2) will suffer from surface diffusion effects during this heating is a critical question that has yet to be answered. To address this issue we have performed a number of high temperature studies of a clean Si(001)2×1 surface. Figure 3(a) shows a series of consecutive images of Si(001)2×1 surface heated to 750 K, *in-situ*. These images show that the step edges are mobile at this temperature, i.e., the contours of the step edges are fluctuating with time, in agreement with previous studies where such step edge "meandering" has been observed.[6] Step edge meandering occurs when the thermal energy of the surface is sufficient to detach dimers from step edges, which then surface diffuse and subsequently re-attach at other step edge sites. Additionally, in Fig. 3(a) it can be seen that the positions of missing dimer defects change with time. This is due to dimers detaching from the edge of a missing dimer defect and diffusing the very short distance to the other side of the defect. However, while both step edge meandering, and small missing dimer defect motion can be seen during the $\sim$13 minute period that the images in Fig. 3(a) were acquired, overall the surface remains largely unchanged. Specifically, there is no net motion of surface dimers, other than at step edges. At a surface temperature of 800 K, however, the surface became mobile, and large "holes" several nanometres in diameter and one or t wo monolayers deep, appeared in the flat terrace regions (not shown). Additionally, regions of $c(4\times4)$ periodicity could be seen in these high temperature images. Figure 3(b) is a room temperature image of a surface heated to 800 K and then cooled, and large regions of the Si(001)$c(4\times4)$ reconstruction are present.

As stated earlier, hydrogen on the Si(001) surface in the monohydride phase desorbs from the Si(001) surface at 825 K; thus hydrogen will not desorb from a monohydride surface heated to 750 K (cf., Fig. 3(a)). However, hydrogen is known to desorb from adsorbed phosphine molecules at this temperature.[7] In a detailed study combining photoemission and STM techniques, Lin *et al.*[7] have investigated the behaviour of PH_3 saturated Si(001)2×1 surfaces that were annealed at high temperatures. They found that at a substrate temperature of 720 K, adsorbed phosphine molecules completely dissociate into P and H. The P

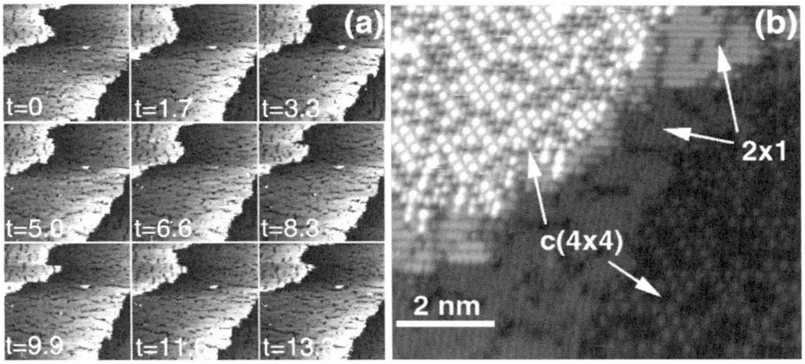

Figure 3. The nine images shown in (a) are consecutive STM images of a clean Si(001)2×1 surface heated *in-situ* to 750 K. Each frame is 80×80 nm^2 in size and took ~ 1.7 minutes to acquire. Tunnelling parameters were -2 V and 0.3 nA. The image shown in (b) is of the same surface after annealing to 800 K. This image was acquired at room temperature with -1.6 sample bias and 0.5 nA tunnelling current. Regions of the metastable Si(001)c(4×4) reconstruction can be seen.

atom displaces a substrate Si atom to become incorporated into the surface (i.e., to form a dimer bond to a surface Si atom and two backbonds to second layer Si atoms). The H atoms either desorb from the surface, or form monohydride dimers with the displaced Si atoms. In conclusion therefore, we expect that the phosphine array fabricated in stage (iii) of our strategy can be incorporated into the top surface layer by heating the surface to between 720–750 K. Additionally, the epitaxial growth of high quality Si layers using MBE can be achieved at and below 750 K,[8] and can also occur on a hydrogen terminated Si(001) surface.[9] As discussed above, Fig. 3(a) shows there is no significant diffusion of surface silicon dimers (except at step edges) for a clean Si(001)2×1 surface heated to 750 K. Thus, the process of incorporating the P atoms into the surface and subsequently encapsulating them with epitaxial silicon can be achieved without raising the sample temperature above 750 K. The results presented here therefore indicate that there will be no significant diffusion of the P atoms in the array at the temperatures required for silicon encapsulation.

4 Summary

In summary, we have outlined a fabrication strategy to construct the Kane quantum computer using STM and MBE. We have demonstrated a high level of control in hydrogen passivating a clean Si(001)2×1 surface. We have presented STM images of *in-situ* heated Si(001)2×1 surfaces that suggest MBE growth of Si overlayers can be performed without significant diffusion of the phosphorus array, provided the substrate is not heated above 750 K. Below this temperature, the localised diffusion of terrace atoms is of order 1 nm, but only in the vicinity of a missing dimer defect. Since the required P separation within the P array of the Kane quantum architecture is ~20 nm, this localised atom diffusion is well within the required fabrication tolerance. In a companion letter[2] we report hydrogen lithography and phosphine adsorption to form an array of phosphine molecules on the surface.

Acknowledgments

This work was funded by the Australian Research Council, and the US National Security Agency through the Army Research Office.

References

1. B. E. Kane, Nature **393**, 133 (1998).
2. J. L. O'Brien *et al.*, IQC01 conference proceedings (Th2), Sydney Australia 2001
3. K. Hata, T. Kimura, S. Ozawa, and H. Shigekawa, J. Vac. Sci. Technol. A **18**, 1933 (2000).
4. D. R. Bowler *et al.*, J. Phys.: Condens. Mat. **12**, 7655 (2000).
5. M. Suemitsu, H. Nakazawa, and N. Miyamoto, Appl. Surf. Sci. **82/83**, 449 (1994).
6. H. J. W. Zandvliet *et al.*, Phys. Rev. Lett. **70**, 2122 (1993).
7. D. S. Lin, T. S. Ku, and T. J. Sheu, Surf. Sci. **424**, 7 (1999).
8. D. J. Eaglesham, H. J. Gossmann, and M. Cerullo, Phys. Rev. Lett. **65**, 1227 (1990).
9. M. Copel and R. M. Tromp, Phys. Rev. Lett. **72**, 1236 (1994).

SINGLE-ELECTRON TRANSISTOR ARCHITECTURES FOR SIMULATION OF SOLID-STATE QUANTUM COMPUTER READOUT

ROLF BRENNER[1,2], TILO M. BÜHLER[1,2], RITA P. McKINNON[1,2], DAVID J. REILLY[1,2], A. R. HAMILTON[1,2], ANDREW S. DZURAK[1,3], ROBERT G. CLARK[1,2], NANCY E. LUMPKIN[1,2], AND GERARD J. MILBURN[4]

[1] *Centre for Quantum Computer Technology, University of New South Wales, Sydney 2052, Australia*

[2] *School of Physics, University of New South Wales, Sydney 2052, Australia*

[3] *School of Electrical Engineering and Telecommunications, University of New South Wales, Sydney 2052, Australia*

[4] *Centre for Quantum Computer Technology, The University of Queensland, St Lucia, QLD 4072, Australia*

Single-spin detection will be crucial for solid-state quantum computer architectures in which the quantum information is stored in the spin-state of single nuclear or electron spins. The difficult problem of single-spin detection can be mapped to a more tractable problem of single-charge detection through spin-dependent electron transfer. Here we describe a readout architecture that can detect the charge-state of two coupled metal dots, which simulates a two-qubit system. This architecture significantly reduces the effect of random charge noise from the signal, reducing the probability of readout errors in the quantum computer.

1 Introduction

It has been shown that algorithms based on quantum mechanics can vastly outperform their classical counterparts for many difficult problems[1,2]. Thus, there is intense interest in the realisation of quantum computers (QCs), which can be scaled up to include many quantum bits (qubits). Recent proposals for solid-state QCs suggest that the qubits can be formed as an array of nuclear or electron spins in a bulk semiconductor material[3,4]. The ability to detect the spin-state of a single qubit is therefore crucial for QC readout. Figure 1 shows a readout architecture based upon the Kane design. In this two-donor system, the electron clouds can only overlap if the two electron spins are anti-parallel, due to the Pauli exclusion principle. A-gates above the two-donor system can be used to set up an electric field, which causes an electron to transfer from one donor to the other if the spins are anti-parallel. Single-electron transistors (SETs) close to the donors detect this charge

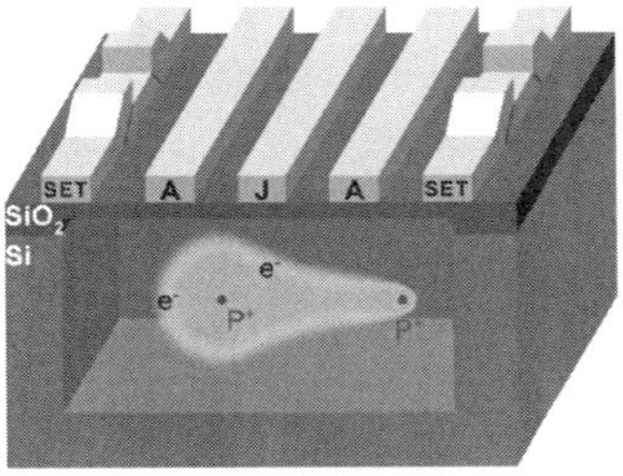

Figure 1. Schematic of a device to detect the spin state of two ^{31}P donors in ^{28}Si.

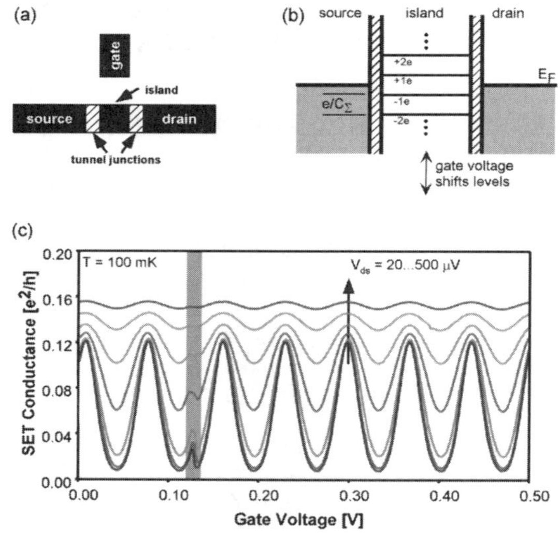

Figure 2. (a) Schematic of a SET and (b) the Coulomb levels on the island (C_I is the total capacitance of the island). (c) Conductance of a SET as a function of gate voltage, showing periodic oscillations due to single-electron c harging, for different source-drain biases.

motion, and thereby determine the relative spin-state of the t wo electrons.

2 Single-electron transistors and charge noise

A SET consists of source and drain leads which connect to an electrically isolated island through tunnel junctions (Fig. 2a). Addition/removal of single electrons to/from the island changes its electrostatic potential, resulting in equally spaced Coulomb levels on the island. Current can only flow if the Fermi levels of source and drain are aligned to these levels (Coulomb blockade). By applying a voltage to the gate, the Coulomb levels move through the Fermi levels of source and drain (Fig. 2b), leading to a periodically oscillating conductance characteristic (Fig. 2c). Random charging of traps in the vicinity of the SET (charge noise) cause discontinuous changes in the SET conductance. Such unwanted events have been observed in previously fabricated SETs (highlighted in grey in Fig. 2c), and they are very similar to the signal expected during quantum-state readout.

3 Twin-SET architecture

To simulate the device shown in Figure 1, we have designed and fabricated an all metal t win-SET/double-dot system (Fig. 3). In this design, the t wo P atoms are replaced by t wo metal islands connected by a tunnel junction (double-dot). Applying a voltage to the t wo control gates A1 and A2 causes single electrons to tunnel from one metal island to the other, just as electrons are transferred from one P donor to the other in Figure 1. Two SETs - one on either side of the double-dot - can detect this charge transfer[5]. When an electron tunnels from the left to the right dot, SET1 senses the departure of an electron, while SET2

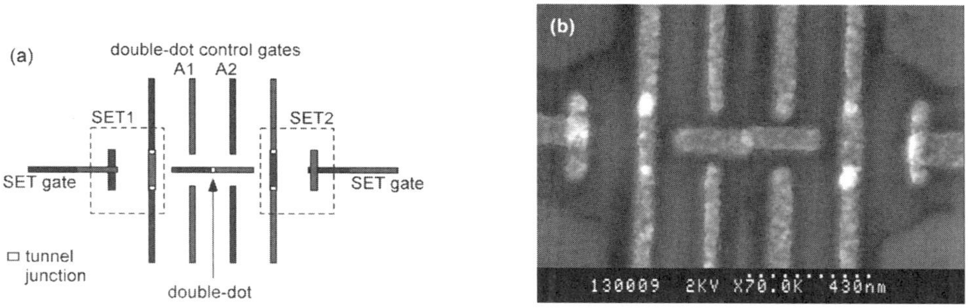

Figure 3. (a) Schematic and (b) SEM image of a twin-SET device, fabricated by double-angle evaporation of Al with intermediate oxidation to form tunnel junctions.

simultaneously senses the arrival of an electron. A significant reduction of spurious signals can therefore be achieved by correlating the outputs of the two SETs.

4 Simulation of quantum computer readout

Applying a differential voltage, $+V_A$ and $-V_A$ to the two control gates allows electrons to tunnel from one metal dot to the other. The charges $Q_1 = -Q_2$ on the dots thus change in discrete steps as V_A increases (Fig. 4a). Biasing the SET-gates appropriately eliminates the effect of the control gates on the SETs, and the SET conductances G change in a step-like manner (Fig. 4b). This can be seen more clearly in Figure 4c, where the transconductance $\partial G/\partial V_A$ shows peaks corresponding to single-electron transfers. Additional biasing of the SET gates tunes the SETs to operate at their most sensitive point (i.e. maximum peak amplitude). Spurious signal rejection is accomplished by using both SETs as shown in Figure 5. Charge transfer in the double-dot is detected by simultaneous and opposite peaks on the twin-SET signals. Charge noise can cause either a peak in just one SET output (I) or simultaneous but equidirectional peaks in both SETs (II). Correlating the outputs of both SETs thus allows significant reduction in unwanted charge noise, which could otherwise lead

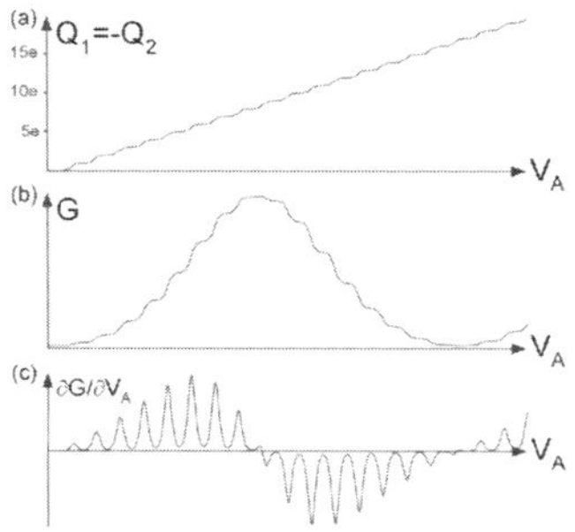

Figure 4. (a) Charge $Q_1 = -Q_2$ on the dots, (b) resulting conductance G of nearest SET, and (c) $\partial G/\partial V_A$ as a function of control gate voltage V_A.

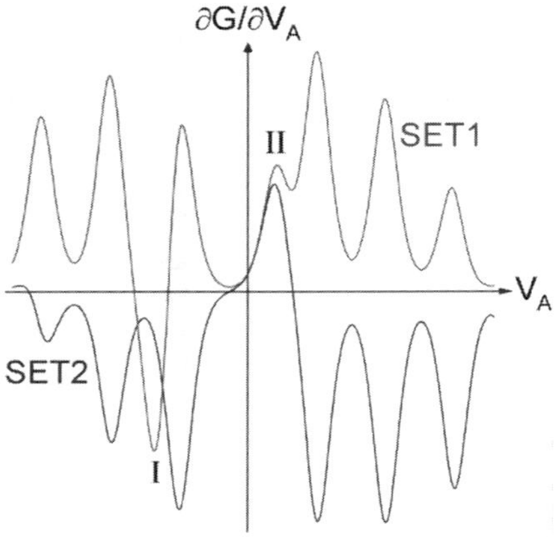

Figure 5. Calculated signal outputs of both SETs. Peaks represent charging events. I and II indicate random charge noise.

to readout errors in a QC.

5 Conclusion

We have fabricated twin-SET devices that are able to detect the charge state of a double-dot. Furthermore, we propose a measurement technique that can discriminate intended charge transfers in the double-dot from random charging of traps in the vicinity of the device. This is crucial for quantum computer read-out where the occurrence of a single charging event has to be detected. These devices will be measured in a low-noise, ultra-low-temperature environment to demonstrate the proposed measurement technique in practice. Future experiments will apply this method to atomic systems.

Acknowledgments

This work was funded by the Australian Research Council and the US National Security Agency through the Army Research Office.

References

1. P.W. Shor in *Proceedings of the 35th Annual Symposium on Foundations of Computer Science, Santa Fe, NM, USA*, ed. S. Goldwasser (IEEE Computer Society Press, Los Alamitos, 1994) 124.
2. L.K. Grover, *Phys. Rev. Lett.* **79**, 325 (1997).
3. B.E. Kane, *Nature* **393**, 133 (1998).
4. R. Vrijen, E. Yablonovitch, K. Wang, H.W. Jiang, A. Balandin, V. Roychowdhury, T. Mor and D. DiVincenzo, *Phys. Rev.* A **62**, 012306 (2000).
5. I. Amlani, A.O. Orlov, R.K. Kummamuru, G.H. Bernstein, C.S. Lent and G.L. Snider, *Appl. Phys. Lett.* **77**, 738 (2000).

NANOFABRICATION OF A MULTI-QUBIT SOLID STATE QUANTUM COMPUTER DEVICE

T.M. BUEHLER[1,2], R.P. MC KINNON[1,2], N.E. LUMPKIN[1,2], R. BRENNER[1,2], A. S. DZURAK[1,3], R.G. CLARK[1,2], D. JAMIESON[1,4] AND J. MC CALLUM[1,4]

[1] *Centre for Quantum Computer Te chnology, University of New South Wales, Sydney 2052, Australia*

[2] *School of Physics, University of New South Wales, Sydney 2052, Australia*

[3] *School of Electrical Engineering and Telecommunications, University of New South Wales, Sydney 2052, Australia*

[4] *School of Physics, University of Melbourne, Parkville 3052, Australia*

We present a fabrication strategy for a silicon-based solid state quantum computer device using high resolution electron beam lithography (EBL) and ion implantation employing a multi-layer resist technology and selective development. The major advantage of our strategy is the self alignment of the quantum bits (qubits) to the overlaying control gate and single electron transistor (SET) readout structures, which are formed by triple angle shadow evaporation. Fabrication of a prototype self-aligned gate and readout SET array for a four-qubit device is reported.

1 Introduction

Silicon-based solid state quantum computation (SSQC) is particularly attractive due to it's compatibility with existing semiconductor integrated circuit technology. In a design proposed by Kane[1], the nuclear spin of ^{31}P atoms embedded in ^{28}Si form the qubits, and operations on the qubits are performed using gates located on the substrate, as depicted in figure 1. An electric field induced by the surface A and J gates is used to polarize a t wo electron system associated with t wo P donors near a charge sensitive SET detector. Charge motion is only possible for a spin-singlet state thereby allowing to readout the P nuclear spin[1,2].

Fabrication of the architecture shown in figure 1 requires the ability to place ^{31}P atoms in ordered arrays in ^{28}Si, and to precisely align these atoms to their control gates and readout SETs. There are essentially t wo approaches to the fabrication of such a structure.

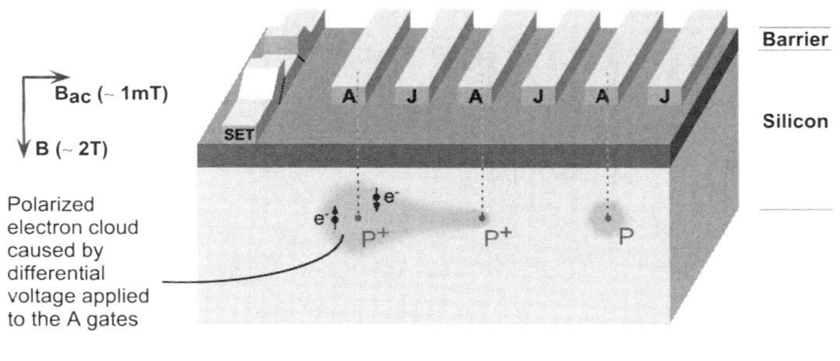

Figure 1. Schematic of a SSQC device as proposed by Kane. The vertical magnetic field polarizes the electron spins which can be brought into resonance with B_{ac} depending on the A gate voltage.

A "bottom up" strategy involves the construction of the computer at the atomic level using scanned probe lithography and epitaxial growth technologies. This will allow for the fabrication of large-scale qubit arrays but will require sophisticated techniques to align the qubits to the control gate architecture. The "top down" strategy reported here, while only designed for small-scale device fabrication at this stage, allows for the precise self-alignment of the control gates to the readout SETs, as well as self-alignment of the control gates to the underlying qubits. This is an essential requirement of several SSQC proposals involving both nuclear-spin qubits[1,4] and electron-spin qubits[3].

2 Top-Down Fabrication Strategy

Figure 2(a) is a schematic of a six donor, four-qubit SSQC in which the two SETs and the outermost ^{31}P atoms are used for state readout. In order to fabricate this structure, the black pattern shown in figure 2(b) is exposed in a bilayer resist of PMMA on top of a copolymer (of MMA and PMAA) using a high resolution EBL. This exposure is used later to form the SETs and the control gates during a triple angle evaporation process as shown in figure 3(b). A third layer of resist which requires a different developer solution than the bilayer is then spun onto the bilayer and the grey horizontal line shown in figure 2(b) is written. A suitable resist for this purpose is PMGI which can be developed using a water-based developer. The aim is to tune the sensitivity of the top layer resist in a way so that the exposure of the second EBL pattern doesn't affect the latent pattern in the PMMA layer.

A linear array of qubits is achieved by partially developing both patterns so that holes for ion implantation are opened only where the two patterns overlap (dots in Fig. 2a). Firstly the cross pattern in the top layer resist is developed, and then the bilayer is partially developed using an organic solution that doesn't affect the PMGI layer. Careful characterization of this second development step allows the opening of holes down to the substrate as shown in figure 3(a).

Ion implantation is performed next, with an area dose such that on average one P ion lands in each hole. Typical ion energies would be of order 20 keV. The probability that a single ion lands in N holes is 0.367^N by Poisson statistics[3], which restricts this fabrication

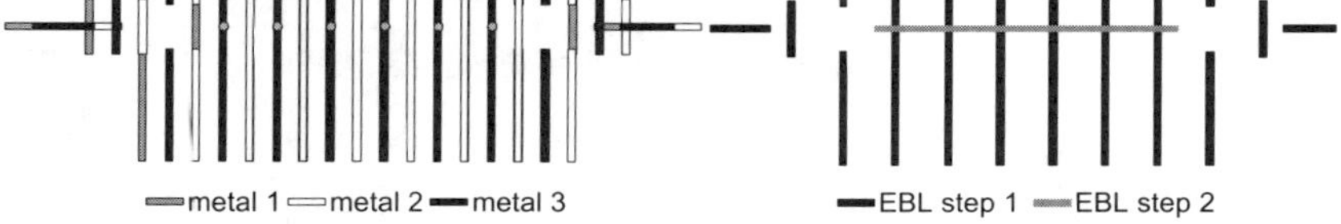

Figure 2. a) Schematic of gate and SET structure as obtained by parallel projection of the EBL pattern in the PMMA layer. The phosphorus donors are indicated by dots in the A-gates. b) The black structure is written into the PMMA layer and the crossing line (grey) is written subsequently into the overlaying resist.

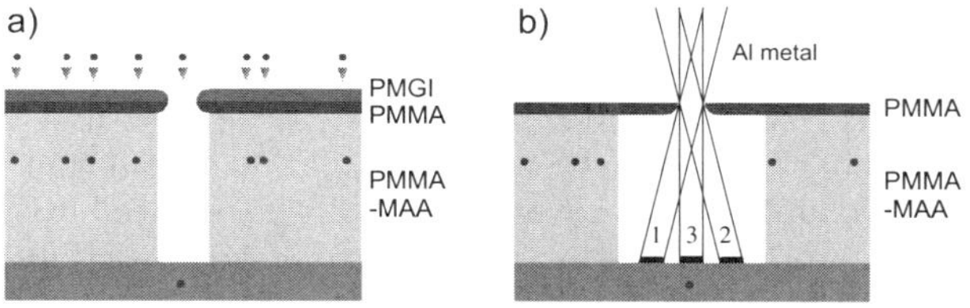

Figure 3. a) Schematic of the triple layer resist structure after development of the holes which form a mask for ion implantation. b) After stripping of the PMGI layer and full development of the remaining bilayer the black pattern of figure 2b) is opened in the PMMA. This structure is the mask for the metal evaporation.

technology to small qubit devices. Damage to the Si substrate caused by the implant process has been modelled[5]. It is anticipated that this damage can be removed by rapid thermal annealing although the exact details still need to be determined.

Stripping of the PMGI layer allows the full development of the latent pattern in the bilayer forming the black pattern in figure 2(b) in the PMMA with a cavity as shown in figure 3(b) in the more sensitive copolymer. A triple angle aluminum shadow evaporation is then performed with an *in-situ* oxidation between the first and second angle evaporations in order to form tunnel junctions that are required for the operation of the Al/Al_2O_3 SETs[6]. The third evaporation at normal incidence forms the A-gate directly above the qubit. Removing the resist leaves the structure shown in figure 2(a).

A second possibility is to have the PMGI as the bottom layer with the copolymer/PMMA bilayer on top. In this case the PMGI is exposed and developed prior to deposition of the copolymer/PMMA bilayer. After exposure of the black pattern shown in figure 2(b) partial development with an IPA based developer that doesn't affect the PMGI results in openings at the substrate only where the t wo EBL patterns cross.

3 Nanofabrication Results

We have fabricated the gate and readout SET array using a bilayer resist structure and triple angle shadow evaporation. Figure 4(a) shows the profile of a cavity formed in a

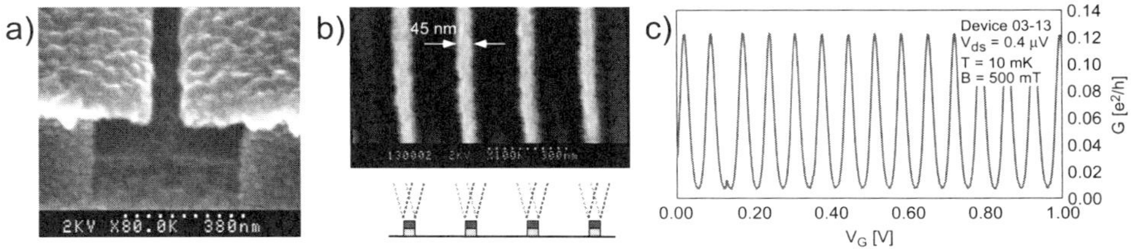

Figure 4. a) SEM image undercut profile formed in a bilayer resist. b) High accuracy overlay of two narrow metal lines as required for the SSQC gate structure. SEM images courtesy UNSW Electron Microscope Unit. c) Coulomb blockade oscillations of an Al/Al_2O_3 SET fabricated using double angle shadow evaporation.

bilayer resist after full development of both layers. The opening in the PMMA is 60 nm and the cavity below is 450 nm wide. Figure 4(b) shows the results obtained when double angle evaporation was performed using the pattern shown in figure 2(b). The resulting lines are only 45nm wide consisting of two lines on top of each other. Using the same shadow evaporation technology we have successfully fabricated SETs which show clear Coulomb blockade oscillations (Fig. 4c). Coulomb blockade oscillations are due to single electron tunneling processes[7]. Their high transconductance provide the high sensitivity of the SETs to charge motion, necessary for SSQC read-out[1,2].

In figure 5 a device is shown on which the triple angle evaporation was performed. The narrow lines are the A-gates which will lie directly above the implanted ^{31}P atoms as shown in Figure 2(a), while the J-gates are slightly wider due to imperfect overlaying of the lines deposited on top of one another. This structure required the formation of a self-supporting PMMA mask over a distance of more than two microns. The limit of the cavity can be seen as the transition between the darker and lighter regions near the corners of the image.

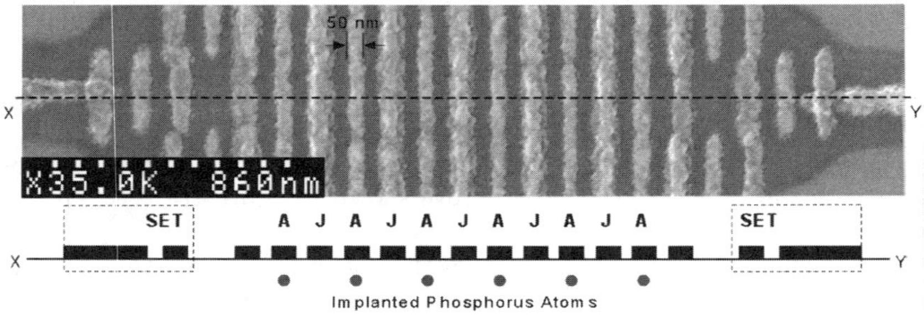

Figure 5. Metal gate and SET structure formed by triple angle shadow evaporation using an EBL exposure pattern as shown in figure 2(b). The schematic below shows the planned location of ^{31}P donors, to be achieved following an ion implantation step which is currently being developed.

4 Conclusion

We have demonstrated the ability to form a self-aligned control gate and readout SET structure for a small-scale SSQC. Extension of the fabrication process to include the third layer of resist and self-aligned single ion implantation is underway. A key advantage of this process is the high alignment accuracy achieved between the qubits and the control gates, together with the production of a regular, linear array of qubits. The registration accuracy made possible with this technique is critical for all silicon-based SSQC architectures.

Acknowledgments

This work was funded by the Australian Research Council, and the US National Security Agency through the Army Research Office.

References

1. B. E. Kane, Nature **393**, 133 (1998).
2. B. E. Kane, N.S. McAlpine, A.S. Dzurak, R.G. Clark, G.J. Milburn, He Bi Sun, and H. Wiseman, Phys. Rev. B **61**, 2961 (2000).
3. R. Vrijen, E. Yablonovitch, K. Wang, H. W. Jiang, A. Balandin, V. Roychowdhury, T. Mor, D. DiVincenzo, Phys. Rev. A **62**, 012306 (2000).
4. B. E. Kane, Fortschr. Phys. **48**, 1023 (2000).
5. N.A. Marks, K. Hee Lee and D.R. McKenzie, in these IQC01 proceedings.
6. T. Fulton and G. Dolan, Phys. Rev. Lett. **59**, 109 (1987).
7. P. Delsing, *Single Charge Tunneling*, edited by H. Grabert and M. Devoret (Plenum, New York, 1992).

CONDITIONAL QUANTUM EVOLUTION INDUCED BY CONTINUOUS MEASUREMENT FOR A MESOSCOPIC QUBIT

HSI-SHENG GOAN AND GERARD J. MILBURN

Center for Quantum Computer Technology and Department of Physics, University of Queensland, Brisbane, Qld 4072, Australia

We consider the problem of an electron tunneling between two coupled quantum dots, a two-state quantum system (qubit), using a low-transparency point contact (PC) or tunnel junction as a detector continually measuring the position of the electron. We focus on the qubit dynamics conditioned on a particular realization of the actual measured current through the PC device. We illustrate the conditional evolutions by numerical simulations. The different behaviors between unconditional and conditional evolutions are demonstrated. The conditional qubit dynamics evolving from quantum jumps to quantum diffusion is presented.

1 Introduction

In condensed matter physics, usually many identical quantum systems are prepared at the same time when a measurement is made upon the systems. For example, in nuclear or electron magnetic resonance experiments, generally an ensemble of systems of nuclei and electrons are probed to obtain the resonance signals. This implies that the measurement result in this case is an average response of the ensemble. On the other hand, for various proposed condensed-matter quantum computer architectures, the need to readout physical properties of a single electronic qubit, such as charge or spin at a single electron level, is demanding. It is particularly important to take account of the decoherence introduced by the measurements on the qubit as well as to understand how the quantum state of the qubit, conditioned on a particular single realization of measurement, evolves in time for the purpose of quantum computing.

We consider, in this paper, the problem of an electron tunneling between two coupled quantum dots (CQDs), a two-state quantum system (qubit), using a low-transparency point contact (PC) or tunnel junction as a detector (environment) continuously measuring the position of the electron. This problem has been extensively studied in Refs. [1,2,3,4]. The master equation (or rate equations for all the reduced-density-matrix elements) for the CQD system (qubit) has been derived and analyzed in Refs. [1,4]. This (unconditional) master equation is obtained when the results of all measurement records (electron current records in this case) are completely ignored or averaged over, and describes only the ensemble average property for the qubit. However, if a measurement is made on the system and the results are available, the state or density matrix is a conditional state conditioned on the measurement results. Hence the deterministic, unconditional master equation cannot describe the conditional dynamics of the qubit in a single realization of continuous measurements which reflects the stochastic nature of an electron tunneling through the PC barrier.

Korotkov [2] has obtained the Langevin rate equations for the CQD system measured by a ideal PC detector in the quantum-diffusive limit[4]. These rate equations describe the random evolution of the density matrix that both conditions, and is conditioned by, the PC detector output. Recently, we [4] presented a *quantum trajectory* measurement analysis of the same system. We found that the conditional dynamics of the qubit can be described by the stochastic Schrödinger equation for the conditioned state vector, provided that the information carried away from the qubit by the PC reservoirs can be recovered by the perfect detection of the measurements. We also analyzed the localization rates at which the qubit becomes localized in one of the two states when the coupling frequency Ω between the states is zero. We showed that the localization time discussed there is slightly different from the measurement time defined in Refs. [3]. The mixing rate at which the two possible states of the qubit become mixed when $\Omega \neq 0$ was calculated as well and found in agreement with the result in Ref. [3]. In this paper, we focus on the qubit dynamics conditioned on a particular realization of the actual measured current through the PC device. We illustrate the conditional evolutions by numerical simulations. The different behavior between unconditional and conditional evolutions are demonstrated. Furthermore, the conditional qubit dynamics evolving from quantum jumps to quantum diffusion [4] is presented.

2 Conditional quantum evolution under continuous measurement

The whole CQD and PC model is described in Refs. [1,2,4]. Basically, when the electron in the CQD system is near the PC (i.e., dot 1 is occupied), there is a change in the PC tunneling barrier, which then results in a change in the effective tunneling amplitude. As a consequence, the current through the PC is also modified. This changed current can be detected, and thus a measurement of the location of the electron in the CQD system (qubit) is effected. We describe parameters used in this paper in the following. We denote the logical qubit states as $|a\rangle$ (i.e., dot 1 is occupied) and $|b\rangle$ (i.e., dot 2 is occupied). The coupling and energy mismatch between the qubit states are $\hbar\Omega$, and $\hbar\mathcal{E}$ respectively. $\Gamma_d = |\mathcal{X}|^2/2$ is the decoherence (dephasing) rate generated by the PC reservoirs in the unconditional dynamics. The parameters $\mathcal{X}$ and $\mathcal{T}$ are given by: $|\mathcal{T}+\mathcal{X}|^2 = D'$, and $|\mathcal{T}|^2 = D$, where D' and D are the average electron tunneling rates through the PC barrier with and without the presence of the electron in dot 1 respectively. Physically, the presence of the electron in dot 1 raises the effective tunneling barrier of the PC due to electrostatic repulsion. As a consequence, the effective tunneling amplitude becomes lower, i.e., $D' = |\mathcal{T}+\mathcal{X}|^2 < D = |\mathcal{T}|^2$. This sets a condition on the relative phase θ between $\mathcal{X}$ and $\mathcal{T}$: $\cos\theta < -|\mathcal{X}|/(2|\mathcal{T}|)$. For simplicity, in this paper we consider the case $\theta = \pi$ as in Refs. [1,2].

The unconditional and conditional master equations of the qubit density matrix, $\rho(t)$, have been obtained and written respectively as sets of coupled stochastic differential equations in terms of the Bloch sphere variables, $x(t)$, $y(t)$, and $z(t)$, in Ref. [4]. We will use these equations to describe the qubit dynamics. In the Bloch sphere representation,

$$\rho(t) = \frac{1}{2}\begin{pmatrix} 1+z(t) & x(t)-iy(t) \\ x(t)+iy(t) & 1-z(t) \end{pmatrix}. \tag{1}$$

Here the variable $z(t)$ represents the population difference between the two dots. Especially, $z(t) = 1$ and $z(t) = -1$ indicate that the electron is localized in dot 2 and dot 1 respectively. The value $z(t) = 0$ corresponds to an equal probability for the electron to be in each dot.

Fig. 1(a) shows the unconditional (ensemble average) time evolution of the population difference $z(t)$ with the initial qubit state being in state $|a\rangle$. The unconditional population difference $z(t)$, rises from -1, undergoing some oscillations, and then tends towards zero, a steady (maximally mixed) state. On the other hand, the conditional time evolution, conditioned on one possible individual realization of the sequence of measurement results,

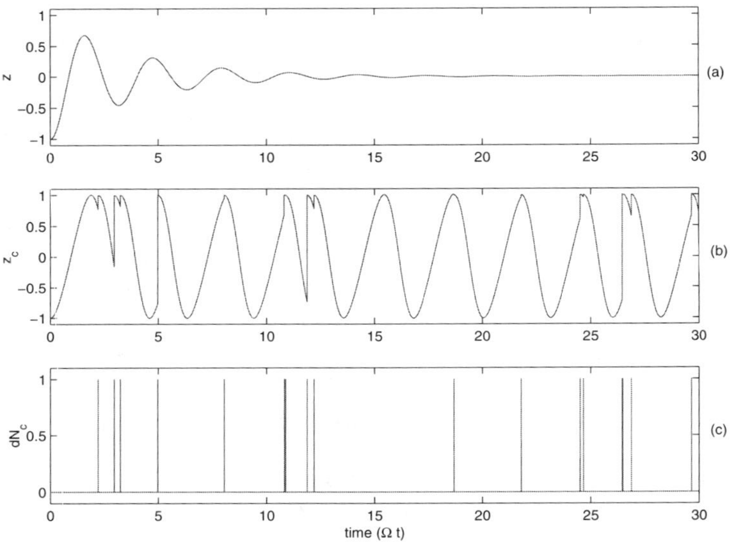

Figure 1. Illustration for different behaviors between unconditional and conditional evolutions. The initial qubit state is $|a\rangle$. The parameters are: $\mathcal{E} = 0$, $\theta = \pi$, $|\mathcal{T}|^2 = |\mathcal{X}|^2 = 2\Gamma_d = \Omega$, and time is in unit of Ω^{-1}. (a) unconditional, ensemble-averaged time evolution of $z(t)$. (b) conditional evolution of $z_c(t)$. (c) randomly distributed moments of detections, which correspond to the quantum jumps in (b).

behaves quite differently. We consider first the situation, where $D' = |\mathcal{T}+\mathcal{X}|^2 = 0$, discussed in Ref. [1]. In this case, due to the electrostatic repulsion generated by the electron, the PC is blocked (no electron passing through) when dot 1 is occupied. As a consequence, whenever there is a detection of an electron tunneling through the PC barrier, the qubit state is collapsed into state $|b\rangle$, i.e., dot 2 is occupied. The conditional, quantum-jump evolution of $z_c(t)$ shown in Fig. 1(b) (using the same parameters and initial condition as in Fig. 1(a)) is quite strikingly different from the unconditional one in that the time evolution is not smooth, but exhibits jumps, and it does not tend towards a steady state. One can see that initially the system starts to undergo a oscillation. As the population difference grows in time so does the probability for an electron tunneling through the PC barrier. This oscillation is then interrupted by the detection of an electron tunneling through the PC barrier, which bring $z_c(t)$ to the value 1, i.e., the qubit state is collapsed into state $|b\rangle$. Then the whole process starts again. The randomly distributed moments of detections, $dN_c(t)$, corresponding to the quantum jumps in Fig. 1(b) is illustrated in Fig. 1(c). The time evolutions in Fig. 1(a) and (b) show little similarity; however, averaging over many individual realizations shown in (b) leads to a closer and closer approximation of the ensemble average in (a).

Next we illustrate how the transition from the quantum-jump picture to the quantum-diffusive picture takes place. In Ref. [4], we have seen that the quantum-diffusive equations can be obtained from the quantum-jump description under the assumption of $|\mathcal{T}| \gg |\mathcal{X}|$. In Fig. 2(a)–(d) we plot conditional, quantum-jump evolution of $z_c(t)$ and the corresponding moments of detections $dN_c(t)$, with different $(|\mathcal{T}|/|\mathcal{X}|)|$ ratios. Each jump (discontinuity) in the $z_c(t)$ curves corresponds to the detection of an electron through the PC barrier. One

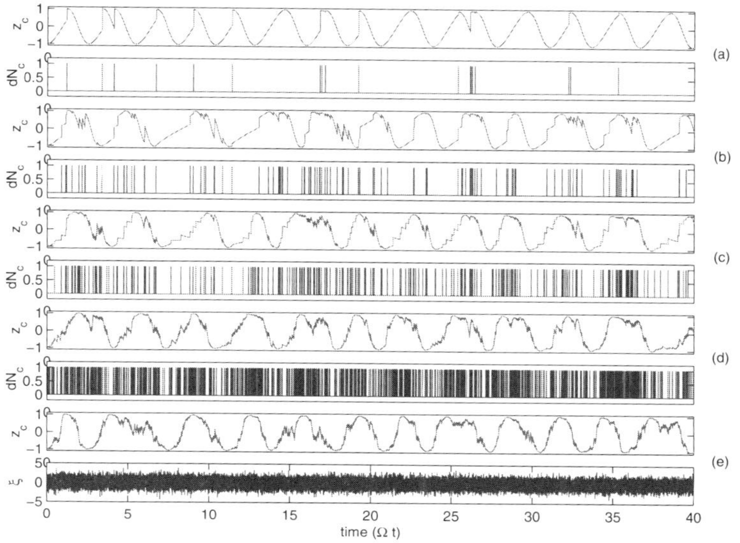

Figure 2. Transition from quantum jumps to quantum diffusion. The initial qubit state is $|a\rangle$. The parameters are: $\mathcal{E} = 0$, $\theta = \pi$, $|\mathcal{X}|^2 = 2\Gamma_d = \Omega$, and time is in unit of Ω^{-1}. (a)–(d) are the quantum-jump, conditional evolutions of $z_c(t)$ and corresponding detection moments with different $|\mathcal{T}|/|\mathcal{X}|$ ratios: (a) 1, (b) 2, (c) 3, (d) 5. With increasing $|\mathcal{T}|/|\mathcal{X}|$ ratio, jumps become more frequent but smaller in amplitude. (e) represents the conditional evolutions of $z_c(t)$ in quantum-diffusive limit. $\xi(t)$, appearing in the expression of current through PC in quantum-diffusive limit, is a Gaussian white noise with zero mean and unit variance.

can clearly observes that with increasing ($|\mathcal{T}|/|\mathcal{X}|$) ratio, the number of jumps increases. The amplitudes of the jumps of $z_c(t)$, however, decreases from $D' = 0$ with the certaint yof the qubit being in state $|b\rangle$ to the case of $(D - D') << (D + D')$ with a smaller probability of finding the qubit in state $|b\rangle$. Nevertheless, the population difference $z_c(t)$ always jumps up since $D = |\mathcal{T}|^2 > D' = |\mathcal{T} + \mathcal{X}|^2$. In other words, whenever there is a detection of an electron passing through PC, dot 2 is more likely occupied than dot 1. The case using quantum-diffusive equations is plotted in Fig. 2(e). We can see that the behavior of $z_c(t)$ for $|\mathcal{T}| = 5|\mathcal{X}|$ in the quantum-jump case shown in Fig. 2(d) is already very similar to that of quantum diffusion shown in Fig. 2(e).

In conclusion, we have discussed the qubit dynamics and illustrated the conditional evolutions by numerical simulations. The difference between unconditional and conditional evolutions is demonstrated. Furthermore, the conditional qubit dynamics evolving from quantum jumps to quantum diffusion is presented. HSG is grateful to useful discussions with A. N. Korotkov, H. M. Wiseman and H. B. Sun.

References

1. S. A. Gurvitz, Phys. Rev. B **56**, 15215 (1997); quant-ph/9808058.
2. A. N. Korotkov, Phys. Rev. B **60**, 5737 (1999); cond-mat/0008461.
3. Y. Makhlin *et al*, cond-mat/9811029; Phys. Rev. Lett. **85**, 4578 (2000).
4. H.-S. Goan *et al*, cond-mat/0006333, to appear in Phys. Rev. B.

QUANTUM ALGORITHMS AND THE MINIMIZATION PROBLEM

LLOYD C.L. HOLLENBERG

School of Physics, University of Melbourne, Victoria 3010, AUSTRALIA.

The global minimization problem occurs at the heart of many computer simulations of physical (and non-physical) systems. It is therefore of great interest to consider how one might perform minimization tasks on a quantum computer. A quantum algorithm framework is described which uses assumed information about the spectrum to reduce the search space to the neighbourhood of the minimum.

1 Introduction

In recent years we have witnessed an explosion of interest in quantum computing on both experimental and theoretical fronts, and everything between. Here we focus our attention on the use of quantum computation in the minimization problem which is important in many areas of study. Quantum algorithms for minimization have been developed since Grover discovered the fast searching algorithm[1]. Most algorithms are thus based on gradual amplitude amplification which is quadratically faster than exhaustive search[2]. A heuristic quantum algorithm was developed by Hogg and Portnov[3] based on problem-tuned phase amplification. There also exists ideas employing adiabatic evolution[5]and quantum annealing[4].

If an algorithm were found which could find the global minimum efficiently (i.e. is polynomial bounded) then all **NP**-complete problems would be solvable. However, since it has been proved that Grover's algorithm[1] is optimal on unstructured search problems[6,7,8], it would seem that this is unlikely to be possible. On the other hand, it is not clear that an exploitation of structure may not lead to faster algorithms for certain problems. In this paper a quantum algorithm framework for minimization is presented which attempts to use "global" structure of the problem in a hybrid sense.

2 A hybrid algorithm

The problem is introduced as the minimization of an energy function $E(x)$ over variables x, with $N = \dim\{x\}$: the task is to find the global minimum $E_{\min}(x_{\min})$. Assume a classical calculation has roughly mapped the density of states with the result $\rho_C(E)$. From $\rho_C(E)$ we can compute the cumulative states function, $n_C(E)$:

$$n_C(E) \equiv \sum_{E'}^{E} \rho_C(E')\, dE'. \tag{1}$$

We store the variables x in register R_x, and the energy values in a second register, R_E. To simplify the argument we assume the x variables are integers: $x \in [0, N-1]$ where $N \equiv 2^{Q_x}$ defines Q_x as the number of qubits in R_x. The state of the computer is initialized as:

$$|\Psi_0\rangle = \frac{1}{\sqrt{N}} \sum_{x=0}^{N-1} |x\rangle\, |0\rangle. \tag{2}$$

The evolution of the computer in the first instance designed to compute and store in R_E not $E(x)$, but rather the bin-function, $E_{\text{bin}}(x)$:

$$E_{\text{bin}}(x) \equiv \begin{cases} E_l, & E(x) \le E_l \\ E_h, & E(x) > E_l, \end{cases} \tag{3}$$

where the "low" and "high" values E_l and E_h may be adjusted externally. The bin position E_l is chosen using $n_C(E)$ so that $n_l \equiv n_C(E_l) \simeq N/4$ (and therefore $n_h \simeq 3N/4$). In the binning process the total quantum state of the two registers can now be re-partitioned into two orthonormal states:

$$|\Psi\rangle = \frac{1}{\sqrt{N}} \sum_{x=0}^{N-1} |x\rangle \otimes |E_{\text{bin}}(x)\rangle = \sum_{b=l,h} c_b\, |\phi_b\rangle \otimes |E_b\rangle, \tag{4}$$

where, for $b = l, h$ (n_b is the number of members of the set $\{x_j^b : E_{\text{bin}}(x_j^b) = E_b\}$),

$$|\phi_b\rangle = \frac{1}{\sqrt{n_b}} \sum_{E_{\text{bin}}(x_j^b)=E_b} |x_j^b\rangle, \quad c_b = \sqrt{\frac{n_b}{N}}. \tag{5}$$

Creating the histogram state with approximately $N/4$ states in the lower bin of interest allows us to take advantage of the special one-step search property of Grover's operator,

$$U_G = -\left(1 - 2\,|\Psi\rangle\langle\Psi|\right)\left(1 - 2|E_l\rangle\langle E_l|\right). \tag{6}$$

Evolving the state according to one application of U_G gives:

$$|\Psi'\rangle = U_G|\Psi\rangle = \sum_{b=l,h} c_b'\, |\phi_b\rangle \otimes |E_b\rangle, \tag{7}$$

with

$$c_b' = \begin{cases} c_l\,(3 - 4c_l^2), & b = l \\ c_h(1 - 4c_l^2), & b = h. \end{cases} \tag{8}$$

Perfect partitioning of $n(E)$ into two bins with $c_l = 1/2$ results in exact projection of the lower quarter of states, $|\phi_l\rangle \otimes |E_l\rangle$. This is the goal we wish to achieve – systematic projection of the lowest quarter of the spectrum – but errors in $n_C(E_l)$ will thwart any attempt. As is obvious from Eq.8, the most crucial bin positioning is the target bin, E_l. To analyse the effect of errors in $n_C(E_l)$ on the bin positions, and hence the deviation of c_l from the ideal, the bin-error, ϵ, is introduced as:

$$n_l \equiv \frac{N}{4}(1+\epsilon) \rightarrow c_l = \frac{1}{2}\sqrt{1+\epsilon}. \tag{9}$$

The bin-error is related to the error in $n_C(E)$. For $\epsilon << 1$, the probabilities of measuring states $|E_l\rangle$ in the energy register are:

$$p_l \equiv c_l'^2 = 1 - \frac{3\epsilon^2}{4} + O(\epsilon^3), \qquad p_h \equiv c_h'^2 = \frac{3\epsilon^2}{4} + O(\epsilon^3). \tag{10}$$

These equations show explicitly how we can tolerate small errors in the position of E_l. The state $|E_l\rangle$ would be found with probability $p_l = 1 - 3\epsilon^2/4$. A measured value of $E \neq E_l$

would tell us immediately that we have not succeeded in removing the higher states (or that the lowest bin was empty). After a successful measurement of $E = E_l$, the x states in the R_x register (which was undisturbed) has been reduced to the subset with $E(x) \leq E_l$. The R_x register may now be made to evolve once more with the energy register, but with a new set of bin parameters, which will again lead to the removal of all but the lower quarter of energy states. The evolution-measurement process corresponds to one step of the state reduction hybrid algorithm. With exact positioning of the $E_l^{(i)}$ (no repeats) the number of steps, k, required to reduce from $E_{\max}$ to $E_l^{(k)} \simeq \Delta E$ is $k = \log_4(N/n(\Delta E))$.

Errors associated with the bin positions that would mean repeating the sequence from the start when one measures $E \neq E_l^{(i)}$ at some iteration. Obviously, for the algorithm to proceed with virtually no repeats requires $|\epsilon| << \sqrt{2/3k}$. However, one can show that the more useful condition $\prod_i^k p_l^{(i)} \geq \frac{1}{M}$ $(M > 1)$ can be achieved with $|\epsilon| \simeq \sqrt{\frac{\log M}{3k}}$ with $O(\frac{1}{k})$ corrections.

3 A non-hybrid algorithm

In any analysis of the overall efficiency of the hybrid algorithm the computational complexity of the classical input must somehow be included in the reckoning (hidden quantum costs are yet another issue). The strategy we will follow is to incorporate a search for the correct $E_l^{(i)}$ position, a process which we assume requires $R^{(i)}$ "bin-search" steps. As we move the value of the lower bin position, let us denote by $E_l^{(i)}(j)$ the jth approximation in the bin-search process. For each j we need to know the positioning error so that the next selected value $E_l^{(i)}(j+1)$ will be closer to the true $E_l^{(i)}$. One has access to the positioning error, $\epsilon_l^{(i)}(j)$ (or equivalently the number of states in the lower bin, $n_l^{(i)}(j)$), indirectly through the probability $p_l^{(i)}(j)$ of measuring $|E_l^{(i)}(j)\rangle$. For each step j we take a sample consisting of $M^{(i)}(j)$ measurements of the energy register, and thus determine an approximation to $p_l^{(i)}(j)$. That this procedure is possible to carry out efficiently is due to the fortunate fact that the evolution of $p_l^{(i)}(j)$ always has the same form (from the expression for c_l').

At the ith iteration it suffices to start the bin-search at $E_l^{(i-1)}$ and ensure that $M^{(i)}(j)$ is large enough to detect the evolution of $p_l^{(i)}(j)$ as $E_l^{(i)}(j)$ is changed. Near the exact $E_l^{(i)}$ the number of sampling points can be increased to achieve a certain accuracy, $\epsilon^{(i)}$. The systematic binning process has ensured that we are no longer dealing with probabilities of the order of $1/N$. Each measurement in the current iteration involves the measurement of the lower bin of each preceding iteration – a fast "cascade" of projections down from the $|x\rangle$ initialization – the previous bin positions $E_l^{(i)}$ have all been determined to high accuracy and so the cascading process will only rarely lead to a repeat.

The quantum minimization algorithm is as follows:

1. ith iteration:
   - (a) $j = 0$: set $E_l^{(i)}(0) = E_l^{(i-1)}$
   - (b) jth move: shift to $E_l^{(i)}(j) = E_l^{(i)}(j-1) - \delta E(j)$
   - (c) Make $M^{(i)}(j)$ measurements, each requiring i steps from the previous iterations:
     - (i) $|x\rangle$ initialization
     - (ii) Cascade: for $p = 1...i-1$

• $E_{\text{bin}}^{(p)}(x)$ evaluation
• $U_G^{(p)}$ evolution
• $|E_l^{(p)}\rangle$ measurement ($p_l^{(p)} \simeq 1$)
• If $E \neq E_l^{(p)} \rightarrow$ go to i.

(iii) Measurement: $|E_b^{(p)}\rangle \rightarrow$ add to probability data sample.

Determine $p_l^{(i)}(j)$ from data sample and hence the next move $\delta E(j+1)$ accordingly.

(e) Repeat from (b) until $p_l^{(i)}(j) \simeq 1$ within tolerance

2. Set $i \rightarrow i+1$ and go to 1, while $i \leq k$.
3. Stop when $E_l^{(k)} \simeq \Delta E$.
4. Use Grover search, or continue moving $E_l^{(i>k)}$ down, to find the global minimum.

The algorithm is initiated with $E_l^{(1)}(0)$ arbitrary, but preferably in the region of the maximum energy. Factoring in all the previous reduction steps, the total number of measurement steps, S_{tot}, required to reduce the superposition to the neighbourhood of the global minimum is $S_{\text{tot}} = \sum_{i=1}^{k} i\, S(i)$, where the total measurement steps count for each iteration is given by $S(i) = \sum_{j=1}^{R^{(i)}} M^{(i)}(j)$. An analysis gives $S_{\text{tot}} \simeq \frac{\beta}{2}\, k^3$, where $\beta >> 1$ and depends on the problem in question.

4 Conclusion

The complexity of the non-hybrid algorithm implies that exploitation of density of states information can be absorbed into a polynomial correction, which means that the Grover limit must assert itself somewhere in the algorithm. Most likely this occurs in the dynamic implementation of the Grover operator, $U_G^{(i)}$. However, the basic algorithm framework may still prove to be useful for specific problems for which structure might possibly be helpful in constructing $U_G^{(i)}$. Work on problems along these lines is in progress.

The author thanks M. Nielson, R. Clark, G. Milburn, B. McKellar and C. Wellard for providing critical feedback and support. This work was supported by the Special Research Centre for Quantum Computer Technology.

References

1. L.K. Grover, *Phys. Rev. Lett.* **79**, 325 (1997).
2. C. Dürr and P. Høyer, "A Quantum Algorithm for finding the Minimum", quant-ph/9607014.
3. T. Hogg and D. Portnov, "Quantum Optimization", quant-ph/0006090.
4. J. Brooke, D. Bitko, T.F. Rosenbaum and G. Aeppli, Science **284** (1999) 779.
5. E. Farhi, J. Goldstone and S. Gutman, "Quantum Computation by Adiabatic Evolution", quant-ph/0001106.
6. C.H. Bennett, E. Bernstein, G. Brassard, and U. Vazirani, quant-ph/9701001.
7. M. Boyer, G. Brassard, P. Høyer and A. Tapp, Fortschr. Phys. **46** (1998), 493.
8. C. Zalka, "Grover's Qusntum Searching Algorithm is Optimal", quant-ph/9711070.

STEP AND REPEAT ION IMPLANTATION INTO ELECTRICALLY ACTIVE SUBSTRATES FOR QUANTUM COMPUTER FABRICATION

C. YANG, D.N. JAMIESON, S.M. HEARNE, P. SPIZZIRRI AND S. PRAWER

Centre for Quantum Computer Technology, Microanalytical Research Centre, School of Physics, University of Melbourne, Victoria 3010, Australia

In the Kane quantum computer proposal,[1] an array of phosphorus ions must be constructed in a silicon substrate with a period of 20-100 nm. We propose a method for constructing this array based on energetic ion implantation into an electrically active substrate through a step and repeat nano-manipulated aperture. The passage of an energetic ion into the substrate creates electron-hole pairs, which can be collected with appropriate electrodes. For example, the substrate could contain an optimised diode structure with a shallow Indium implanted surface layer forming the top contact of a PN-junction. The top contact must be very thin, but less than 20 nm appears to be possible in the-state-of-art technology. The single ion-induced charge will produce an electrical transient in the electrodes, which would be registered by a conventional instrumentation. Implantation would occur through a 20 nm aperture in a mask mounted on a readily available stage capable of the required position accuracy. This paper discusses the technical and practical aspects of the design of such a system.

Keywords: quantum computer, ion implantation, single ion-induced charge and electrical transient.

1 Introduction

For the construction of the Kane solid-state quantum computer, ^{31}P atoms are proposed to be incorporated in silicon substrate for the fabrication of a large scale qubit array.[1] The ion implantation method is a promising technique for the fabrication of such array of ^{31}P atoms in silicon substrate[2]. However, it is highly challenging to implant single ^{31}P ions in each designed site accurately, since the ion emission from ion source is random with time and the ability to focus the ion beam to the required precision does not exist. We propose a system for the online detection and manipulation of a single-ion-implantation process. The key idea is to utilise the silicon substrate, which is to be implanted, as a detector for the measurement of the single ion implantation event. The energy deposited by the single ion in the electrically active silicon substrate produces a number of electron-hole (e-h) pairs. The e-h pairs can be separated by a build-in electric field and their charge can be collected and converted to a voltage pulse, known as the energy pulse. The energy pulse contains rich information on the nature of the single ion implantation. It provides a control signal, which can be subject to online energy discrimination for feedback to an ion beam switch. Hence single ion implantation can be performed with a high accuracy.

To demonstrate the principle, we have performed experiments with well-established MeV microbeam technology where the technique is known as Ion Beam Induced Charge (IBIC)[3]. MeV He ions were implanted into a commercially available pin-diode. This allows online detection and control of the single ion implantation process. In this pilot work, the demand on the energy resolution is not very high. However, for further work with the envisaged 15-20 keV ^{31}P single ion implantation to construct the Kane computer, the energy resolution of the present system needs to be improved by at least by 5-10 times. The limit of the intrinsic energy resolution for the silicon-based detectors and technique options for improving the present system are now discussed.

2 The detection of a single energetic ion by silicon semiconductors

The detection of single ions arriving at random time intervals and implanting into an electrically active substrate is routinely done with 100% efficiency. Several options exist for fabrication of the electrically active silicon substrate. These include deposition of a thin metal surface layer to form a shallow Schottky barrier junction, fabrication of a surface PN-junction or using a piece of intrinsic silicon with an external electrical field applied from surface electrodes. Each ion implanted into the substrate yields a large number of e-h pairs that can be separated by an electric field built-in or applied between two separated electrodes attached to the surface. After the separation of the e-h pairs, the charge is collected and converted to the energy pulse through dedicated charge sensitive instrumentation. The energy pulse height linearly depends on the energy deposited in the silicon substrate by the single ion[4]. The energy spectrum of the implanted ions usually appears as a Gaussian shape.

An example of the PN junction approach is a solid-state nuclear particle detector for MeV ions. For keV ions, an optimised diode structure could be constructed similarly with a shallow indium implanted surface layer forming the top contact of the PN-junction. In order to reduce the energy loss for the incident ions, the top contact must be very thin (less than 20 nm). The energy pulse height depends linearly on the energy deposited in the electrical active substrate by each implanted ion. The pulse can be registered online and analysed and to reveal information on the depth of the incident ion and the nature of the substrate such as defects.

3 The system for online detection and control of the single-ion-implantation events

The ion emission from ion sources is a random process. This is one of major difficulties for the fabrication of a large array of ^{31}P ions in silicon. To overcome this problem, a system with the online control of the single ion implantation is required.

Figure 1 displays the layout of the proposed system capable of online detecting and manipulating the single ion implantation. The system consists of four major parts: ^{31}P ion source, ion beam on-demand controller, positioning controller and detection system for single ion implantation events. When the positioning controller is ready, computer sends a "start" signal to the ion beam on-demand controller to make ion beam to pass through. If the detection system registers an energy pulse high enough above the "analyser" rejection threshold, a "stop" signal will be sent to the ion beam on-demand controller to close the ion beam path and this ends the control loop for a single ion implantation. The computer sends another command to the positioning controller so that a new site for the next cycle of the single ion implantation will be prepared. When a new site is ready, a new cycle of the single ion implantation starts. In this way, a large array of ^{31}P ions may be constructed with pre-programming and automation.

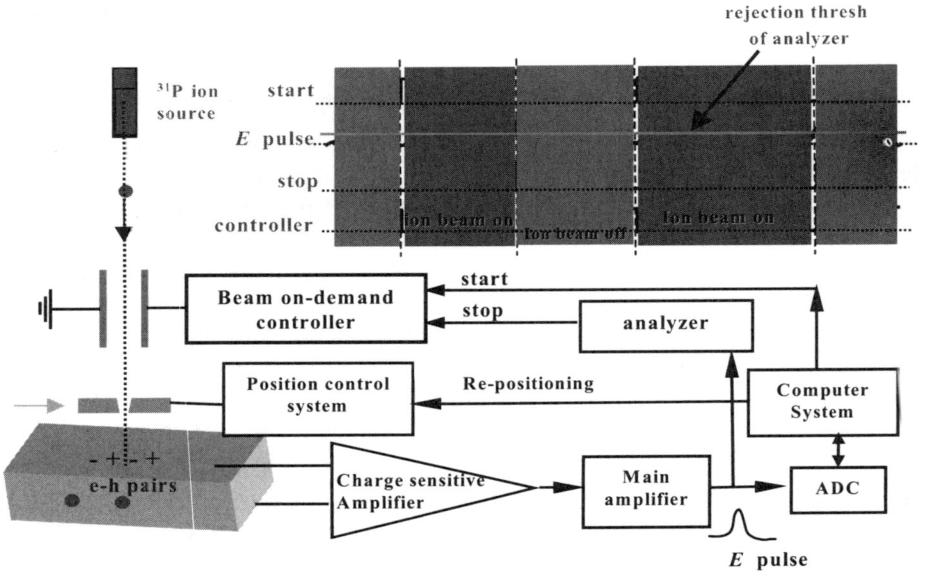

Figure 1. The system for online detection and control of single ion implantation events.

4 Experiments with a focused MeV He ions

An integrated system with a low energy phosphorous ion source is still under construction in our lab. In order to demonstrate the principle of the proposed operation system described in last section, an experiment was carried out with well-established MeV microbeam technology where the technique is known as Ion Beam Induced Charge (IBIC)[3].

In the experiment, a 2 MeV focused helium ion beam was used for the single ion implantation into a Hamamatsu pin-diode. At the first stage of the experiment, a large numbers of ions were implanted into the pin-diode at a 4 by 4 spot array. Each spot in the array received 3.75×10^5 ions, which induced a high density of defects. The spot size of the implantation was the focused beam size, which was about 3 microns. At the second stage of the experiment, a larger two-dimensional scan with 256 by 256 pixels was made over the previously implanted area with dose of one single ion at each pixel. The energy pulse measured at each pixel was stored together with the X-Y position information in the computer. An energy pulse contrast map, termed a detection efficiency contrast map, was constructed to reveal the hot spots where reduced detection efficiency was shown against the undamaged area where 100% detection collection efficiency was remained. The experiment result was displayed in figure 2. For the construction of this energy pulse contrast map, only a single ion implantation per pixel is required. It is evident that the localized damage at this dosage level does not reduce the detection efficiency at the other undamaged area of the silicon substrate.

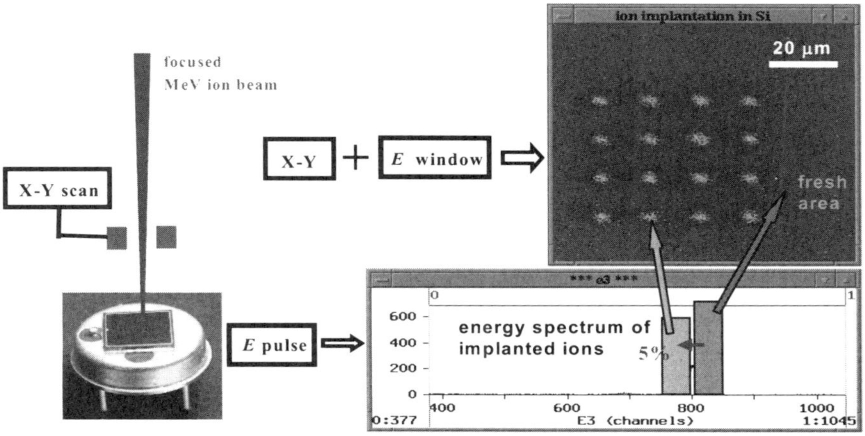

Fig. 2. Single ion imaging MeV ion implantation into a pin-diode.

5 Improvement of the detection system for 15-20 keV phorsphorous-31 ions

In the measurement of the MeV ion implantation event in silicon substrate, the present standard nuclear instrumentation is readily available for the normal energy resolution about 10-20 keV. This energy resolution is enough for resolving the detailed information on the MeV ion implantation. However, for the low energy phosphorous ions, efforts have to be made in the improvement of the energy resolution for the present system. Usually, the total energy resolution consists the following two components:[4]

$$\Delta E=(\Delta E)_{intrinsic}+(\Delta E)_{electronic} \quad (1)$$

If the electronic noise can be reduced close to zero, the ultimate energy resolution is determined by the intrinsic energy resolution. The relative energy resolution:[4]

$$R= \Delta E/E=2.35(F\bullet w/E)^{1/2} \quad (2)$$

Where F is the Fano Factor (~ 0.12), w is the average energy for e-h pair creation (in Si 3.62 eV at 300K, 3.81 eV at 77K). For the measurement of 20 keV ^{31}P ions, this gives $R\sim 2\%$, and $\Delta E \sim 200eV$, that is sufficient for the detection of 15-20 keV ^{31}P ions. For the large reduction of the electronic noise $(\Delta E)_{electronic}$, following approaches will be considered: optimizing the charge sensitive amplifier; cooling the electronics and electromagnetic shielding.

Conclusions

The detection of the single-ion-implantation event in silicon through the charge collection of the induced e-h pairs is a promising method for the on-line control of the single-ion-implantation for the massive production of qubits. The electrical registration of the single-ion-implantation in silicon substrate has been developed and demonstrated successfully for MeV high-energy-ion implantation. The energy resolution for the detection system utilized in the present ion implantation should be improved toward the ultimate limit of the intrinsic energy resolution (200 eV) of silicon detectors for 15-20 keV phosphorus ion implantation. This will allow large arrays of single phosphorous ions to be constructed.

References

1. B.E. Kane, Nature, 393 (1998) 133-7.
2. D.N. Jamieson, "A role for ion implantation in quantum computing" IQC01
3. M.B.H. Breese, D.N. Jamieson and P.J.C. King, Materials Analysis with a Nuclear Microprobe, John Wiley and Sons, New York, 1996.
4. W.R.Leo, Techniques for Nuclear and Particle Physics Experiments, Springer-Verlag, 1987,1994, p159.

THE STUDY OF C_{60} AS A THIN FILM RESIST IN THE KANE QUANTUM COMPUTER

V.MILLAR, I.ANDRIENKO, A.CIMMINO AND S.PRAWER

Spe cialResearch Centre for Quantum Computer Technology and The Microanalytical Research Centre, School of Physics, The University of Melbourne, Parkville, 3010, Victoria, Australia

T.C.RALPH

Department of Physics, University of Queensland, QLD 4072, Australia

The Kane proposal for a solid state quantum computer, based on an array of phosphorus nuclear spins constructed in silicon, offers a practical method of realising a quantum computer. One proposal for the construction of a phosphorus spin array is based on ion implantation of P ions directly into the silicon matrix. The location of each P atom is revealed by using the damage created in thin film resists which have been deposited onto the silicon substrates prior to implantation. Results are presented in this paper of the use of a novel resist which is very sensitive to the passage of ions. The novel resist is a monolayer of C_{60}. This is normally an insulator, but ion implantation can result in the formation of conducting regions in the film which can then be detected by Scanning Probe Microscopy techniques. Experiments described here involved in-situ Ar ion irradiation of C_{60} films on silicon substrates under UHV conditions and subsequent use of the STM to perform current mapping of the surface. The resulting images revealed considerable changes to the C_{60} layer for doses above 10^{14}ions/cm^2. Areas corresponding to ion impact sites showed an increase in surface conductivity by an order of magnitude. However close analysis of the STM images show that the C_{60} films do not form uniform monolayers and are highly defective. Consequently the changes in topography and conductivity created by low doses (less than 10^{13} ions/cm^2) were not distinguishable from nanometre scale spatial inhomogeneity in the as-deposited C_{60} films, making it difficult to observe the effect of single ion impacts.

1 Introduction

Practical implementation of a quantum computer poses a formidable challenge. The Kane proposal for a solid state quantum computer assumes single P atoms can be placed into a Si crystal with near atomic precision [1]. In Si, significant coupling will occur when $4J \approx \mu_B B$, where μ_B is the Bohr magneton, B is the applied magnetic field and $4J$ is the exchange energy which depends on the overlap of the electron wavefunctions. This condition approximates a separation between P donors of 100 - 200Å. This separation is near the limit of current electronics fabrication technology and so new techniques of locating single ions need to be explored.

One approach for construction of a P spin array is to randomly introduce P atoms into Si through a thin film resist that will register the passage of a single ion. A resist layer is often chosen such that the damage created by ion implantation causes a measurable change in some property of the resist under ion impact.The location of each implanted P atom is then revealed by using the latent damage created in the thin film resists.

In the present work we explore the use of C_{60} as a resist layer for registering the passage of a single ion. The most abundant fullerene cluster is C_{60} which has a closed "soccerball"-type structure comprised of 20 hexagons and 12 pentagons in which each carbon atom is sp^2 bonded to its neighbours [2]. In thin film form the material as a whole is electrically

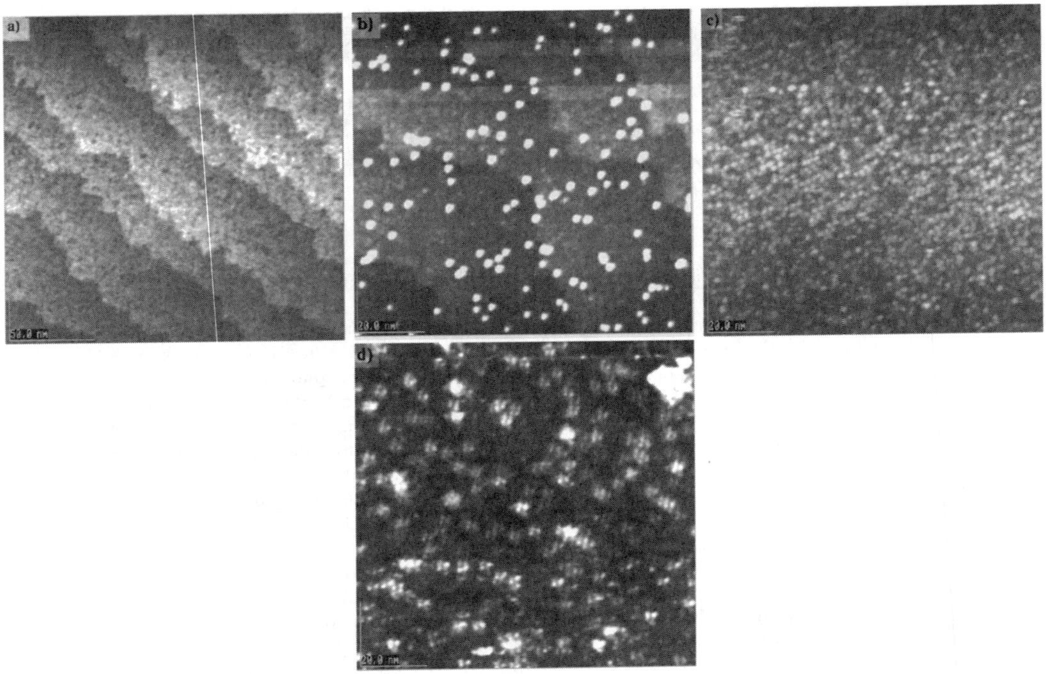

Figure 1. STM images showing the various stages of the evaporation of C_{60} onto the Si substrate a)clean Si(100)2 $\times$ 1 surface showing monoatomic steps and some defects b)C_{60} absorbed on the Si surface with a coverage of approximately 0.08 monolayers c)full monolayer coverage of the Si substrate before irradiation d)the C_{60} surface after irradiation (10^{14} ions/cm^2)

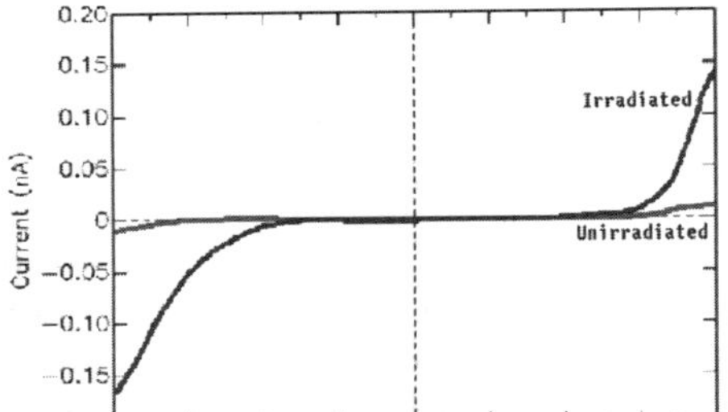

Figure 2. Tunnelling Current versus voltage for irradiated and unirradiated ares of the C_{60} film.

insulating [3]. However conducting regions can be created by ion implantation. The ion beam-fullerene interaction is such that each C_{60} molecule is destroyed in a single event; no smaller fullerene clusters can be detected. The disintegration occurs when C atoms are knocked out of the molecule either directly by the incident ion or by an energetic knock-on C atom within the damage cascade [4]. Upon disintegration each C_{60} molecule releases

60 carbon atoms which disperse amongst the remaining intact C_{60} spheres giving rise to hopping conductivity in the film.

2 Experiment

Phosphorus-doped n-type Si wafers of (100) orientation were used for all studies in this report. The main UHV chamber in the JEOL JAFM 4500XT is pumped by a 200 l/s custom design ion pump and the base pressure remains below 3×10^{-10} Torr. The pressure rise during the final annealing of the Si sample does not go beyond 1×10^{-9} Torr in this system. The Si substrates were mounted on the scanning stage of the microscope and baked at 150°C for 10 hours to remove impurities from the surface. The clean Si(100)2 × 1 reconstructed surface was obtained by repeatedly annealing the Si(100) wafers up to 1200°C followed by rapid cooling to 900°C and gradual cooling to room temperature over a period of 3 minutes [5].

The C_{60} was evaporated in-situ onto the Si substrate, held at room temperature, using a UHV tungsten filament evaporator located in the direct line of the sample. The in-situ evaporation allowed detailed STM imaging and analysis of the C_{60} as a function of evaporation time. To obtain the best possible stability for scanning, electrochemically etched tungsten tips were used.

For convenience, ion irradiation was performed not with ^{31}P but with the Ar^+ ion gun that is located in the UHV chamber. Although ^{31}P is required for the Kane proposal, Ar^+ ions are suitable for this work as we are only interested in registering ion impacts. The samples were transferred under UHV conditions to the preparation chamber where they were irradiated with 1keV Ar^+ ions with doses between 10^{12} and greater than 10^{14} ions/cm^2.

3 Results

During the evaporation of C_{60} onto the substrate and following irradiation STM images were periodically taken so the surface topography could be monitored. Figure 1a) shows the clean(100)2 × 1 reconstructed surface 1b) shows approximately 0.08 monolayers of C_{60} absorbed on the Si surface 1c)shows a complete monolayer coverage of C_{60} and 1d)shows the C_{60} surface following irradiation with a dose of 10^{14} ions/cm^2. Large changes in topography and conductivity (see Figure 2) were seen following irradiation with doses $\geq 10^{14}$ ions/cm^2.

However the ion dose was sufficiently high that the ion tracks are expected to overlap and so it cannot be concluded that each depression in the C_{60} film corresponds to a single ion impact. As the detection of single ions is necessary for the Kane proposal the experiment was repeated with lower doses of 10^{12} and 10^{13} ions/cm^2. The variations in topography and conductivity created by these doses were indistinguishable from those seen in the film prior to irradiation. For example, after irradiation the surface roughness (0.73 ± 0.22nm) was comparable to the surface roughness prior to irradiation (0.62 ± 0.24nm). Variations in the conductivity following irradiation were of the same order of magnitude as those observed prior to irradiation.

4 Discussion

The variation in the topography and conductivity seen in the C_{60} films prior to irradiation suggests that the as-deposited C_{60} films do not form perfect monolayers. Work by Wang.*et.al* suggests that the bonding between C_{60} and the Si substrate is much stronger than that among the C_{60} molecules. As a result the C_{60} monolayer does show local ordering but no large scale ordering [6]. Depressions in the STM images may therefore correspond to the Si substrate or a highly defective film which produces higher current at defective points. Protrusions may correspond to a well ordered film or multiple C_{60} layers which have low conductivity. As the damage caused by low ion doses is indistinguishable from the natural variations already present in the C_{60} film the detection of single ion tracks was not possible. Our group is currently pursuing organic self-assembled monolayers, which show large scale ordering, as another possibility for single ion registration.

5 Conclusion

For large ion doses ($\geq 10^{14}$ ions/cm^2) the effects of irradiation are clearly visible. At this dose, irradiation of the C_{60} film results in large changes to the surface topography and an increase in conductivity over areas where single or multiple ion impacts have occured. Changes induced by doses lower than 10^{14} ions/cm^2 are not distinguishable from the innate spatial inhomogeneity of the unirradiated C_{60} film. This is a result of C_{60} not forming a perfect monolayer. For single ion impacts to be registered a well-ordered defect free film needs to be created one example is organic self-assembled monolayers.

References

1. B.E Kane, *Nature*, **393**, 133, (1998).
2. H.W.Kroto, J.R.Heath, S.C.O'Brien, R.F.Curl and R.E.Smalley,*Nature*, **318**, 162, (1985).
3. R.C.Haddon, *Acc.Chem.Res.*, 25, (1992).
4. S.Prawer, K.W.Nugent, S.Biggs, D.G.McCulloch, W.H.Leomg, A.Hoffman and R.Kalish, *Phys.Rev.B.*, **52**, 841, (1995).
5. R.J.Hamers,R.M.Tromp adn J.E.Demuth,*Phys.Rev.B.*, **34**, 5343 (1986).
6. X.-D.Wang,T.Hashizume,H.Shinohara,Y.Saito,Y.Nishina and T.Sakurai, *Phys.Rev.B.*, **47**, 15923 (1993).

THE FABRICATION OF TEST STRUCTURES FOR PHOSPHORUS SPIN DETECTION USING *MEV* ION-IMPLANTATION.

A.C.Y. LIU AND J.C. MCCALLUM

Centre for Quantum Computing Tehcnology, Micro-Analytical Research Centre, School of Physics, University of Melbourne, Parkville, Melbourne, Victoria 3010, Australia

J. WONG-LEUNG

Department of Electronic Materials Engineering, Research School of Science and Engineering, Australian National University, Australian Capital Territory, 0200, Australia

The crystallisation of deep amorphous silicon volumes with lateral dimensions of 1 μm constrained by single crystal silicon is investigated. These volumes, denoted amorphous silicon wells have been formed using MeV ion implantation through a mask. During annealing the amorphous phase is transformed to the crystalline phase through the process of solid phase epitaxy, which involves the organisation of atoms at the interface between the phases. The dynamics of this process and the secondary structures that are incorporated into the crystal have been c haracterised using atomic force microscopy and transmission electron microscopy . The same mask and a different implantation regime may provide test structures for phosphorus spin detection using magnetic resonance force microscopy .

1 Introduction

Solid phase epitaxy describes the process by which the amorphous phase of some materials may be transformed to the crystalline via layer-by-layer organisation of the amorphous volume using a crystalline seed as a template and energy, added to the system in the form of heat or ion bombardment. We examine this phenomenon in deep amorphous volumes which are constrained in t wo dimensions by the seed material. Interfaces occur both parallel and perpendicular to the surface normal of the wafer and crystallisation may proceed vertically and laterally. Interest in these deep amorphous structures and their behaviour upon heat treatment is motivated by the use of high energy implants to create uniform doped regions with a higher aspect ratio than is possible through diffusion techniques[1].

2 Experiment

(100) silicon wafers are subjected at liquid nitrogen temperature to a multi-energy self implant through a nickel grid. Eight energies in the range 100 $keV - 8\ MeV$ are sufficient to amorphise the regions exposed to the impinging beam to a depth of 5 μm. The sample thus consists of an array of these amorphous structures with lateral dimensions of 1 μm and a period consistent with that of the apertures in the grid (12 μm). At a temperature of 620 ^oC crystallisation proceeds inwards at both the lateral and vertical interfaces. This process is sumarised in Fig. 1. The structures are characterised at stages in this processing to articulate the dynamics of crystal recovery and describe the defects that are incorporated into the recovered crystal. This is perfomed with atomic force microscope (AFM) topographies and with transmission electron microscopy (TEM) analyses.

Annealing was perfomed on a hot stage with a vacuum chuck in air. It is known that the infiltration of hydrogen from the native oxide supresses epitaxy, however, in this sample

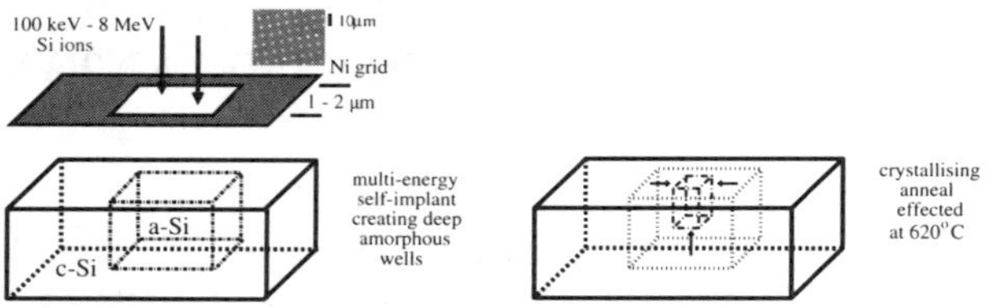

Figure 1. The process of creating an amorphous silicon well. An optical image of the nickel grid used as a mask is shown as an inset.

geometry enough material remains unperturbed by this effect for the intrinsic behaviour to be studied[2]. AFM surveys were performed in air. Preparation of electron transparent regions for the TEM studies followed conventional thinning techniques [3]. All the TEM images were obtained on a Philips EM 430 microscope operated at 300 keV.

3 Results and Analysis

3.1 Atomic Force Microscope Topography

The topology of the samples was mapped using an AFM. The density of amorphous silicon is 1.8 % less dense than that of the crystalline phase[4]. Investigations with the AFM discovered that the amorphous volumes plastically deformed to equilibrate this density difference giving rise to a sample topology of regular surface swellings. Fig. 2 displays topographies of the as-implanted case (top) and also the sample after it has undergone an anneal at 620 ^oC for 20 $mins$ (bottom). Several things may be noted from these topographies. Firstly, the implanted region corresponds well to the footprint of the grid apertures. Secondly, the amorphous volume protruding from the surface diminishes upon annealing consistent with the fact that the amorphous material is being incorporated into the crystalline phase. However, this protrusion does not disappear, even after subsequent prolonged annealing suggesting that the atoms in the surface feature are not all transported down to participate in the crystal growth.

3.2 Transmission Electron Microscopy

TEM analases were performed on these samples to gain insight into the dynamics of the crystallisation and to observe the quality of the regrown material. Fig. 3 displays a zone axis bright field image of a partially crystallised amorphous volume treated for 7 $mins$ at 620 ^oC and studied in plan-view. The original lateral amorphous crystalline interface is evident as a band of dislocations. Linear faults grow in from this to the current lateral interface which marks the perimeter of the remaining amorphous volume. The crystallisation demonstrates favoured growth directions and concomitant faceting to accommadate these. Comparison with the selected area diffration pattern allows these directions to be identified as [010] directions. This is in accordance with earlier studies of lateral solid phase epitaxy [5].

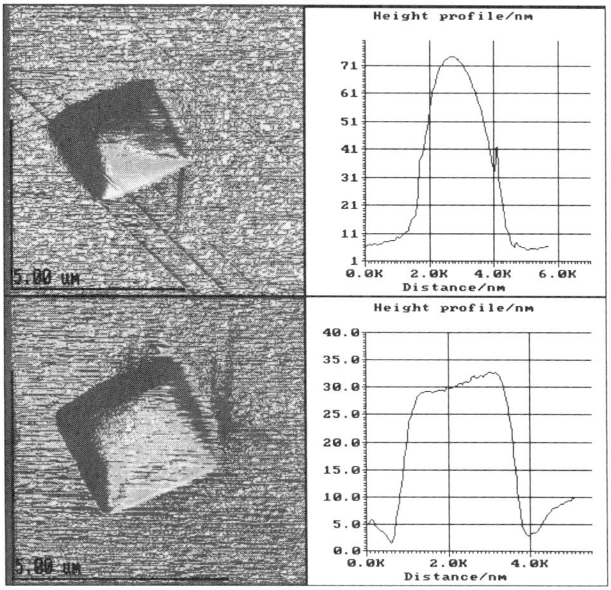

Figure 2. AFM topographies of an as-implanted (top) and annealed (bottom) well with the height profiles taken through the centre of each well displayed to the right.

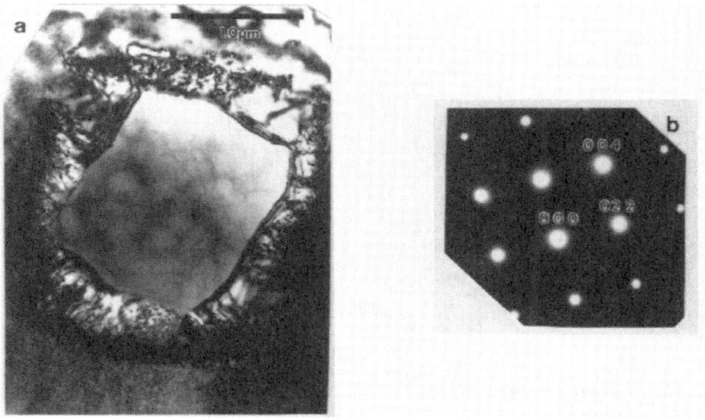

Figure 3. Zone axis bright field of a partially crystallised amorphous well. Many defective structures may be identified. Comparison with the selected area diffraction pattern shown in part b) of the figure allows the prominent growth directions to be identified as [010].

4 Test Structures For Phophorus Spin Detection

It is proposed that local crystalline volumes bounded by amorphous silicon may be obtained using the same mask and a different implantation regime. This regime is outlined in Fig. 4. A large area multi-energy self-implant amorphises the entire area. A multi-energy boron implant through the grid creates select volumes of nearly constant boron concentration.

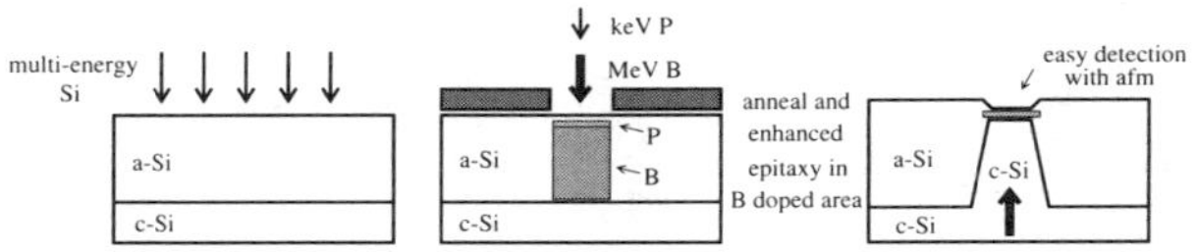

Figure 4. Schematic illustrating the production of test structures for phosphorus spin detection.

A low energy phosphorus implant through the grid creates regions of sparse phophorus contamination for study. As the system is taken to crystallising temperatures crystal growth occurs seeded by the underlying crystal. Enhanced growth due to the presence of boron [6] will result in pillars of crystalline material embedded in the amorphous phase. The density difference between the phases implies that a depression is created atop the crystalline regions which will be straightforward to locate with magnetic resonance force microscopy.

5 Conclusions

The crystallisation of deep amorphous silicon structures constrained by the single crystal seed material proceeds according to fast growth along [010] directions. Many defective structures are incorporated into the resulting crystal. A method has been outlined for the construction of test structures for phosphorus spin detection. These rely on the creation of crystalline silicon volumes embedded in the amorphous phase.

Acknowledgements

The author would like to acknowledge the expertise of Alberto Cimmino in generating the AFM topographies and the use of the Philips EM 430 microscope at the EM unit in ANU.

References

1. L. Rubin and W. Morris, Semiconductor International (1997), April, p. 77.
2. G. Olson and J. Roth, Mat. Sci. Rep. bf 3 1 (1998).
3. PhD thesis, Jennifer Wong-Leung, ANU, 1997.
4. Olson
5. H. Yamamoto, H Ishiwara and S. Furukawa, Jap. J. Appl. Phys. **25** 667 (1986).
6. Olson

STUDIES OF PMMA FOR THE ION BEAM ROUTE INTO QUANTUM COMPUTING

DAVID A. BRETT, DAVID N. JAMIESON, ALBERTO CIMMINO, IGOR ANDRIENKO AND STEVEN PRAWER

Special Research Centre for Quantum Computer Te chnology, Microanalytical Research Centre, School of Physics, University of Melbourne, Victoria 3010, Australia

Several parallel routes have been proposed for construction of a phosphorus spin array for a solid state quantum computer. These include nanofabricaton based on hydrogen resists and implantation of energetic ions. We have investigated techniques for the use of ion implantation. In our proposal we employ single ions to create latent damage in a thin layer of PMMA deposited on a silicon substrate. Development and possible metalisation of the latent damage could result in a self-aligned structure with metal dots above buried phosphorus atoms. This is a necessary requirement for the A gates in a Kane quantum computer. This paper investigates the use of PMMA as a suitable resist for registering the passage of a single ion. Issues that are addressed include the sensitivity of the resist to ion impact and evaluation of schemes to produce the required spacing of the phosphorus atoms for creation of a few qubit quantum computer. Initial experiments are reported based on pilot studies of alpha particle implants into 50–100 nm PMMA resist films using the Melbourne nuclear microprobe that show considerable promise for this method.

1 Introduction

A few qubit quantum computer could be constructed using randomly implanted phosphorus ions. These ions would be implanted into a substrate containing an array of pre-fabricated structures, called cells, for control of the qubits. One of the key requirements is a material which is capable of registering the passage of a single ion into the substrate and thereby marking its position. Usefully positioned ions in cells can then be identified by an appropriate surface scan to locate the registration marks.

Polymethyl Methacrylate (PMMA) is capable of fulfilling this role. PMMA is a commonly used positive resist for nanolithography, has a high resolution and is reasonably sensitive. Previous work has indicated that PMMA is capable of detecting single heavy ions,[1,2] which is promising for our proposal to detect the passage of single phosphorus ions.

Pilot experiments were performed with thin PMMA films and MeV helium ions conveniently available in our laboratory. Numerical simulations with the code "The Stopping and Range of Ions in Matter 2000.39" (SRIM)[3] showed that MeV helium could be used as an approximation to phosphorus. Furthermore, an evaluation of the expected yields of usefully placed ions randomly implanted into arrays of cells prefabricated in the substrate was also conducted.

2 Theoretical Considerations

2.1 Yields

For an array of cells with a given period on the substrate, the dose of the randomly implanted ions can be optimised to maximise the probability of getting one ion per cell. At the optimum dose, Poisson statistics show that the probability a given cell is occupied is 36.8%. In the Kane quantum computer, a one qubit device requires two adjacent phosphorus ions. The probability that two adjacent cells are occupied is 13.5%. However, by careful arrangement of the cell configuration, it will be possible to raise the number of adjacent occupied cells. Geometrically, the optimum configuration is that of an equilateral hexagon, as it allows the maximum connectivity between cells. If this configuration is considered, then the probability of getting at least two adjacent cells with one phosphorus atom in each is 65.7%. If this configuration were to be tessellated, then the probability is further increased. However, this configuration may not be compatible with the fabrication techniques for adjacent control circuitry.

It should be noted that additional problems will reduce the yield of useful qubits. These include range straggling of the ions, damage to the silicon crystal and contamination of the crystal with other elements. These problems are discussed further below.

We propose the implantation of 10 keV P^+ ions through a mask into a bi-layer resist some 50 nm thick, the bottom layer of which is sensitive to single ion passage. This thickness is much greater than the range of 10 keV P^+ ions and therefore few ions reach the substrate. The advantage of this is that the ions that reach the substrate generally have straight paths, as scattering from the straight path virtually eliminates the chance that the ion will reach the substrate. SRIM simulations indicate that the ions do not penetrate very far into the substrate, but it is proposed that by orienting the crystal such that a low order crystal axis is parallel with the incident beam, then the final depth of the ions may be improved by ion channelling.

It should be noted that SRIM uses the binary collision approximation, which is not valid for energies below one keV, thus the information obtained for the end-of-range distribution of the ions may not be accurate. For our purposes, we will assume that 90% of phosphorus atoms arriving in the silicon substrate are at an appropriate depth.

Damage to the crystal caused by ion implantation may also reduce the yield of useful cells. However, most of this damage can be removed by rapid thermal annealing without causing the implanted ions to migrate very far. We assume that 10% of cells are inoperable due to crystal damage.

When implanting through a resist such as PMMA, atoms in the resist are knocked into the substrate. These atoms can alter the magnetic environment in which the phosphorus atoms find themselves and prevent a cell from functioning. SRIM simulations indicate that for each phosphorus atom that reaches the substrate, we could expect it to be accompanied by one oxygen, four carbon and, more importantly, six hydrogen atoms, which possess non-zero nuclear moments. For our purposes, we will assume that 90% of cells are uncontaminated.

Using our assumptions for yield reduction factors discussed above, we obtain a success probability of 12.4% per pair of holes, or 62.6% for the equilateral hexagon for a two qubit.

2.2 PMMA

Ion irradiation of PMMA produces latent damage along the ion tracks. Development of this latent damage makes the ion track visible. Single ion tracks have previously been noted in PMMA for high energy heavy ions, however there has been no prior work regarding keV ion tracks in this material. For our pilot experiments helium ions were chosen for this purpose, as SRIM indicates that the electronic stopping for helium ions at three to four MeV was approximately 11 eV/Åcompared with that of 9 eV/Åfor phosphorus at 10 keV. Phosphorus possesses a much higher nuclear stopping power than helium, but considering the fact that the longest range phosphorus ions, which we require, undergo little or no nuclear stopping, it was felt that experiments with helium were still useful. Future experiments will be done with phosphorus when our new phosphorus implanter is commissioned.

Combining the known dose required to expose PMMA[4] with values for the range of electrons emitted by passage of a three MeV Helium ion,[5] a diameter of 25–30 nm was obtained for the latent damage caused by single ion passage. Therefore the damage caused by single Helium ion passage in PMMA should be detectable.

3 Experiment

Silicon substrates were coated with PMMA at the Semiconductor Nanofabrication Facility at the University of New South Wales (UNSW) and implanted with three and four MeV Helium ions using the nuclear microprobe attached to the University of Melbourne's 5U Pelletron. Following implantation, the samples were returned to UNSW for chemical etching and the returned to the University of Melbourne for imaging with the School of Physics' Jeol JSTM–4200A Scanning Probe Microscope.

4 Results

Imaging of irradiated regions after developing revealed holes in the resist consistent with single ion passage, once it is noted that prolonged development gradually removes unexposed PMMA as well. Representative images are shown in figure 1. The images shown in figure 1 are from one implant site, where the ion flux varied from complete exposure of the PMMA (top left) through to no exposure (bottom right). All images show an area of one square micron. One of the problems encountered with the experiment was that the ion flux was non-uniform. Consequently, it is impossible to accurately determine a local ion dose and confirm that the areal density of the observed holes was the same as the ion dose. Nevertheless, the structures are consistent with the dose used. Also evident in the images are balls of PMMA, which have been noted previously and are believed to be balls of polymer chain.[6]

5 Conclusion

On the basis if the pilot experiments it appears that PMMA is capable of detecting the passage of single ions, however further work is required to optimise the PMMA film thickness and development process. When used in conjunction with hexagonal arrays of cells, the developed PMMA surface layer imaged by SPM can identify the locations of adjacent cells loaded with qubits for further study.

Acknowledgements

We would like to acknowledge the invaluable assistance of Andrew Dzurak and Nancy Lumpkin at the University of New South Wales.

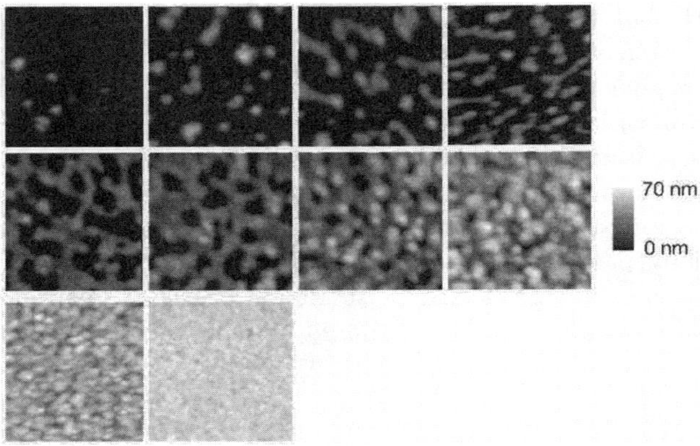

captionDeveloped PMMA showing a range of exposures. Each image covers an area of 1 square micron.

References

1. R.M. Papaléo, L.S. Farenzena, M.A. de Araújo, R.P. Livi, Nucl. Instr. and Meth. B 151 (1999) pp. 135–139.
2. R.M. Papal´eo,L.S. Farenzena, M.A. de Araújo, R.P. Livi, M. Alluralde, G. Bermudez, Nucl. Instr. and Meth. B 148 (1999) pp. 126–131.
3. J.F. Ziegler, J.P. Biersack, U. Littmark, The Stopping and Range of Ions in Solids (Pergamon Press, New York, 1985)
4. J.A. Van Kan, J.L. Sanchez, B. Xu, T. Osipowicz, F. Watt, Nucl. Instr. and Meth. B 158 (1999) pp. 179–184
5. A. Bettiol, PhD thesis, University of Melbourne, 1999.
6. F. Schrempl, W. Witthuhn, Nucl. Instr. and Meth. B 132 (1997) pp. 430–438

DC SQUID DEVICES FOR NMR IN PHOSPHOROUS-IMPLANTED SILICON

C.I. PAKES, D.N. JAMIESON, S. PRAWER

Centre for Quantum Computer Technology, School of Physics, University of Melbourne, Victoria 3010, Australia

P.W. JOSEPHS-FRANKS, R.P. REED

Fundamental Metrology Group, National Physical Laboratory, Queens Road, Teddington, Middlesex TW11 0LW, United Kingdom

The Kane solid state quantum computer[1] aims to exploit as qubits phosphorous-31 nuclear spins embedded in a silicon matrix. One proposal for the fabrication of an array of phosphorous spins is by the ion implantation of P ions directly into silicon. Little is known about the subsequent damage this may cause within the environment of the embedded ion and the influence of damage on its operation as a qubit. We propose an experiment to investigate this issue by performing nuclear magnetic resonance for a low population ($\sim 10^6$) of phosphorous donors using a miniature dc superconducting quantum interference device (SQUID) based system. We present preliminary results which examine the scaling of flux sensitivity of our devices with SQUID loop dimension and demonstrate that our smallest device, with loop dimension 3 micron, has a spin sensitivity of about 40 spins in a 1 Hz bandwidth. We address the task of coupling our SQUID sensor to an ion-implanted sample situated within the loop of a pick-up coil.

Keywords: Quantum Computer, Ion implantation, SQUID, NMR

1 Introduction

Nuclear spins have been incorporated as qubits into several proposals for the development of a quantum computer.[1-3]. We are particularly interested in the solid state quantum computer proposed by Kane[1], which is based on an array of $I = \frac{1}{2}$ phosphorous-31 ions embedded within a silicon matrix. One proposal for the experimental realisation of such an array is by the implantation of phosphorous ions directly into silicon.[4]. A major issue relating to the effectiveness of this technique is the extent of damage imposed on the silicon matrix by the implanted ion and the influence of damage on its operation as a qubit. We propose an experiment to measure nuclear magnetic resonance (NMR) at low temperature, using a dc superconducting quantum interference device (SQUID), for a low dose of phosphorous ions implanted in intrinsic silicon. This capability would allow us to explore the potential benefit of both implantation along channel planes in silicon and post-implantation annealing in attempting to create phosphorous donor sites in a defect-free silicon environment.

The nuclear magnetization of phosphorous donors in silicon has been studied in two ways. In the first method a dc SQUID was used to measure the static nuclear magnetization in samples containing $\sim 10^{17}$ donor atoms.[5]. In the second method the moment of the substitutional phosphorous nucleus was probed through a hyperfine interaction with its bound donor electron, in a sample of $\sim 10^{16}$ donor atoms, by Electron-Nuclear Double Resonance.[6]. Combining the extreme sensitivity of a miniature thin-film dc SQUID sensor[7,8] with the selectivity of NMR may be useful in probing the nuclear magnetization of a significantly smaller population[9] of implanted phosphorous donors. A major consideration in achieving the required sensitivity of a SQUID sensor is the scaling of the internal dimension of the SQUID loop. Several niobium dc SQUID devices, with varying loop dimension, have been designed and commercially fabricated by HYPRES. In an earlier publication we outlined the measurement of the flux sensitivity for our device with smallest loop size (3μm × 3μm).[10]. Here we examine the variation of flux sensitivity with the scaling of the loop dimension. Our smallest SQUID should be capable of detecting about 40 spins in a 1 Hz bandwidth. To utilise the full sensitivity of the SQUID, the sensor must be shielded from an applied magnetic field and the signal to be measured must be brought to the sensor via a flux transformer. We outline a scheme for achieving this arrangement and show that the expected spin sensitivity of our complete system incorporating a flux transformer is ~5×10^4 spins in a 1 Hz bandwidth.

2 Squid Sensitivity Measurements

Reducing the loop area of the SQUID, which reduces its inductance, can increase the energy sensitivity of the device to near quantum limit operation. Furthermore, a device incorporating a SQUID of small loop area has reduced sensitivity to external magnetic fields. In the thermal limit, the energy sensitivity of a SQUID, of capacitance C and inductance L, operating at a temperature T is given by

$$\varepsilon = 16k_B T(LC)^{1/2}, \tag{1}$$

where k_B is the Boltzmann constant. The electronic spin sensitivity is given by

$$S_n = a\frac{\phi_{ns}}{2\pi\mu_B\mu_0}, \tag{2}$$

in units of spins (of moment μ_B) per √Hz, where ϕ_{ns} is the flux noise density, related to the energy sensitivity by $\phi_{ns} = (2\varepsilon L)^{1/2}$, a is the dimension of the SQUID loop. The SQUID loop inductance $L \sim 1.25\mu_0 a$.

We have measured and made a comparison of the sensitivity of three devices, of loop dimension 3μm, 10μm and 30μm. All incorporate dc SQUIDs with junction size 3μm × 0.2μm, dielectric thickness 3μm, junction capacitance 10 fF and measured critical current density ~10^3 Acm^{-2}. Each junction is resistively shunted (R = 10Ω). The devices were mounted in a helium cryostat, which was magnetically shielded using a double walled μ-metal shield to a level of 10 nT, while external field fluctuations were less than the expected flux sensitivity of the SQUIDs. A coil inside the shield was employed to permit the application of both a constant and time-dependent magnetic field.

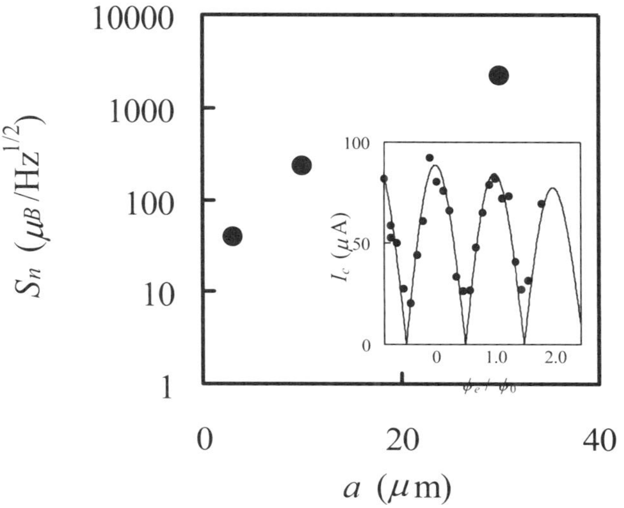

Fig. 1. Variation of spin sensitivity with SQUID loop dimension. Inset: Variation of I_c with applied magnetic flux, ϕ_e/ϕ_0, for the 3μm × 3μm SQUID ($\phi_0 = h/4\pi$)

Measurement of the magnetic field dependence of the I-V characteristics of each device have been carried out at a chosen operating temperature of 8 K, for which no hysteresis was observed in the transport measurements of any of our SQUIDs. Typical results, showing the observed variation of critical current, I_c, with applied magnetic flux, ϕ_e, are shown in Fig. 1(inset). The data are for the 3μm × 3μm SQUID. The solid lines are a fit to theory.[10]. The sensitivity was evaluated by determining the peak voltage response, $R\partial I_c/\partial\phi_e$, obtained from Fig. 1 (inset). Noise measurements demonstrate a white noise floor level of 1 nV/√Hz, which is limited by the head amplifier of our commercial SQUID electronics (Oxford Instruments DC SQUID Controller). This corresponds to a flux noise density of 5×10^{-7} ϕ_0/√Hz for our smallest SQUID. Evaluating $R\partial I_c/\partial\phi_e$ for all devices yields a scaling of spin sensitivity (2) with SQUID loop dimension as shown in Fig. 1, where we have taken the head amplifier noise as limiting the performance of each device. The 3μm × 3μm device has a measured spin sensitivity of 38.7 spins/√Hz. In the thermal noise limit (1), a spin sensitivity of 2.5 spins/√Hz is anticipated at our chosen operating temperature of 8 K.

3 Expected sensitivity with flux transformer

The proposed arrangement of signal and pick-up coils to form a flux transformer is as follows. The 3μm × 3μm SQUID chip is to be housed in a niobium can. The signal to be measured is brought to the SQUID by a single-turn niobium signal coil, of inductance L_S~5 pH, fabricated upon the same chip. The signal coil is connected by means of a twisted pair of niobium leads, of inductance

L_L~0.1 μH, to a pair of counterwound niobium pick-up coils, each of N_P=10 turns, inner diameter 50 μm and inductance L_P~1 fH, fabricated on an intrinsic silicon wafer. The sample is to be formed by implanting 20 keV phosphorous ions directly within one of the split pair, through a 50 μm pinhole mask. The pick-up coil, along with RF field coils, are to be contained within a lead tube to trap a steady external magnetizing field.

A flux change ϕ_P through the pick-up coil will couple to the SQUID a flux

$$\phi_{SQ} = -\frac{M_S N_P}{L_S + 2L_P + L_L}\phi_P, \tag{3}$$

where $M_S \sim 0.2$ fH is the mutual inductance between the signal coil and the SQUID. For our 3μm × 3μm SQUID, the sensitivity limit corresponds to a flux noise density at the pick-up coil of 3×10^{-5} $\phi_0/\sqrt{\text{Hz}}$. The corresponding spin sensitivity of our complete system, incorporating the flux transformer, is evaluated to be 5×10^4 spins/√Hz.

5 Conclusions

We have demonstrated the scaling of spin sensitivity with loop dimension of commercial dc SQUIDs. Our smallest device is found to have a spin sensitivity of about 40 spins in a 1 Hz bandwidth. A scheme for employing such devices as sensors for the detection of NMR of phosphorous donors in silicon has been outlined, based upon the implantation of phosphorous ions within the pick-up loop of a flux transformer. The spin sensitivity of our complete system is estimated to be of order 5×10^4 spins/√Hz, which represents potentially an improvement in sensitivity over previous experiments of several orders of magnitude.

Acknowledgements

We would like to thank A.R. Hamilton (UNSW) for useful discussion.

References

1. B.E. Kane, Nature **393** (1998) 133
2. D.P. DiVincenzo, Science **270** (1995) 255
3. N.A. Gershenfeld and I.L. Chuang, Science **275** (1997) 350
4. D.N. Jamieson, "A role for ion implantation in quantum computing" IQC01
5. K. Andres, R.N. Bhatt, P. Goalwin, T.M. Rice and R.E. Walstedt, Phys. Rev. B **24** (1981) 244
6. G. Feher, Phys. Rev. **114** (1959) 1219
7. J.C. Gallop and W.J. Radcliffe, IEEE Trans. Magn. **21** (1985) 602
8. M.B. Ketchen, D.D. Awschalom, W.J. Gallagher, A.W. Kieinsasser, R.L. Sandstrom, J.R. Rozen and B. Bumble, IEEE Trans. Magn. **25** (1989) 1212
9. L.R. Narasimhan, M. Takigawa and M.B. Ketchen, Appl. Phys. Lett. **65** (1994) 1305
10. P.W. Josephs-Franks, R.P. Reed and C.I. Pakes, Physica B: Condens. Matter **284** (2000) 540

ASPECTS OF TRANSPORT AND NOISE FOR MESOSCOPIC CHARGE DETECTORS

FREDERICK GREEN
Centre for Quantum Computer Technology, School of Physics, University of New South Wales, Sydney NSW 2052, Australia
and
Department of Theoretical Physics, Institute of Advanced Studies, The Australian National University, Canberra ACT 0200, Australia

MUKUNDA P. DAS
Department of Theoretical Physics, Institute of Advanced Studies, The Australian National University, Canberra ACT 0200, Australia

A structured array of coherently interacting spins may provide the best strategy for a scalable quantum computing engine. Communication with the outside world will take place via active mesoscopic devices, operating as electrometers for the electron states within the spin array. One candidate for a data transducer is the single-electron transistor. We give a brief survey of certain features of charge control and noise in such devices, broadly determined by the physics of the electron gas. It may be feasible to alter the form of these features and thus to mitigate a potential source of noise: charge fluctuations in the source and drain regions.

1 Introduction

We focus on some issues of the *open-boundary connection* between a mesoscopic conductor and its macroscopic leads. The physics of a channel's connection to its open reservoirs must influence the formulation of good predictive device models, and thus also the design of read-out devices for a solid-state quantum-spin computer [1,2]. Mesoscopic conductors – be they simple wires or active multi-terminal structures such as single-electron transistors (SETs) – derive part of their observable characteristics from the details of the interfaces with the leads. A well-known example is the contact resistance [3].

Non-equilibrium transport and noise in open systems has a substantial literature [4,5,6,7,8]. We recall two basic aspects: *global charge balance* across a driven open channel and its constraint on the non-equilibrium thermal fluctuations (noise), and *Coulomb suppression* of such fluctuations when the channel is strongly non-uniform.

2 Global Charge Balance: A Constraint on Fluctuations

Here we consider a SET-like structure as simply a two-terminal mesoscopic contact linked to large, metallic source and drain leads; the explicit details of transport in the central gate-control island do not enter at this level. No matter how high the channel current may be, the conductor remains overall charge-neutral. The *buffer layers* at the channel-lead interfaces screen out any electric fields, confining them to the active region [6,8].

Let $f_{\mathbf{k}}(\mathbf{r}, t)$ be the time-dependent electron distribution for wave vector $\mathbf{k}$, at point $\mathbf{r}$ in the channel; spin and sub-band labels are suppressed for brevity. If N is the total number of carriers within the active region Ω (*including* the buffer layers), summation over local sub-cells $\Omega(\mathbf{r})$ making up the active volume yields

$$N = \sum_{\mathbf{r}} \Omega(\mathbf{r}) \left\{ \frac{1}{\Omega(\mathbf{r})} \sum_{\mathbf{k}} f_{\mathbf{k}}(\mathbf{r}, t) \right\} = \sum_{\mathbf{r}} \Omega(\mathbf{r}) \left\{ \frac{1}{\Omega(\mathbf{r})} \sum_{\mathbf{k}} f_{\mathbf{k}}^{\mathrm{eq}}(\mathbf{r}) \right\} \tag{1}$$

where f^{eq} is the distribution at equilibrium; the bracketed expressions are locally defined densities. N is constant and remains fully compensated by the non-participating positive background, integrated over Ω. Gauss' theorem implies, unconditionally, the global neutrality of the device for all driving fields. This is true if and only if the inner active region is indeed screened by the electron gas in the macroscopic leads. Degeneracy and screening guarantee that the leads are *locally equilibrated*, and *locally neutral*, at all times

Random external perturbations induce a fluctuation background that displaces the instantaneous distribution $f_{\mathbf{k}}(\mathbf{r}, t)$ from its ensemble average. The *same* external stochastics will act on the channel both at equilibrium and when it is driven by a given applied voltage. Let ΔN be the mean-square thermal fluctuation in N. Then Gauss' theorem, constraining Eq. (1) as a dynamical relation, implies that

$$\Delta N = \sum_{\alpha} \Delta f_{\alpha}(t) = \sum_{\alpha} \Delta f_{\alpha}^{\mathrm{eq}} \tag{2}$$

where we write $\alpha \equiv (\mathbf{k}, \mathbf{r})$ and $\Delta f_{\alpha}(t)$ is the mean-square thermal fluctuation of the (microscopically determined) distribution function.

This is a rigorous, non-equilibrium, kinetic-theoretical relation [9] which holds within the limit of the local-gradient expansion for the electron gas. It governs the physics over length scales greater than the Fermi wavelength, itself short compared to typical device sizes.

The outcomes of Eq. (2) may surprise. The first is that even the *non-equilibrium* thermal fluctuations in a degenerate electron system necessarily scale with the thermal energy $k_{\mathrm{B}}T$, for any value of the applied potential. The microscopic form of the distribution $\Delta f_{\alpha}(t)$ is known explicitly [9]; here we recall a milder result, the global sum rule

$$\sum_{\alpha} \Delta f_{\alpha}(t) = \sum_{\alpha} \Delta f_{\alpha}^{\mathrm{eq}} \to k_{\mathrm{B}}T \left\{ \sum_{\mathbf{r}} \Omega(\mathbf{r}) \mathcal{N}[\varepsilon_{\mathrm{F}}(\mathbf{r})] \right\}, \tag{3}$$

in which the Fermi-Dirac limit of Δf^{eq} has been used to introduce the density of states $\mathcal{N}$ at the local Fermi energy $\varepsilon_{\mathrm{F}}(\mathbf{r})$:

$$\Omega(\mathbf{r})^{-1} \sum_{\mathbf{k}} \Delta f_{\alpha}^{\mathrm{eq}} = \Omega(\mathbf{r})^{-1} \sum_{\mathbf{k}} k_{\mathrm{B}}T \frac{\partial f_{\alpha}^{\mathrm{eq}}}{\partial \varepsilon_{\mathrm{F}}}(\mathbf{r}) \to k_{\mathrm{B}}T \mathcal{N}[\varepsilon_{\mathrm{F}}(\mathbf{r})].$$

There is an immediate corollary for the *current auto-correlation function*, which gives shape to the noise spectrum for the structure. The thermal current correlations scale with Δf. From Eq. (3), it is evident that the thermal component of the noise must have a persistent proportionality to the ambient temperature T, even well away from the linear low-field regime for the device, where the Johnson-Nyquist formula [10] enforces T-scaling.

3 Inhomogenity and Noise Suppression

The second outcome of the microscopic theory leading to Eq. (2), is that a strongly non-uniform mesoscopic conductor manifests *suppression* of charge fluctuations below those of a uniform but otherwise equivalent medium [9]. The other condition necessary for suppression is carrier degeneracy. Carriers in a non-uniform metallic channel will, to some extent, be localized. The confining potential can come about through modulation doping, band-structure engineering, or both as for a GaAs/AlGaAs heterojunction quantum-well channel [11]. Taking the latter as our example, we ask for the effect of the large self-consistent Coulomb energy on the fluctuations of the two-dimensional electron gas (2DEG). A channel of density n_s in the plane, contains $N = \Omega n_s$ carriers in area Ω:

$$N = \Omega \mathcal{D} k_B T \ln\{1 + \exp[(\mu - \varepsilon_0(n_s))/k_B T]\}. \tag{4}$$

Here $\mathcal{D} = m^*/\pi\hbar^2$ is the 2DEG density of states and μ is the chemical potential, determined asymptotically in the leads. For simplicity we assume ground-state occupation only, at energy $\varepsilon_0(n_s)$. The density dependence of sub-band energies reflects the strong Coulomb repulsion within the confined 2DEG.

Equation (4) can be varied in two ways to arrive at the charge-fluctuation strength over the channel. If the internal potential is frozen, $\varepsilon_0(n_s)$ remains at a *fixed value*. With this restriction, the 2DEG form of Eq. (2) for the driven channel becomes [9]

$$\sum_\alpha \Delta f_\alpha(t) = \Delta N = k_B T \left[\frac{\partial N}{\partial \mu}\right]_{\varepsilon_0(n_s)} = \frac{\Omega \mathcal{D} k_B T}{1 + \exp[(\varepsilon_0(n_s) - \mu)/k_B T]}. \tag{5}$$

Alternatively, if one allows for the natural *relaxation* of charge fluctuations via the density dependence of $\varepsilon_0(n_s(\mu))$, the result is self-induced screening:

$$\sum_\alpha \widetilde{\Delta} f_\alpha(t) = \widetilde{\Delta} N = k_B T \left(1 - \frac{\delta \varepsilon_0}{\delta \mu}\right) \left[\frac{\partial N}{\partial \mu}\right]_{\varepsilon_0(n_s)} = \frac{\Delta N}{1 + \left(\dfrac{\Delta N}{\Omega k_B T}\right) \dfrac{d\varepsilon_0}{dn_s}}, \tag{6}$$

where $\widetilde{\Delta} f_\alpha(t)$ denotes the dynamical non-equilibrium distribution of fluctuations for the confined channel, with full self-consistency. Eq. (6), like Eq. (2) before it, is an exact relation with a rigorous kinetic-theoretical basis [9].

Just as Eq. (2) entails the T-scaling of non-equilibrium thermal fluctuations, so does Eq. (6) entail, in an inhomogeneous mesoscopic contact, the persistent scaling of non-equilibrium fluctuations with Coulomb suppression, *as prescribed in the equilibrium state.* This should have definite and observable physical consequences [9]. One might also be able to maximise the degree of suppression by careful engineering of the band structure.

We cannot cover here the microscopic relation between suppression and the fluctuation-dissipation theorem [10]; sufficient to state that the standard form of that theorem is recovered within the inhomogeneous kinetic description, in a straightforward way [9]. It takes its place alongside the other conserving *sum rules*, all of them vital to any theory of mesoscopic transport and noise that calls upon the physics of degenerate electrons [12].

4 Summary, with a Possible Application

Global neutrality dominates the physics of a SET channel. That is the consequence of the non-equilibrium fields' total confinement to the active region. Everywhere beyond, the fields are evanescent. This is due to screening by mobile charges in the buffer zones at the device-lead interfaces. Two properties follow for current noise in a mesoscopic channel:

- Thermally induced fluctuations of *degenerate* carriers must continue to scale with ambient temperature even when they are strongly driven.
- When the degenerate carriers are also *strongly inhomogeneous*, their thermal fluctuations are Coulomb-suppressed.

In the literature, a major issue is the degree to which external noise propagates to a quantum-spin array and influences computational coherence [2,13,14,15,16]. The broad-scale issues that we have discussed do not explicitly address the dynamic response within a SET's island. On the other hand, one knows that the noise performance of a transistor also depends on the current fluctuations within its *access regions* [17]. For the SET, these are the transition zones between the central island and the outer circuit.

If a SET were fabricated on a heterojunction substrate, Coulomb suppression of fluctuations would act throughout the leads [9], reducing charge-fluctuation noise in the access regions below levels in a uniform device with the same conductance, fabricated by three-dimensional metallization. This may lower the overall noise figure (defined in terms of the current signal-to-noise ratios and referred to the SET gate terminal).

Acknowledgments

We thank the members of the Centre for Quantum Computer Technology for valuable discussions. FG also thanks the Centre, and the Institute of Advanced Studies at the Australian National University, for their warm hospitality. This work was funded by the Australian Research Council, and the US National Security Agency through the Army Research Office.

References

1. B E Kane, *Nature* **393**, 133 (1998).
2. S Das Sarma, J Fabian, X Hu, and I Žutić, *Preprint* cond-mat/0006369.
3. Y Imry and R Landauer, *Rev. Mod. Phys.* **71**, S306 (1999).
4. W R Frensley, *Rev. Mod. Phys.* **62**, 45 (1990).
5. F Sols, *Phys. Rev. Lett.* **67**, 2874 (1991).
6. E W Fenton, *Phys. Rev.* B **46**, 3754 (1992).
7. M D Johnson and O Heinonen, *Phys. Rev.* B **51**, 14421 (1995).
8. W Magnus and W Schoenmaker, *Phys. Rev.* B **61**, 10883 (2000).
9. F Green and M P Das, *J. Phys: Condens. Matter* **12**, 5251 (2000).
10. C Kittel, *Elementary Statistical Physics* (Wiley, New York, 1958).
11. C Weisbuch and B Vinter, *Quantum Semiconductor Structures: Fundamentals and Applications* (Academic Press, San Diego, 1991).
12. D Pines and P Nozières, *The Theory of Quantum Liquids* (Benjamin, New York, 1966).
13. W G Unruh, *Phys. Rev.* A **51**, 992 (1995).
14. B E Kane *et al*, *Phys. Rev.* B **61**, 2961 (2000).
15. Y Makhlin, G Schön, and A Schnirman, *Phys. Rev. Lett.* **85**, 4578 (2000).

16. D V Averin, *Preprint* cond-mat/0010052.
17. A Cappy, in *Compound Semiconductor Device Modelling*, ed. C M Snowden and R E Miles (Springer, London, 1993).

THE STUDY OF DIAMOND NV CENTRES AS A POTENTIAL SINGLE PHOTON SOURCE

ELIZABETH TRAJKOV AND STEVEN PRAWER

Special Research Centre for Quantum Computing Technology and Micro Analytical Research Centre, School of Physics, The University of Melbourne, Parkville, Victoria 3010, Australia

The nitrogen vacancy (NV) centre in diamond has been investigated as a potential single photon source. The intensity of the NV centre fluorescence saturates with increasing laser power, which indicates that NV centres might show antibunching.

1 Introduction

In quantum cryptography, perfect security depends upon each bit of information to be coded on a single quantum object such as a single photon. Fundamentally, it is impossible to replicate the entire quantum state of a single object, hence this deters any potential intruder from disrupting the message without the receiver realising. [1]

An ideal single photon source emits light such that only one of two detectors behind a semitransparent beam splitter registers an event. A single fluorescing two-level atom is a potential single photon source. [2,3,4] Since excitation and subsequent decay to the initial state of a two-level atom takes a finite time, only one photon is emitted at a time. The emitted light is thus antibunched and tends to consist of single photons separated by a time interval determined by the excited state lifetime and the pumping rate. [2,3,4]

One such two-level system is individual nitrogen-vacancy (NV) centres in synthetic diamond, with a zero phonon line at a wavelength of 637 nm. [5] They are formed by a substitutional nitrogen atom with a vacancy trapped at an adjacent lattice position. NV centres are suitable for single photon generation because of their high radiative quantum efficiency of close to one as well as a short decay time of the excited state. [6]

To demonstrate the antibunching properties of the fluorescence from single NV centres, the intensity correlation function $g^{(2)}(\tau)$ is measured using a Hanbury-Brown-Twiss experimental set-up as depicted in Fig. 1. This correlation function is the joint probability of detecting a photon in two successive time intervals, normalised by the probability of recording two statistically independent photons.

If antibunching occurs for a single NV centre, then we expect the intensity correlation function $g^{(2)}(\tau)$ to be minimum at $\tau = 0$, where τ is the detection time difference, as shown in the curve below.

This curve indicates that the probability of detecting two photons in coincidence is less than the probability of detecting two photons that have a nonzero separation time.

2 NV Centre Fluorescence

The NV centres were produced in HPHT diamond via 2 MeV He^+ irradiation (1.63 x 10^{16} ions/cm^2) and subsequent annealing at 900 °C for 2 hours. Only a small portion of the sample was irradiated (1x1 mm).

The sample fluorescence was excited using the 514.5 line from an Argon ion laser and collected using a DILOR XY confocal micro-Raman spectrometer operating at room tem-

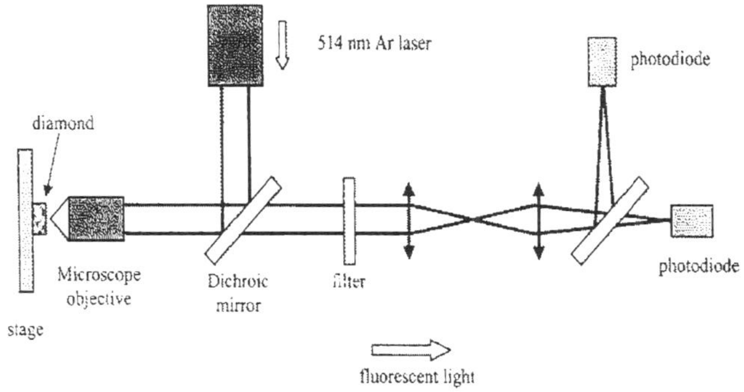

Figure 1. Hanbury-Brown-Twiss Configuration

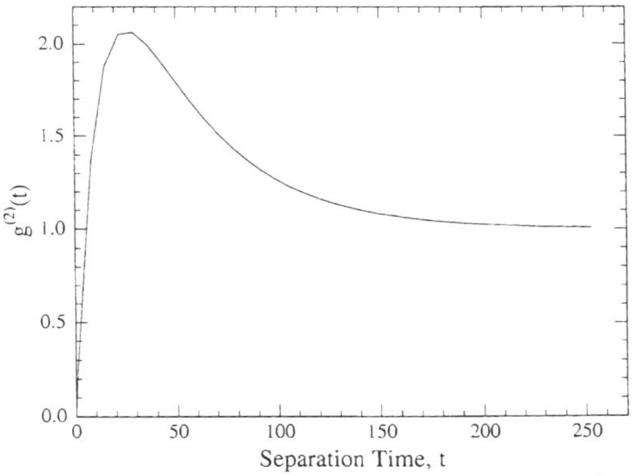

Figure 2. Intensity Correlation Function Exhibiting Antibunching

perature. The excitation power was 15 mW and the beam was focussed using the x80 Olympus microscope objective.

Prior to irradiation no NV centres were present in our sample. The irradiation and annealing produced a notable concentration of NV centres, which is indicated by the increase in the NV centre fluorescence intensity as shown in Fig. 3.

The NV centre fluorescence from bulk diamond was investigated as a function of excitation power, which is presented in Fig. 4.

We found that the fluorescence signal began to saturate as the laser power increased. In experiments by Brouri *et al.* [2] and Kurtsiefer *et al.* [3] this phenomenon was also observed for single NV centres which exhibited antibunching properties. This saturation behaviour of the NV centre fluorescence might be essential for antibunching.

From lateral scans of the irradiated diamond we found that there was no spatial variation

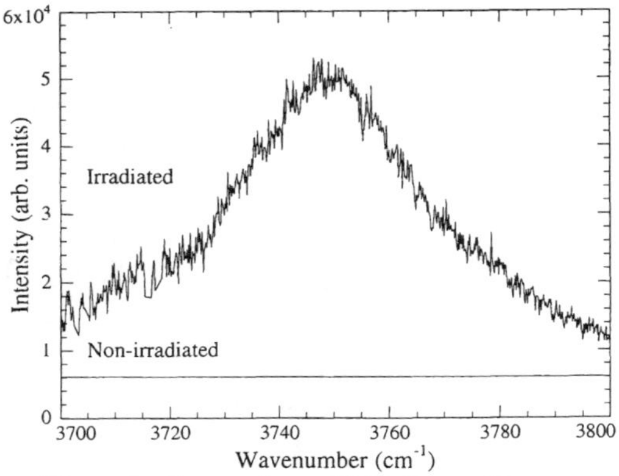

Figure 3. Comparison of NV Centre Fluoresence

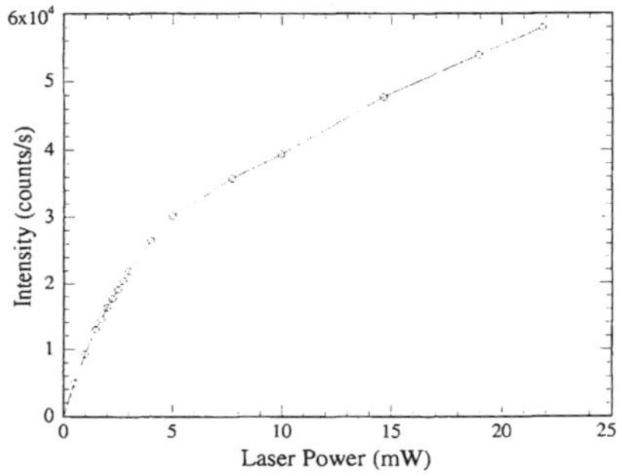

Figure 4. Behaviour of NV Centre Fluorescence

between individual NV centres, hence it is unclear whether the above saturation curve is from a single NV centre. To investigate antibunching we will need to vary the irradiation dose to determine which concentration will allow us to resolve individual centres.

3 Conclusion

The saturation of the NV centre fluorescence indicates that NV centres might show antibunching behaviour. We hope to validate the antibunching behaviour of NV centres and also explore the practicability of diamond as a single photon source by investigating such things as stability and efficiency of the system.

References

1. W. Tittle, G. Ribordy and N. Gisin, Phys. World **11** 41 (1998).
2. R. Brouri, A. Beveratos, JP. Poizat and P. Grangier, Optics Letters **25** 1294 (2000).
3. C. Kurtsiefer, S. Mayer, P. Zarda and H. Weinfurter, Phys. Rev. Lett. **85** 290 (2000).
4. R. Vyas and S. Singh, J. Opt. Soc. Am. B **17** 634 (2000).

5. A. Gruber, A. Drabenstedt, C. Tietz, L. Fleury, J. Wrachtrup and C. von Borczyskowski, Science **276** 2012 (1997).
6. A.T. Collins, M.F. Thomaz and M.I.B. Jorge, J. Phys. C **16** 2177 (1983).

QUANTUM COMPUTATION WITH HARMONIC OSCILLATORS

STEPHEN D. BARTLETT AND BARRY C. SANDERS
Department of Physics, Macquarie University, Sydney, New South Wales 2109, Australia
E-mail: Stephen.Bartlett@mq.edu.au; Barry.Sanders@mq.edu.au

BENJAMIN T. H. VARCOE
Max-Planck-Institut für Quantenoptik, 85748 Garching, Germany
E-mail: B.Varc œ@mpq.mpg.de

HUBERT DE GUISE
Facult 'eSaint-Jean, University of Alberta, 8406 rue Marie-Anne Gaboury, Edmonton, Alberta, Canada T6C 4G9
E-mail: hdeguise@gpu.srv.ualberta.ca

By encoding a qudit in a harmonic oscillator and investigating the $d \to \infty$ limit, we give an entirely new realization of continuous-variable quantum computation. The generalized Pauli group is generated by number and phase operators for harmonic oscillators. We describe a physical realization in terms of modes in a microwave cavity ,coupled via a standard Kerr nonlinearity .

1 Introduction

The use of continuous-variable (CV) quantum computing allows information to be encoded and processed much more compactly and efficiently than with discrete-variable (qubit) computing. With CV realizations, one can perform quantum information processes using fewer coupled quantum systems: a considerable advantage for the experimental realization of quantum computing. The rapidly developing field of CV quantum information theory has applications to quantum error correction [1], quantum cryptography [2] and quantum teleportation [3], including an experimental realization of CV quantum teleportation [4].

The most common proposed realization of CV quantum computation at present employs position eigenstates as a computational basis [1]; these states are approximated experimentally using highly squeezed states [3]. Other CV realizations are described by a generalized Pauli group generated by the number operator $\hat{N}$ and a phase operator $\hat{\theta}$, and computational bases given by either harmonic oscillator number eigenstates or phase eigenstates [5]. These realizations are obtained formally by taking the $d \to \infty$ limit of the qudit, the d-dimensional generalization of the qubit, and are important for five key reasons: (i) these CV realizations are entirely distinct from the position eigenstate computational basis realization, both in terms of the computational basis and in terms of the SUM gate; (ii) the SUM gate employs a standard Kerr optical nonlinearity to couple two modes; (iii) these realizations give natural extensions of the qubit-based (discrete-variable) Pauli group, with a well-defined limiting procedure; (iv) this CV quantum computation presents an appealing realization in terms of coupled modes in a microwave cavity using the powerful methods of state preparation [6] in such cavities; and (v) these realizations give a new implementation of the well-studied phase operator [7].

2 CV Quantum Computation using Harmonic Oscillators

To describe CV quantum computation, one requires an infinite–dimensional Hilbert space $\mathcal{H}$ and a representation of the generalized Pauli group [1] that serves as a group of unitary transformations on $\mathcal{H}$. By defining $\mathcal{H}$ to be the Hilbert space for a harmonic oscillator, one can define specific representations of the generalized Pauli group in terms of operators that act on the harmonic oscillator Hilbert space.

The natural generalization of the Pauli group [5] is the group generated by the number operator $\hat{N}$ and the (Pegg–Barnett [7]) phase operator $\hat{\theta}$. Note that it is convenient to work with the finite–dimensional Hilbert space $\mathbb{H}_d$ of boson number not greater than $d-1$. On this Hilbert space, the phase operator is well–defined, and the system incorporates a finite energy cutoff and corresponding phase resolution.

With this generalization of the Pauli group, there are two natural computational bases. The first, the *number state basis*, consists of the harmonic oscillator energy eigenstates

$$\hat{N}|n\rangle = n|n\rangle\,, \quad n = 0, 1, \ldots, d-1\,. \tag{1}$$

The second basis, the *phase state basis*, consists of eigenstates of the phase operator

$$\hat{\theta}|\phi\rangle = \phi|\phi\rangle\,, \quad \phi = 0, \frac{2\pi}{d}, \ldots, \frac{2(d-1)\pi}{d}\,. \tag{2}$$

These bases are "dual" in the sense that $\langle\phi|n\rangle = \exp(i\phi n)$.

To perform universal CV quantum computation [8], it is necessary to be able to: (*i*) initially prepare a qudit in an arbitrary state for computation; (*ii*) realize an arbitrary unitary transformation on a single qudit; (*iii*) have a controlled two–qudit interaction gate such as the SUM gate [1]

$$\text{SUM}: \; |s_1\rangle_1 \otimes |s_2\rangle_2 \mapsto |s_1\rangle_1 \otimes |s_1 + s_2\rangle_2\,; \tag{3}$$

and (*iv*) perform PVM measurements in the computational basis.

3 Proposed Realization as Coupled Modes of a Microwave Cavity

One approach to physically realizing CV (qudit–based) quantum computation is by using a multimode microwave cavity. A single qudit is realized as the state of a (longitudinal) mode, and qudit–qudit interactions are performed by appropriately coupling different modes. By using a microwave cavity, one can take advantage of the recent technical advances of the micromaser: a micromaser provides a physical realization of the Jaynes–Cummings model, which describes the coupled system consisting of a single two–level atom and a single mode of the radiation field. Single atom interactions enable state preparation, qudit transformations, two–qudit coupling, and state measurement. Current high–Q microwave cavities allow photon lifetimes up to 0.3 s [9], several orders of magnitude larger than the atom interaction time.

State preparation in a microwave cavity is highly developed, both theoretically and experimentally. Proposals [6] for preparation of an arbitrary state of the radiation field involve injecting a sequence of atoms and are conditional on measurement of the output atoms; a deterministic approach to the preparation of Fock states (not involving conditional measurements) has been experimentally realised [9]. Recently, it has been proposed [10] that any state of the radiation field can be prepared with arbitrarily high fidelity using a sequence of entangled atoms; this technique has the important features that the resulting state of the field is not entangled with that of the atoms, and that the method is independent of the initial state of the field.

An arbitrary unitary transformation on a single qudit (single mode), to any desired precision, can be performed *efficiently* using a combination of phase–space displacements, squeezing, and a nonlinear Kerr effect [8]. By an appropriate combination of these transformations, one can approximate (to arbitrary accuracy) any polynomial Hamiltonian in $\hat{a}^\dagger$ and $\hat{a}$.

Linear transformations of a cavity mode state may be performed by coupling the mode to an external monochromatic coherent field via transmission through one of the mirrors [11]. A strong coherent field is directed into the cavity through a low–transmissivity mirror, and the result on the mode state is that it is displaced; i.e., the field state is acted upon by a unitary Glauber displacement operator. The choice of frequency for the incoming coherent field determines which of the longitudinal field states (each with a different frequency) is displaced.

Squeezing the field state requires a nonlinear process, as does the nonlinear Kerr transformation of the field state. The squeezing transformation may be achieved via a $\chi^{(2)}$ nonlinearity and the latter by a $\chi^{(3)}$ nonlinearity. A nonlinearity is effected by passing atoms through the cavity one at a time: during the time of passage, the field states undergoes a nonlinear evolution. An effective nonlinearity is obtained by adiabatically eliminating the slowly varying atomic degrees of freedom. The field in the cavity may be squeezed by injecting t wo–level atoms into a micromaser cavity [12] or three–level atoms [13]. By choosing atomic levels and motional parameters carefully, the squeezing (Bogoliubov) transformation can be approximated. The Kerr, or $\chi^{(3)}$, nonlinearity is generally smaller, but it is possible to employ a four–level atom, passing through the cavity, to obtain a large Kerr nonlinearity [14]. Thus, by directing atoms through the cavity, unitary $\chi^{(2)}$ and $\chi^{(3)}$ evolutions particular field states can be effected. The Kerr nonlinearity can also effect a cross–phase modulation to perform the micromaser equivalent of the optical cavity quantum electrodynamics conditional phase shift [15].

For quantum computation, we must also realize a gate that performs a t wo–qudit interaction. Consider t wo oscillators coupled by the four–wave mixing interaction Hamiltonian $\chi\hat{N}_1\hat{N}_2 = \chi\hat{a}_1^\dagger\hat{a}_1\hat{a}_2^\dagger\hat{a}_2$. This Hamiltonian for an optical system describes a four–wave mixing process in which χ is proportional to the third–order nonlinear susceptibility [16]. Let oscillator 1 be in a state $|s_1\rangle_1$ encoded in the number state basis, and let oscillator 2 be in a state $|s_2\rangle_2$ encoded in the phase state basis. This interaction Hamiltonian generates the transformation

$$e^{-\mathrm{i}\chi\hat{N}_1\hat{N}_2 t}|s_1\rangle_1 \otimes |s_2\rangle_2 = |s_1\rangle_1 \otimes |\chi t s_1 + s_2\rangle_2 \,. \tag{4}$$

Thus, with time $t = \chi^{-1}$, this Hamiltonian generates the SUM transformation $|s_1\rangle_1 \otimes |s_2\rangle_2 \to |s_1\rangle_1 \otimes |s_1 + s_2\rangle_2$.

Quantum computation with multiple qudits could be performed by coupling several modes in a single cavity; each mode realizes a single qudit. Modes are coupled via a SUM interaction of the time described above. Note that the control qudit for the sum operation must be encoded in the number state basis, and the target qudit must be in the phase state basis

Measurements in the number state computational basis can be performed using a QND measurement of photon number [17]. Atoms passing through the cavity have their momentum coupled to the radiation field; the momentum distribution of outgoing atoms yields a measurement of photon number.

In summary, we have presented a new form of continuous variable computation in terms of number and phase operators, and describe a realization in terms of coupled modes of a microwave cavity using linear transformations, squeezing, and nonlinear Kerr media. This new approach has the advantage over position–eigenstate CV computation in that the computational basis states, for large but finite d, are well–defined and obtainable, and do not require "infinite–squeezing" of Gaussian wavepackets.

Acknowledgments

This project has been supported by an Australian Research Council Large Grant and by a Macquarie University Research Grant. We acknowledge helpful discussions with Samuel Braunstein, Terry Rudolph, and Benjamin Varcoe.

References

1. D. Gottesman, A. Kitaev, and J. Preskill, "Encoding a qudit in an oscillator," quant-ph/0008040.
2. T. C. Ralph, Phys. Rev. A **61**, 010303(R) (2000).
3. S. L. Braunstein and H. J. Kimble, Phys. Rev. Lett. **80**, 869 (1998).
4. A. Furusawa, J. L. Sørensen, S. L. Braunstein, C. A. Fuchs, H. J. Kimble, and E. S. Polzik, Science **282**, 706 (1998).
5. S. D. Bartlett, H. de Guise and B. C. Sanders, "Quantum Computation with Harmonic Oscillators," quant-ph/0011080.
6. K. Vogel, V. M. Akulin, and W. P. Schleich, Phys. Rev. Lett. **71**, 1816 (1993).
7. D. T. Pegg and S. M. Barnett, J. Mod. Optics **44**, 225 (1997).
8. S. Lloyd and S. L. Braunstein, Phys. Rev. Lett. **82**, 1784 (1999).
9. S. Brattke, B. Varcoe and H. Walther, Phys. Rev. Lett. (submitted) (2001); B. T. H. Varcoe, S. Brattke, M. Weidinger, and H. Walther, Nature **403**, 743 (2000); B. T. H. Varcoe, S. Brattke and H. Walther, J. Opt. B: Quantum Semiclass. Opt. **2**, 154 (2000).
10. T. Wellens, A. Buchleitner, B. Kümmerer, and H. Maassen, Phys. Rev. Lett. **85**, 3361 (2000).
11. C. T. Bodendorf, G. Antesberger, M. S. Kim, and H. Walther, Phys. Rev. A **57**, 1371 (1998).
12. F. le Kim, Phys. Rev. A **44**, 3282 (1991).
13. M. Orszag, R. Ramírez, J. C. Retamal and L. Roa, Phys. Rev. A **45**, 6717 (1992); R. R. Puri, C. K. Law and J. H. Eberly, Phys. Rev. A **50**, 4212 (1994)
14. S. Rebi´c,S. M. Tan, A. S. Parkins, and D. F. Walls, J. Opt. B: Quantum Semiclass. Opt. **1**, 490 (1999).
15. Q. A. Turchette, C. J. Hood, W. Lange, H. Mabuchi, and H. J. Kimble, Phys. Rev. Lett. **75**, 4710 (1995).
16. G. J. Milburn and D. F. Walls, Phys. Rev. A **28**, 2065 (1983).
17. M. J. Holland, D. F. Walls, and P. Zoller, Phys. Rev. Lett. **67**, 1716 (1991).

UNITARITY AND QUANTUM COMPUTATION

TIEN D KIEU

School of Biophysical Sciences and Electrical Engineering, Swinburne University of Te chnology, Hawthorn 3122, Australia

E-mail: kieu@swin.edu.au

The requirement of unitarity imposes severe constraints not only on the type of errors recoverable by quantum error correction codes, but also on the halting of a universal quantum computer.

1 Introduction

The unitarity of quantum evolution is known to impose severe constraints on the dynamics — as in the case of scattering matrix in quantum field theory, for instance. In this short note, we illustrate the restrictive nature of unitarity as constraints on the quantum error correction, which is essential and even vital for the physical realisation of a quantum computing device of considerable size operated for a usefully long time. We then follow with some restriction [1] on the halting of a universal quantum computer.

2 Unitarity and quantum error correction

The discovery of error correcting codes for quantum computation was a pleasant surprised, see [2,3] for a review. Without this possibility, realisation of quantum computation would have been unthinkable as the computers are inevitably and constantly subject to errors induced by either the internal interactions or the environment or both.

We present here the steps of a general recipe for quantum error correction [2]:

- It is necessary to encode the protected qubits in an state $|\phi\rangle$ in a certain way by an (unitary) operation E acting on the original and some auxiliary qubits (needed to store the information of the original qubits to avoid the no-cloning theorem for quantum states)

$$E|\phi\rangle|0\rangle = |\phi; E\rangle$$

- The induced errors are represented by operators acting on the encoded state $\sum_s |e_s\rangle M_s|\phi; E\rangle$, with corresponding environment state $|e_s\rangle$ and the sum is over all t ypes of recoverable errors s.

- The qubits are then padded with yet more extra 'ancilla' qubits in the initial state, say, $|0\rangle$ to be used for identifying the t ypeof error s through some syndrome extraction operator A

$$A\left(\sum_s |e_s\rangle M_s|\phi; E\rangle|0\rangle\right) = \sum_s |e_s\rangle M_s|\phi; E\rangle|s\rangle$$

- These error-labelling qubits $|s\rangle$ are then measured to yield one particular state with which the error could be identified.

- Accordingly, the identified error on the originally encoded qubits could be corrected,
- The error-labelling qubits are then reset for the next round of error detection and correction.

Now with A being unitary as a part of the quantum dynamics, we have

$$\langle \phi; E|M_{s'}^{\dagger} M_s|\phi; E\rangle(1 - \langle s'|s\rangle) = 0$$

From which, as $\langle s'|s\rangle = 0$ for unambiguous error identification, the unitarity constraint follows

$$\langle \phi; E|M_{s'}^{\dagger} M_s|\phi; E\rangle = 0$$

This presents a severe constraint and a necessary condition for error correction. In the case of one qubit, examples of errors are the bit-flip represented by the Pauli matrix $M_s = \sigma_x$, flipping the qubit between its two states, and the phase error represented by $M_s = \sigma_z$. As $M_{s'} M_s = \sigma_x \sigma_z \neq 0$ but the unitarity constraint requires that the operator product has zero expectation value in the encoded state, these errors can thus only be recovered by appropriate encoding on the original state padded with extra qubits to enlarge the Hilbert space. The encoding has to be done in such a way that errors of the extra encoding qubits are also correctable. In general, and especially for the types of errors due to the correlations and interactions among the many qubits, the unitarity constraint above imposes severe constraints on the type of quantum mechanically recoverable errors and demands highly non-trivial choices of the the number of extra qubits used for encoding, their initial state and the encoding E.

3 Halting a universal quantum computer

For some algorithms when the nature and the number of computational steps are known in advance, as with the Shor's factorisation [4], in which case one might be able to build a purpose machine just to carry out those particular functions. This is an example of the hardwired quantum network [5]. However, not always the number of computational steps is fixed for all input data for a general algorithm. In addition, we might not want a hardwired black box but might want a programmable computer.

The concept of (classical) universal Turing machine not only provides such a programmable machine, but has also proved to be invaluable in the theoretical study of classical computer science – particularly in the complexity classification of various computational problems. Deutsch [6] has thus generalised the concept of universal Turing machine to that of a universal quantum computer (shortened as qucomputer), where the input contains both the initial data and the algorithm to be carried out on this set of data.

While universal Turing machine has the well-known halting problem where we cannot always tell whether an algorithm with some arbitrary input will halt, qucomputer also has its own halting problem of a different kind. Here, the problem is when to stop the quantum computation, even with some computation task which is known to finish after finite but *a priori* unknown number of executation steps. We can not just simply check the result by the most direct measurement as this action, being nonunitary and dissipative, would destroy the quantum evolution of the qucomputer, and thus stop the computation at that point.

Let us define precisely a qucomputer and a general mechanism for halting for our discussion purposes. A qucomputer is a closed quantum system governed by a time-independent, unitary dynamics $\mathcal{U}$ for each time step between the admission of the initial input, which is some vector, representing both the programs and data, in the Hilbert space of $\mathcal{U}$ and the final reading-out measurements. To check whether the computation has finished, an extra qubit, the halt qubit, is added [6]. The halt qubit is set to $|0_{\rm h}\rangle$ initially, and to $|1_{\rm h}\rangle$ when the computed result is obtained, upon which time one could halt the computation. By measuring this halt qubit one would not disturb the quantum calculation even if the computation is not yet finished. An immediate problem with this scheme was pointed out in [7]. In this paper we point out even a more fundamental problem with the halt qubit: *it in fact cannot be switched at all, with finite dimensional Hilbert space* [8]*, and thus cannot be used in this case to flag whether the computation is finished or not.*

For halting, it is desirable of the dynamics to be able to store the output, which is finite in terms of qubit resources, *invariantly* (that is, unchanged under the unitary evolution) after some finite time when the desirable output has been computed. This reserved space, from which the output can be read out anytime afterwards, is mathematically an invariant subspace V which could be defined by the input state itself –so that different input problems can have different outputs. This is a generalisation of the proposal with the augmented halt qubit, where the final halt state $|1_{\rm h}\rangle$ establishes, in fact defines, an invariant subspace.

A qucomputer can hardly be useful without such an invariant subspace. Unless it is known in advance when to stop the qcomputer just in time, the hard-earned results of computation will be lost in the ever-changing evolution of states under the unitary dynamics. Indeed, it is not always possible to predetermine the necessary number of calculational steps involved in solving general computable functions –except for a small class of problems or where the completion of the calculation need not be interrogated.

We now show below that the switching of the halt qubit (or its suitable generalisation) is not compatible with the desirable properties of

1. The orthogonality of the two states of the halt qubit (in order to yield unambigious measurement results);

2. The invariance of the halt subspace under the qucomputer dynamics;

3. The unitarity of the qucomputer dynamics.

Let the initial input state be $|\psi_0\rangle = |x_0\rangle \otimes |0_{\rm h}\rangle$, where the halt state $|1_{\rm h}\rangle$ is not restricted to be a single qubit state but can be general and as large as necessary, but still finite, to store the output. Let $|\psi_N\rangle$ be the state at the time step N. And let N_0 be the switching time when $|\psi_{N_0}\rangle$, having the halt qubit in state $|0_{\rm h}\rangle$, is turned into a vector in the invariant subspace V that has the halt qubit in the state $|1_{\rm h}\rangle$.

- Since $|\psi_{N_0}\rangle$ is orthogonal to V, then $|\psi_{N_0+1}\rangle = \mathcal{U}|\psi_{N_0}\rangle$ is orthogonal to the unitary transformation $\mathcal{U}$ of any vector in V.

- In particular, $|\psi_{N_0+1}\rangle$ is orthogonal to the unitary transformation of every vector of some (at most infinitely countable) orthonormal basis of V.

- Such a situation can only happens if $|\psi_{N_0+1}\rangle = 0$ since $|\psi_{N_0+1}\rangle$ is in the invariant subspace V by assumption and the unitary transformation of an orthonormal basis also constitutes another orthonormal basis.

Because $|\psi_{N_0}\rangle \neq 0 \neq |\psi_0\rangle$ for any non-trivial computation, we thus arrive at a contradiction proving that the halt state cannot be switched.

This chain of reasoning can easily be generalised to cover the more general initial state for *any* halting protocol which requires some invariant subspace in particular, and the above desirable properties in general.

4 Concluding remarks

The restricting power of unitarity of quantum evolution has been demonstrated for the quantum error correction and for the general halting of a universal quantum computer. It should be noted that our arguments and conclusions herein do not cover the case of *probabilistic/statistical* halting.

Acknowledgments

The author wishes to acknowledge the collaboration with Michael Danos on the halting problem for universal quantum computers, and to thank Alan Head, Richard Josza and Robert Lee for discussions. In particular, the author is grateful to Sam Gutman for discussions on the inifnite dimensional Hilbert space.

References

1. T.D. Kieu and M. Danos, *e-print quant-ph/9811001*;
   T.D. Kieu and M. Danos, to appear in *Acta Physica Hungarica.*
2. A. Steane, *Rept. Prog. Phys.* **61**, 117-173 (1998).
3. E. Knill and A. Laflamme, *Phys. Rev.* **A55**, 900 (1997).
4. P.W. Shor, *e-print quant-ph/9508027.*
5. D. Deutsch, *Proc. R. Soc. Lond.* **A425**, 73-90 (1989).
6. D. Deutsch, *Proc. R. Soc. Lond.* **A400**, 97-117 (1985).
7. J.M. Myers, *Phys. Rev. Lett.* **78**, 1823-4 (1997).
8. For infinite dimensional case, Sam Gutman in a private communication has given an argument –similar to that of the Hilbert Hotel having an inifinite number of rooms– to circumvent the unitarity restriction given here on the halting of a qucomputer.

OPTIMAL PROBABILISTIC GUESS OF QUANTUM ORDER-FINDING COMPUTATION BY LINEAR PROGRAMMING THEORY

RYUICHI SAWAE, YUSUKE MATUDA, KENJI FUKUDA, YOSHIYUKI MORI, MINARU TEI, KENICHI TAKARABE

Okayama University of Science, Ridai-cho, Okayama City, Japan 700-0005

TOSHIO SAKATA

Kyushu Institute of Design, Minami-ku Fukuoka City, Japan 815-8540

Linear programming theory is applied to estimate the optimal probabilistic guess for the order-finding problem. The order-finding problem for a permutation π with the 2^n members is to find the least number r for any member y in the 2^n members by the r iterations of which the member y returns to the original position. This problem is known to be hard for classical computation, because the number of oracle queries needed to find the order r increases exponentially in n of π. The quantum computation overcomes this difficulty by its inherent parallel processing and the number of oracle queries increases only in n. However, to show the advantage of the quantum computation for the order-finding problem, it is needed to know whether the quantum computation could give the probability to get the correct answer larger than the classical probabilistic guess. The probabilistic guess problem for the quantum order-finding computation is considered by the liner programming theory.

1 Introduction

The quantum computation is able to reduce the computational cost to polynomial time by its inherent parallel computation. The file search problem among the N files needs the N steps at worst by the classical computation. Grover, however, showed that the file search is able to be solved by the $O(\sqrt{N})$ steps by the quantum computation[1]. The Grover's algorithm increases the amplitude of the quantum state of the target file by the composed unitary operation of Walsh-Hadamard and inverse (sign change) transformation of the target quantum state. The amplitude enhancement of this way is clearly explained by the rotation of the group $SU(2)$ with the geometrical consideration by Jozsa, Yu and Sun[2,3]. Bennett and Zalka also showed that the Grover's algorithm is optimal[4,5]. The Grover's algorithm is further extended for the generalization of the initial state of the quantum computaion[6,7], the use of the general phase change other than the sign change[8] and the replacement of the Walsh-Hadamard transformation with the more general unitary transformation[9]. All these extensions are shown to work properly for the Grover's algorithm. Most recently, Brasard et al[9]. and Galindo[10] independently proposed the Quantum Amplitude Amplification method and Buhrman et al. applied this method to find a triangle in a general graph[10]. Vadersypin et al. realized very recently the order-finding of the permutation π of the four member by NMR experiment[11]. In this paper we extend the quantum simulation to more qubits to explore the behavior of the optimal probabilistic guess of the order-finding when the Discrete Fourier transformation basis is extended. The emphasis will be on the fact that the advantage of the quantum computation for the order-finding problem is understood by the optimal probabilistic guess calculated by the linear programming theory.

2 Order-finding problem

For a natural number n, a permutation $\pi \in S_{2^n}$ replaces the member y in the set $\{1, 2, 3, \ldots, 2^n\}$ with the other member. After the subsequent permutations the member y returns to the original position. We define the order r is the least number for y to return to the original position. The order r depends on both y and π, so the problem is defined as to "find the least number r which satisfies $\pi^r(y) = y$." The number of steps to find r increase as 2^n by the classical algorithm[12]. The quantum computation reduces the steps to find r to n. There is, however, a further complication on this problem. When the distribution of r is uniform on the set $\{1, 2, 3, \ldots, 2^n\}$, the maximization of the probabilistic guess is easy. But, for the arbitrary distribution $\{\alpha_i\}$ of r, the linear programming theory is needed to maximize the probabilistic guess. If the maximum quantum probabilistic guess is larger than the classic guess, it is safely said that the quantum computation is more efficient than the optimal classical calculation. The result for the case of $n = 2-6$ is shown in the section 4.

3 Quantum circuit for order-finding

In this section we discuss the quantum circuit which was used for the order-finding problem and simulated by an old-fashioned classical computer.

Like in the paper of Vandersypen et al[11], the quantum algorithm and the quantum circuit are as follows.

Step 1. Input $\overbrace{0 \otimes 0 \otimes \cdots \otimes 0}^{l} \otimes y_0 \otimes y_1 \otimes \cdots \otimes y_{n-1}$ as a data.
Hereafter, we may write the above input as $0 \otimes y$.

Step 2. Apply the Walsh-Hadamard transformation to the first l qubits.
The state will be entangled such that $\sum_{j=0}^{j=2^l-1} j$ without the normalization constant.

Step 3. Apply the unitary transformation $x \otimes y \to x \otimes \pi^x(y)$.
This transformation related to the permutation π is the oracle query.

Step 4. Apply the discrete Fourier transformation (DFT) on the first l qubits.

Step 5. Measure the first l qubits.

4 Application of the linear programming theory for the order-finding problem

We take the quantum circuit (see Figure 1) for the order-finding problem. This $l + n$ qubits quantum circuit consists of two parts, one of which represents the iterations of a permutation π, the other of which represents a member y and the permutation π.

Suppose the probability function $P(r' = r)$ gives the value of probability when the guess value r' is equal to the order r. $P(r' = r)$ is rewritten as follows.

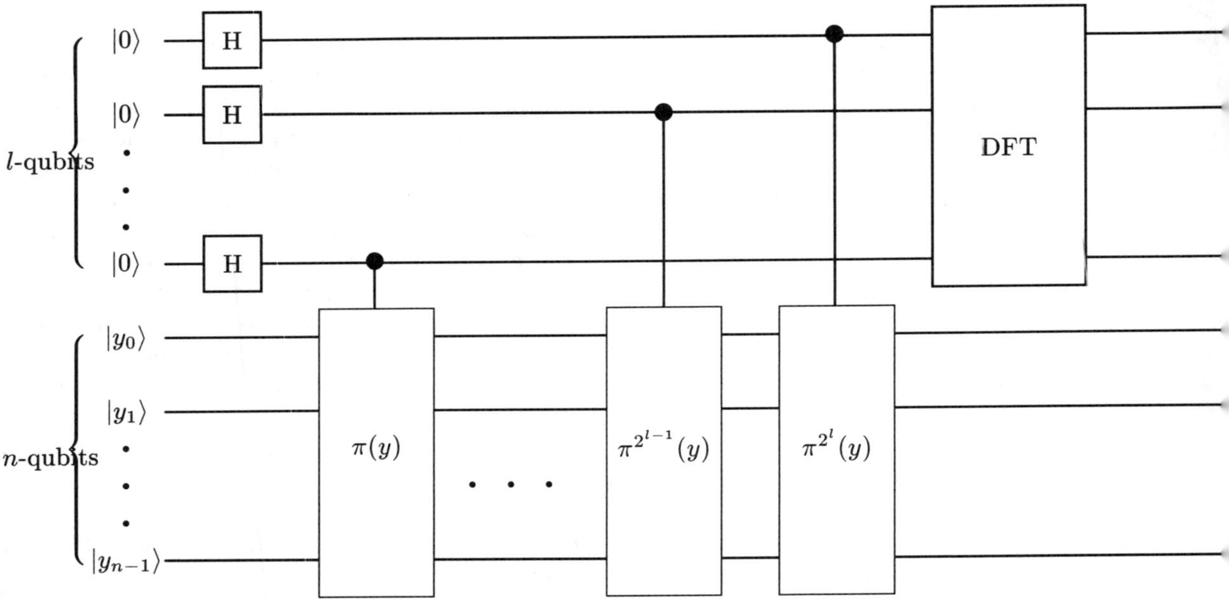

$$P(r' = r) = \sum_{j=0}^{2^l-1} P(r' = r|m = j)P(m = j) \tag{1}$$

$$= \sum_{j=0}^{2^l-1} \sum_{i=1}^{2^n} P(r' = i, r = i|m = j)P(m = j) \tag{2}$$

$$= \sum_{j=0}^{2^l-1} \sum_{i=1}^{2^n} P(r' = i|m = j)P(r = i|m = j)P(m = j) \tag{3}$$

$$= \sum_{j=0}^{2^l-1} \sum_{i=1}^{2^n} P(r' = i)P(m = j|r = i)P(r = i) \tag{4}$$

Put $\alpha_i = P(r = i)$ and $\theta_{ij} = P(r' = i|m = j)$.
Then, we get

$$P(r' = r) = \sum_{j=0}^{2^l-1} \sum_{i=1}^{2^n} \theta_{ij} P(m = j|r = i)\alpha_i \tag{5}$$

$$= \sum_{i=1}^{2^n} \alpha_i \sum_{j=0}^{2^l-1} \theta_{ij} P(m = j|r = i) \tag{6}$$

To maximize the probability $P(r' = r)$ for the arbitrary distribution $\{\alpha_i\}$ of r, it is to find $\{\theta_{ij}\}$ under the condition of $P_1(\theta) = P_2(\theta) = \cdots = P_{2^n}(\theta)$, where $P_i(\theta) = \sum_{j=0}^{2^l-1} \theta_{ij} P(m = j|r = i)$.

Since the probability of $P(m|r = 1)$ in the quantum circuit for all order-finding problems becomes $P(m = 0|r = 1) = 1$ and $P(m = j|r = 1) = 0$ for $j > 0$, the objective funtion can be easily found to be θ_{10}.

5 Quantum Computer Simulation

The result for the case of $n = 2$ and $l = 3$ is compiled in Table 1. Here the r is given and the squared amplitude(probability) of each DFT basis is measured by the quantum circuit of Figure 1. The probability for $r = 2$ is concentrated on $m = 0$ and $m = 4$ and that on the other basis is zero. This is simply due to the fact that the $r = 2$ means the iteration occurs every t wo steps. The similar result is seen on the case of $r = 4$. This just matched behavior occurs due to the DFT basis are even. On the other hand, for the odd case of $r = 3$, this just matched behavior does not occur because the even DFT basis are not divided completely by the odd number. So the probability is spread on the all basis, but the enhancement is seen on the 0, 3 and 5 because they are approximately equally separated 1/3. Thus the simulated results are intuitively understood by the iteration of the permutation.

If we measure the output of Figure 1 without the knowledge about the permutation π. We may have the probability of one of m. What can we say about the pearmutation π from this measurement?

For example, if it is $m = 4$Cthe r might be 2 or 3 or 4. The guess method for r is optimized by the before mentioned linear programming theory for an arbitrary distribution of r and the result is compiled in Table 2. If the measurement is $m = 4$Cwe can say that the permutation is $r' = 4$ with 100% accuracy. The 100% accuracy guess is insured for the measurement except $m = 0$. If the measurement is $m = 0$, we can give the optimal guess with 0.55046 for $r' = 1$.

The result for the case of $n = 3$ and $l = 4$ is compiled in Table 3, in which the measured probability with the given r is simulated. The clear iteration of the concentration of the probability is seen on the DFT basis and the approximately equally separated iteration is also seen for the odd r. It is naturally expected that the iteration behavior for the odd r might be improved as expanding the DFT basisD

The optimal probabilistic guess for this case is compiled in Table 4. The 100% accurate guess is insured for the seven cases out of the 16 cases. The complexity again occurs for the case of $m = 0$. From Table 3 the simulated probability of $m = 0$ is maximum for the case of $r = 1$ but the optimal probabilistic guess is $r = 3$ in the case of no knowledge about r distribution.

The behavior of optimal probabilistic guess on $P(r' = r)$ is surveyed as a function of the DFT basis and the result is compiled in Table 5. The asymptotic improvement of $P(r' = r)$ is seen as increasing the DFT basis. From all the cases investigated here it is known that the optimal probabilistic guess becomes larger than the classical guess if the condition $l > n$ is satisfied. This interesting behavior can be proved. However the solution does not exist beyond $l = 6$ for the case of $r = 2$. This curious behavior is under consideration.

Table 1: The probabilities of the measurement m for given r

m	$P(m\|r=1)$	$P(m\|r=2)$	$P(m\|r=3)$	$P(m\|r=4)$
0	1	0.5	0.343750	0.25
1,7	0	0	0.014515	0
2,6	0	0	0.062500	0.25
3,5	0	0	0.235485	0
4	0	0.5	0.031250	0.25

Table 2: The optimal probabilities of a guess r' for given m

r'	$P(r'\|m=0)$	$P(r'\|m=\text{odd})$	$P(r'\|m=2,6)$	$P(r'\|m=4)$
1	0.55046	0	0	0
2	0.10092	0	0	1
3	0.14679	1	0	0
4	0.20184	0	1	0

Table 3:(QC $l=4$, $n=3$) The probabilities of the measurement m for given r

	$P(m\|r=1,2,3,4,5,6,7,8)$							
m	$r=1$	$r=2$	$r=3$	$r=4$	$r=5$	$r=6$	$r=7$	$r=8$
0	1	0.5	0.33594	0.25	0.20313	0.17188	0.14844	0.125
1, 15	0	0	0.00282	0	0.00369	0.00726	0.00859	0
2, 14	0	0	0.00363	0	0.00726	0.03125	0.11222	0.125
3, 13	0	0	0.00597	0	0.17803	0.11774	0.02454	0
4, 12	0	0	0.01563	0.25	0.01563	0.01563	0.04688	0.125
5, 11	0	0	0.22951	0	0.01326	0.11774	0.07836	0
6, 10	0	0	0.05887	0	0.11774	0.03125	0.01278	0.125
7, 9	0	0	0.01170	0	0.05502	0.00726	0.13851	0
8	0	0.5	0.00781	0.25	0.01563	0.17188	0.00781	0.125

Table 4: (QC $l = 4$, $n = 3$) The optimal probabilities of a guess r' for given m

$P(r'|m = 0, 1, 2, 3, 4, 5, 6, 7, 8, 9, 10, 11, 12, 13, 14, 15)$

r'	$m = 0$	$m = 1, 5, 15$	$m = 2$	$m = 3$	$m = 4$	$m = 6$
1	0.345	0	0	0	0	0
2	0	0	0	0	0	0
3	0.655	0	0	0	0	0
4	0	0	0	0	0.379	0
5	0	0	0	0.105	0	0.258
6	0	1	0	0.895	0	0
7	0	0	0	0	0	0
8	0	0	1	0	0.621	0.742

r'	$m = 7, 9$	$m = 8$	$m = 10, 13$	$m = 11$	$m = 12$	$m = 14$
1	0	0	0	0	0	0
2	0	0.690	0	0	0	0
3	0	0	0	0.543	0	0
4	0	0	0	0	1	0
5	0	0	1	0	0	0
6	0	0.310	0	0.457	0	0
7	1	0	0	0	0	0.604
8	0	0	0	0	0	0.396

Table 5: The optimal probabilistic guess as a function of DFT basis

qubits	$n = 2\,(r = 1..4)$	$n = 3\,(r = 1..8)$	$n = 4\,(r = 1..16)$	$n = 5\,(r = 1..32)$
4	0.391			
5	0.550			
6	×	0.221		
7	×	0.345		
8	×	0.464	0.121	
9	×	0.489	0.208	
10	×	0.497	0.340	0.063
11	×	0.499	0.412	0.115
Classical Probability	0.5	0.25	0.125	0.0625

6 Conclusion

The optimal probabilistic guess of the order-finding problem is formulated with the linear programming theory. The optimal probabilistic guess in all the investigated cases of $n = 2$ to 5 exceed the classic probabilistic guess when the DFT basis becomes larger than the size of permutation. The optimal probabilistic guess is improved also as increasing the DFT basis except the case of $n = 2$. These behaviors are partly understood by the fact that the iterative nature of the permutation becomes projected more clearly on the larger DFT basis. However the upper bound of the improvement by expanding the DFT basis and the exceptional behavior of the case $n = 2$ is left for the future problem.

References

1. Lov K. Grover. Quantum Mechanics helps in searching for a needle in a haystack. Phys.Rev.Lett.79,(1997)325.
2. Charles H.Bennett, Ethan Bernstein, Gilles Brassard, and Umesh Vazirani. Strengths and Weaknesses of Quantum Computing. SIAM J. Comput. 26,(1997)1510.
3. Christof Zalka. Grover's quantum searching algorithm is optimal. Phys. Rev. A60,(1999)2746.
4. Richard Jozsa. Searching in Grover's Algorithm,1999. http://xxx.lanl.gov/abs/quant-ph/9901021.
5. Sixia YU and Chang-Pu Sun. Quantum searching's underlying SU(2) structure and its quantum decoherence effects,1998. http://xxx.lanl.gov/abs/quant-ph/9903075.
6. Eli Biham, Ofer Biham, David Biron, Markus Grassl, and Daniel A.Lidar. Grover's Quantum Search Algorithm for an Arbitrary Initial Amplitude Distribution. Phys.Rev.A60,(1999)2742.
7. A. Hosoya and A. Carlini. Quantum Computers and Unstructured Search:Finding and Counting Items with an Arbitrarily Entangled Initial State,1999. http://xxx.lanl.gov/abs/quant-ph/9909089.
8. Lov K.Grover. Quantum computers can search rapidly by using almost any transformation. Phys.Rev.Lett.80, (1998)4329.
9. Gilles Brassard, Peter Hoyer, Michele Mosca, and Alain Tapp. Quantum Amplitude Amplification and Estimation,2000. http://xxx.lanl.gov/abs/quant-ph/0005055.
10. Alberto Galindo and Miguel A. Martin-Delgado. A Family of Grover's Quantum Searching Algorithms,2000. http://xxx.lanl.gov/abs/quant-ph/0009086.
11. Lieven M.K. Vandersypen, Matthias Steffen, Gregory Breyta, Costantino S.Yannoni, Richard Cleve, and Isaac L.Chuang. Experimental realization of order-finding with a quantum computer. To appear in Phys. Rev. Lett.
12. Richard Cleve. The query complexity of order-finding, 1999. http://xxx.lanl.gov/abs/quant-ph/9911124.